Elke Winterhager (Ed.)

Gap Junctions in Development and Disease

Elke Winterhager (Ed.)

Gap Junctions in Development and Disease

With 47 Figures, 10 in Color

Springer

PROFESSOR DR. ELKE WINTERHAGER
Medical Faculty of the
University Duisburg-Essen
Hufelandstr. 55
45122 Essen
Germany
e-mail: e.winterhager@uni-essen.de

Library of Congress Control Number: 2005926503

ISBN-10 3-540-26156-7 Springer-Verlag Berlin Heidelberg New York
ISBN-13 978-3-540-26156-8 Springer-Verlag Berlin Heidelberg New York

Springer-Verlag is a part of Springer Science+Business Media
springeronline.com

Printed in Germany

Cover design: *design&production*, Heidelberg, Germany
Typesetting and production: LE-TEX Jelonek, Schmidt & Vöckler GbR, Leipzig, Germany
31/3150 YL 5 4 3 2 1 0 - Printed on acid-free paper

Preface

These are challenging times for scientists characterizing signaling pathways of their special interest, because the techniques employed to analyze molecular functions or interactions have developed rapidly in recent years. With regard to gap junctions, molecular cloning of the different connexin isoforms, sequencing of several vertebrate genomes, generation of mice carrying mutations of the different connexin genes, and targeting of inherited human diseases to mutated connexin genes have given completely new insights into this research field. The use of modern techniques has contributed to specifying and broadening our knowledge of these intercellular channels. However, at the same time, more sophisticated new questions have arisen. Just to mention some milestones in the history of gap junction research: this special cell–cell contact was identified as an intercellular nexus between adjacent cells by Dewey and Barr in 1962 and further characterization of the structure of this intercellular channel in the following years led to the well-known structure of the two hemichannels, each composed of six connexin subunits which form a water-filled pore for the exchange of molecules. The gap-junction world became complicated with the identification of different connexin proteins, starting with the cloning of connexin32 by David Paul in 1986 and hopefully ending with all the known 20 members in the human and 21 members in the mouse genome. Furthermore, the existence of at least 20 different channels in humans with different electrophysiological properties has been shown by evidence that the channels can be composed of different connexin isoforms, leading to modifications of their channel properties.

The divergency of the channels has evoked the question of whether these channels form a redundant system, or are highly specific in mediating signal cascades. To answer this question numerous connexin knockout mice have been established, starting with the connexin43 knockout mouse by Reaume et al. in 1995 with the unexpected result of a special heart defect.

The specific function of the connexins in development and organ function has been evaluated by systemically knocking out the different connexins and the replacement of one connexin by another isoform, which is reviewed in the first chapter by Klaus Willecke. Since the gap junctions are indeed channels responsible for ion transport across adjacent cell borders and the propagation of membrane electricity and Ca^{2+} waves, an intensive field of gap-junction physiology has developed, defining the different channel properties. However, up to now, it remains an open question as to what types of molecules in what amounts are crossing this channel. The regulation of discrimination between the molecules which could pass with limitations through these channel pores is still under discussions. It has always been puzzling that the channel properties exclusively could govern the specific tissue functions found in the different mouse mutants. In particular, the hypothesis already postulated by Loewenstein in the 1960s, stating that these channels are involved in growth control and thus responsible for tumorigenesis, never really fit the known channel properties. Meanwhile, it has become evident that several connexin functions could be mediated via protein–protein interactions at the C-terminus of the connexin. Thus, there is a need to discriminate between channel and protein function. Impairment of development and diseases need not be based on connexin mutations, but rather on wrong signaling mediated via protein–protein interactions at the C-terminus. Research on gap junctions gained much more social respect when it became obvious with the publication of Bergoffen et al. (1993) that connexin32 mutations are responsible for the X-linked Charcot Marie tooth disease. This discovery was followed by discoveries of several other connexin mutations which lead to specific diseases in humans. The most important ones are described here in this volume.

In the meantime, it has become impossible to cover all aspects of connexin research. In this volume, we have focused on the role of connexin channels in embryonic development and in the development of human diseases.

This volume deals with connexin channel redundancy, specificity, cell signaling, and conductivity, which are responsible for developmental processes and appropriate organ function. Although these contributions cannot completely answer the question why the loss or mutation in gap-junction proteins leads to a specific phenotype, at least they describe the direction of gap-junction research and how research will proceed in future.

I would like to express my profound thanks to all the authors who contributed the various chapters on their special topics. They did a great job and it was a joy to collaborate with my colleagues while editing this book.

Elke Winterhager
Essen, February 2005

Contents

Contributors

Alcoléa, S.
Laboratoire de Génétique et Physiologie du Développement (UMR CNRS 6545), Institut de Biologie du Développement de Marseille (IBDM, IFR 15), Université de la Méditerranée, Marseille, France

Bierhuizen, M.F.A.
Department of Medical Physiology, University Medical Center, Utrecht, The Netherlands

Boucard, N.
Department of Cell Physiology and Metabolism, University of Geneva, Medical School, 1211 Genève 4, Switzerland

Caille, D.
Department of Cell Physiology and Metabolism, University of Geneva, Medical School, 1211 Genève 4, Switzerland

Cancela, J.
Department of Cell Physiology and Metabolism, University of Geneva, Medical School, 1211 Genève 4, Switzerland

Charollais, A.
Department of Cell Physiology and Metabolism, University of Geneva, Medical School, 1211 Genève 4, Switzerland

Charpantier, E.
Department of Cell Physiology and Metabolism, University of Geneva, Medical School, 1211 Genève 4, Switzerland

Chey Sin, W.
Department of Cellular and Physiological Sciences, The University of British Columbia, Vancouver, BC, Canada V6R 1W9

Cohen-Salmon M. (e-mail: mcs@pasteur.fr)
INSERM U587, Unité de Génétique des Déficits Sensoriels, Institut Pasteur, 25, rue du Dr Roux. 75724 Paris Cedex 15, France

DeRosa A.M. (e-mail: aderosa@ic.sunysb.edu)
Program in Genetics, State University of New York, Stony Brook, New York 11794–8661, USA

del Castillo, F.J.
INSERM U587, Unité de Génétique des Déficits Sensoriels, Institut Pasteur, 25, rue du Dr Roux. 75724 Paris Cedex 15, France

Dermietzel, R. (e-mail: rolf.dermietzel@rub.de)
Department of Neuroanatomy and Molecular Brain Research, Ruhr-University Bochum, Universitätsstrasse 150, 44801 Bochum, Germany

Dunk, C.
Samuel Lunenfeld Research Institute, Mount Sinai Hospital, Toronto, Canada

Dupays, L.
Laboratoire de Génétique et Physiologie du Développement (UMR CNRS 6545), Institut de Biologie du Développement de Marseille (IBDM, IFR 15), Université de la Méditerranée, Marseille, France

Dupont, E.
National Heart and Lung Institute, Imperial College London, London, UK

Eiberger, J.
Institut für Genetik, Abt. Molekulargenetik, Römerstr. 164, 53117 Bonn, Germany

Gellhaus, A.
Institute of Anatomy, University of Essen, Essen, Germany

Goldberg, G.S.
Department of Molecular Biology, University of Medicine and Dentistry of New Jersey, Stratford, New Jersey 08084, USA

Gros, D.B. (e-mail: gros@ibdm.univ-mrs.fr)
Laboratoire de Génétique et Physiologie du Développement (UMR CNRS 6545), Institut de Biologie du Développement de Marseille (IBDM, IFR 15), Université de la Méditerranée, Marseille, France

Haefliger, J.A.
Department of Internal Medicine, University of Lausanne, Medical School, 1011 Lausanne, Switzerland

Kaba, R.
National Heart and Lung Institute, Imperial College London, UK

Kibschull, M.
Institute of Anatomy, University of Essen, Essen, Germany

Kidder, G.M. (e-mail: gerald.kidder@fmd.uwo.ca)
Department of Physiology and Pharmacology, The University of Western Ontario, London, Ontario N6A 5C1, Canada

Klee, P.
Department of Cell Physiology and Metabolism, University of Geneva, Medical School, 1211 Genève 4, Switzerland

Lye, S. (e-mail: lye@mshri.on.ca)
Samuel Lunenfeld Research Institute, Mount Sinai Hospital, 600 University Avenue, Toronto, Ontario M5G 1X5, Canada

Maltzahn, J. von
Institut für Genetik, Abt. Molekulargenetik, Römerstr. 164, 53117 Bonn, Germany

Martinez-Wittinghan, F.J.
Department of Physiology and Biophysics, State University of New York, Stony Brook, New York 11794–8661, USA

Mathias, R.T.
Department of Physiology and Biophysics, State University of New York, Stony Brook, New York 11794–8661, USA

Meda, P. (e-mail: Paolo.Meda@medecine.unige.ch)
Department of Cell Physiology and Metabolism, University of Geneva, C.M.U., 1 rue Michel Servet, 1211 Geneva 4, Switzerland

Meier, C.
Department of Neuroanatomy and Molecular Brain Research, Ruhr-University Bochum, Universitätsstrasse 150, 44801 Bochum, Germany

Meysen, S.
Laboratoire de Génétique et Physiologie du Développement (UMR CNRS 6545), Institut de Biologie du Développement de Marseille (IBDM, IFR 15), Université de la Méditerranée, Marseille, France

Michon, L.
Department of Cell Physiology and Metabolism, University of Geneva, Medical School, 1211 Genève 4, Switzerland

Miquerol, L.
Laboratoire de Génétique et Physiologie du Développement (UMR CNRS 6545), Institut de Biologie du Développement de Marseille (IBDM, IFR 15), Université de la Méditerranée, Marseille, France

Naus, C.C.G. (e-mail: christian.naus@ubc.ca)
Department of Cellular and Physiological Sciences, Faculty of Medicine, The University of British Columbia, Vancouver, BC V6R 1W9, Canada

Nlend, R.N.
Department of Cell Physiology and Metabolism, University of Geneva, Medical School, 1211 Genève 4, Switzerland

Petit, C.
INSERM U587, Unité de Génétique des Déficits Sensoriels, Institut Pasteur, 25, rue du Dr Roux. 75724 Paris Cedex 15, France

Peyrou, M.
Department of Cell Physiology and Metabolism, University of Geneva, Medical School, 1211 Genève 4, Switzerland

Populaire, C.
Department of Cell Physiology and Metabolism, University of Geneva, Medical School, 1211 Genève 4, Switzerland

Richard, G. (e-mail: gabi@genedx.com)
GeneDx Inc., 207 Perry Parkway, Gaithersburg, MD 20877, USA

Severs, N.J. (e-mail: n.severs@imperial.ac.uk)
Cardiac Medicine, National Heart and Lung Institute (Imperial College), Guy Scadding Building, Dovehouse Street, London SW3 6LY, UK

Teunissen, B.E.J.
Department of Medical Physiology, University Medical Center, Utrecht, The Netherlands

Théveniau-Ruissy, M.
Laboratoire de Génétique et Physiologie du Développement (UMR CNRS 6545), Institut de Biologie du Développement de Marseille (IBDM, IFR 15), Université de la Méditerranée, Marseille, France

Thomas, N.
National Heart and Lung Institute, Imperial College London, UK

White, T.W. (e-mail: thomas.white@sunysb.edu)
Program in Genetics, Department of Physiology and Biophysics, BST 5-147, State University of New York, Stony Brook, New York 11794-8661, USA

Willecke, K. (e-mail: genetik@uni-bonn.de)
Institut für Genetik, Abt. Molekulargenetik, Römerstr. 164, 53117 Bonn, Germany

Winterhager, E.
Institute of Anatomy, University of Essen, Essen, Germany

Zulianello, L.
Department of Cell Physiology and Metabolism, University of Geneva, Medical School, 1211 Genève 4, Switzerland

1 Connexin and Pannexin Genes in the Mouse and Human Genome

Klaus Willecke, Jürgen Eiberger, Julia von Maltzahn[1]

1.1 Introduction

Up to the present (December 2004), 20 connexin genes have been identified in the mouse genome. For almost all of them, with the exception of connexin33, orthologous genes appear to occur in the human genome. In addition, two connexin genes, i.e. connexin25 and connexin59 appear to be only present in the human genome which thus contains 21 human connexin genes. The criteria for the identification of connexin genes in the genomic data bases and the chromosomal location of these genes have been described in a recent review article (Söhl and Willecke 2003). This article also included a comparison of the connexin nomenclature, where the "Cx" symbol is followed by the theoretical molecular mass (in kDa) of the corresponding connexin proteins. The alternative nomenclature uses the letters Gja to Gjd, followed by a consecutive number assigned to each connexin. The corresponding human connexin gene symbols are GJA to GJD, followed by the same number as the mouse connexin genes (cf. Söhl and Willecke 2003).

For the purpose of this volume, in which the expression and function of many connexin genes are discussed, we thought it appropriate to present an updated list of the mouse and human connexins (see Table 1.1). With the additions discussed below, this table is very similar to the previously published table (Söhl and Willecke 2003). Thus, we refer the reader to the previous table regarding all details or questions not mentioned below.

Five years ago, another class of genes was described which apparently codes for pannexin proteins in higher vertebrates, including mouse and man (Panchin et al. 2000). This class of genes shows high sequence homology to innexins that form gap junction channels in invertebrates (for review, see Phelan 2004). Recently, it has been shown that rat pannexin1 and -2 proteins can form gap junction channels after expression in *Xenopus* oocytes. Therefore, at the end of this short review, we have summarized the

[1]Institut für Genetik, Abt. Molekulargenetik, Römerstr. 164, 53117 Bonn, Germany, E-mail: genetik@uni-bonn.de

Gap Junctions in Development and Disease
(ed. by E. Winterhager)

Table 1.1. Summary of currently known mouse and human connexin genes (orthologues are listed in line) comparing phenotypes of Cx-deficient mice to known hereditary diseases. This table represents an updated version of Table 3 published by Söhl and Willecke (2003) and includes new information on the characterization of several connexin genes discussed in this chapter

Mouse connexin	Major expression	Phenotype(s) of Cx-deficient mice	Human hereditary disease(s)	Human connexin
mCx23	n.a.	n.a.	n.a.	hCx23
	n.a.		n.a.	hCx25
mCx26	Breast, skin, cochlea, liver, placenta	Lethal on ED 11 *OTO-cre*: hearing loss	Sensorineural hearing loss, palmoplantar hyperkeratosis, keratitis-ichthyosis-deafness syndrome (KID), hystrix-like ichthyosis–deafness syndrome (HID), Vohwinkel's syndrome	hCx26
mCx29	Schwann cells, oligodendrocytes	n.a.	n.a.	hCx30.2 (hCx31.3)
mCx30	Skin, brain, cochlea	Hearing loss	Nonsyndromic hearing loss, hydrotic ectodermal dysplasia, keratitis-ichthyosis-deafness syndrome (KID), Clouston's syndrome	hCx30
mCx30.2	Vascular smooth muscle cells, cardiac conduction system	n.a.	n.a.	hCx31.9
mCx30.3	Skin	n.a.	Erythrokeratodermia variabilis (EKV)	hCx30.3

Table 1.1. (*continued*)

Mouse connexin	Major expression	Phenotype(s) of Cx-deficient mice	Human hereditary disease(s)	Human connexin
mCx31	Skin, cochlea, placenta, uterus	Placental dysfunction	Hearing impairment, erythrokerato-dermia variabilis (EKV)	hCx31
mCx31.1	Skin	n.a.	n.a.	hCx31.1
mCx32	Liver, Schwann cells	Decreased glycogen mobilization, increased liver carcinogenesis, CMTX	CMTX (Charcot-Marie-Tooth neuropathy)	hCx32
mCx33	Testis	n.a.		
mCx36	(Inter)-neurons	Visual transmission defects	n.a.	hCx36
mCx37	Endothelium	Female sterility	Association with atherosclerosis	hCx37
mCx39	Developing striated muscle fibers	n.a.	n.a.	hCx40.1
mCx40	Heart, endothelium	Atrial arrythmias	n.a.	hCx40
mCx43	Many cell types	Lethal on P0 heart malformations *MHC-cre*: arrythmias *GFAP-cre*: dysregulation of spreading depression	Visceroatrial heterotaxia oculoden-todigital dysplasia syndrome (ODDD) syndactyly type III	hCx43[a]
mCx45	Heart, smooth muscle, neurons	Lethal on ED 10.5 *Nestin-cre*: visual transmission defects	n.a.	hCx45

Table 1.1. (*continued*)

Mouse connexin	Major expression	Phenotype(s) of Cx-deficient mice	Human hereditary disease(s)	Human connexin
mCx46	Lens fiber cells	Zonular nuclear cataract	n.a.	hCx46
mCx47	Oligodendro-cytes	Myelin deformation	Pelizaeus-Merzbacher-like disease (PMLD)	hCx47
mCx50	Lens fiber cells	Microphthalmia, zonular pulverulent and congenital cataract	Zonular pulverulent cataract	hCx50
	n.a.		n.a.	hCx59
mCx57	Retinal horizontal cells	n. a.	n.a.	hCx62
Σ 20				**Σ 21**

[a] At the time of writing this review (i.e. December 2004), the human genomic data base lists an additional human Cx43 gene, located on human chromosome 5 that was previously described as a pseudogene, based on the absence of an intron and the presence of a short polyA tail (i.e. ten residues) within the genomic sequence. This pseudogene was abbreviated to hpsiCx43 by Söhl and Willecke (2003). The hCx43 gene and the hpsiCx43 pseudogene differ in a few nucleotide alterations, but the hpsiCx43 sequence does not contain any frameshift or nonsense mutation. Thus, both genes hCx43 and hpsiCx43 might have been subjected to selective pressure during evolution. Thus, the current data bases list the hpsiCx43 gene as a hypothetically expressed connexin gene. This hypothesis could be experimentally checked in the future.

Table 1.2. Comparison of mouse and human pannexin proteins and the chromosomal assignments of their genes

Mouse pannexin	Protein (aa)	Protein (kDa)	Chromosome	Human pannexin	Protein (aa)	Protein (kDa)	Chromosome
Panx1	448[a]	48.07[b]	11[a]	PANX1	426[b]	47.6[b]	9[b]
Panx2	607[a]	73.27[b]	22[a]	PANX2	633[b]	69.5[b]	15[b]
Panx3	392[a]	44.98[b]	11[a]	PANX3	392[b]	44.7[b]	9[b]

[a] Baranova et al. (2004); [b] Bruzzone et al. (2003)

Table 1.3. Transcriptional expression of mouse and human pannexin genes

Mouse pannexins	mRNA expression	Human pannexins	mRNA expression
Panx1	Adrenal gland[a], thyroid[a], prostate[a], brain[a], eye[a], bladder[a], spinal cord[a], stomach[a], liver[a], large intestine[a]	PANX1	Heart[b], skeletal muscle[b], testis[b], ovary[b], brain[b], placenta[b], thymus[b], prostate[b], small intestine, kidney[b], lung[b]
Panx2	Eye[a], spinal cord[a], thyroid[a], brain[a], prostate[a], liver[a], kidney[a], heart[a]	PANX2	Adult brain[b], not in spinal cord[b]
Panx3	Skin[a]	PANX3	At low level in brain[b]

[a] Bruzzone et al. (2003); [b] Baranova et al. (2004)

available information on mouse and human pannexin genes in Tables 1.2 and 1.3.

1.2 Connexin Genes

Connexin26

Several human hereditary diseases affecting the skin and in some cases also the inner ear have recently been characterized to be due to mutations in the Cx26 gene, for example, the keratitis–ichthyosis–deafness syndrome (KID), the hystix-like ichthyosis–deafness syndrome (HID) and Vohwinkel's syndrome. Current knowledge on gap junction diseases affecting the skin has been summarized by van Steensel (2004).

Connexin29

Altevogt and Paul (2004) have described a Cx29-deficient mouse in which the Cx29 coding region has been replaced by a lacZ reporter gene, coding for β-galactosidase that can easily be analyzed in tissues of transgenic mice. Analysis of reporter gene expression confirmed that Cx29 is transcribed in myelin forming Schwann cells and oligodendrocytes where it is expressed together with Cx32 (Altevogt et al. 2002; Nagy et al. 2003). In addition, Cx29/lacZ was found to be activated in Bergmann glia cells. When Cx29 was expressed in *Xenopus* oocytes, it did not form gap junction channels, but altered the electrophysiological properties of coexpressed Cx32 channels (Altevogt et al. 2002). Apparently, Cx29 does not form gap junction channels (plaques), but is expressed in a rather diffuse manner in plasma membranes of myelinated cells, perhaps concentrated at adaxonal membranes (Altevogt et al. 2002; Nagy et al. 2003). No phenotypic abnormality of the Cx29 defective mouse has been reported so far by Altevogt and Paul 2004, in contrast to earlier findings on Cx32-defective mice (Nelles et al. 1996) or Cx32 and Cx47 double-deficient mice (Odermatt et al. 2003; Menichella et al. 2003) that exhibited morphological or behavioural abnormalities, respectively. Recently, we have confirmed the results of Altevogt et al. (2004) with an independently generated Cx29/lacZ-deficient mouse line (Eiberger et al. unpublished results). In addition to myelin forming cells, our results show that mouse Cx29 appears to be expressed in the adrenal gland and in chondrocytes.

Connexin30

Certain mutations in the human Cx30 gene cause Clouston's syndrome, a hereditary skin disease (recently reviewed by van Steensel 2004). In addition, the keratitis–ichthyosis–deafness syndrome (KID) can be due to a mutation in the human connexin30 gene (Jan et al. 2004).

Connexin30.2

The mouse Cx30.2 was first characterized by Nielsen et al. (2003), and suggested to be expressed – based on immunoreactivity – in the blood vessel wall, presumably in smooth muscle cells. Recently, we have found that the Cx30.2 protein is expressed in the conduction system of the heart, particularly at the sinus node and atrioventricular node (Kreuzberg et al. 2005, Circul. Research, in press). Thus, Cx30.2 is the fourth mouse connexin, in addition to Cx43, Cx40 and Cx45 that is expressed in cardiomyocytes. The single channel conductance of Cx30.2 gap junction channels in HeLa cell transfectants is the lowest among all known connexin channels (Kreuzberg et al. 2005, Circul. Research, in press). It will be of great interest to define its functional role for the propagation of cardiac action potentials relative to the other "cardiac" connexin proteins (cf. Söhl and Willecke 2004).

Connexin39

Recently, von Maltzahn et al. (2004) have characterized the mouse Cx39 gene and found it to be expressed in striated muscles during development from E13.5 to birth (i.e. in the diaphragm, intercostal muscles etc.) and, in adult muscles, during regeneration after injection of barium chloride into the tibialis anterior muscle. Although no functional intercellular gap junction channels could be detected (by neurobiotin spreading) in HeLa-Cx39 transfected cells, myotubes in neonatal diaphragm (P0) appeared to be coupled after microinjection of Alexa 488, although no other known connexin could be detected in this tissue at this time of development (von Maltzahn et al. 2004). Thus, Cx39 expression appears to fulfill a unique functional role in myogenesis of striated muscles which could be analyzed when Cx39-deficient mice become available.

Connexin45

Cx45-deficient mice were described by two groups (Krüger et al. 2000; Kumai et al. 2000). These mice died during development (ED 9.5–10.5), due to morphological and functional abnormalities in the heart (Kumai et al. 2000), or in the blood vessel system (Krüger et al. 2000). More recently, Cx45 conditional mice were generated (Nishi et al. 2003) in which the Cx45 gene was deleted under control of the CA actin promoter which is active at ED 9.5 in cardiomyocytes. These mice died at about the same time as the germline Cx45-deficient mice (Kumai et al. 2000), thus confirming that the cardiac defect is largely responsible for the death of the Cx45-deficient mice.

We have also generated conditional Cx45-deficient mice in which the Cx45 coding region was deleted under control of the nestin promoter, specific for neurons and smooth muscle cells. These mice are viable and are lacking Cx45 expression in blood vessels and neurons (but retain it in the cardiac conduction system). When Cx45 was deleted in these cells, an EGFP reporter gene was activated instead. We have studied these mice with regard to the function of Cx45 in the retina and found Cx45 expressed in optic nerve (ON) cone bipolar cells. These cells are coupled by gap junctions to AII amacrine cells which express Cx36. Cx36-deficient mice (Güldenagel et al. 2001) showed the same reduction of the b-wave in the electroretinogram as the newly generated conditional Cx45-deficient mice (Maxeiner et al. 2005). These functional analyses, together with the cell type-specific expression of Cx45 and Cx36, strongly suggest that Cx45 and Cx36 hemichannels can form functional heterotypic gap junction channels between ON cone bipolar cells and AII amacrine cells, respectively. This is the first example where functional heterotypic gap junction channels with identified connexins have been found in a mammalian organ (Maxeiner et al. 2005).

Connexin47

Uhlenberg et al. (2004) have identified Cx47 (GJA12) mutations in three human families that show the symptoms of Pelizaeus-Merzbacher-like disease (PMLD). The affected human patients showed severe demyelination and died at an early age. In one family with three patients, a single base mutation was found in both alleles of human Cx47. In another family, a further single base exchange was found in one Cx47 allele, accompanied by nonsense mutations in the other human Cx47 allele (compound heterozygous). The fifth patient inspected carried a frameshift mutation in combination with a point unitation in the presumed second transmembrane domain. The severe phenotype of these patients differs strongly from the phenotype of Cx47-deficient mice (cf. Odermatt et al. 2003 and Menichella et al. 2003) that was rather mild and showed only rare vacuolation in the optic nerve and other parts of the central nervous system. The much more severe phenotype of PMLD (Cx47) patients is reminiscent of the phenotype of Cx47/Cx32 double deficient mice that show an action tremor and die after about 2 months. Thus, it was speculated (Uhlenberg et al. 2004) that in the brain of Cx47-deficient patients, the mutated Cx47 proteins might exert a trans-effect on the Cx32 protein, thus leading to functional inactivation of both proteins. Another difference was that the human Cx47 gene appears to be expressed in Schwann cells and oligodendrocytes, whereas the mouse Cx47 gene is only expressed in oligodendrocytes (cf. Odermatt et al. 2003; Menichella et al. 2003).

Connexin57

Several years ago, the mouse Cx57 gene was characterized by Manthey et al. (1999), and suggested to be transcriptionally expressed in several tissues. With the recently generated Cx57-deficient mice, which express the lacZ reporter gene instead of the Cx57 coding region, we found expression of the lacZ gene only in horizontal cells of the adult mouse retina, in some thymic cells and in the embryonic kidney (Hombach et al. 2004). We also identified a previously not recognized splice site in the coding region of the mouse Cx57 gene. Thus, the last 25 C-terminal amino acids of mouse Cx57 protein, as described by Manthey et al. (1999), need to be replaced by 12 new amino acids encoded by exon3 (Hombach et al. 2004). This leads to a change in the molecular mass of 55.64 kDa instead of 57.12 kDa. For simplicity, we suggest keeping the original designation Cx57. Thus, Cx57 appears to be a very restricted neuronal connexin which is expressed in only one type of neurons. However, the expression pattern of the mouse Cx57 protein needs to be confirmed by specific Cx57 antibodies when they become available. The horizontal cells in the Cx57-deficient retina show less than 1% of neurobiotin coupling, compared to wild-type retina (Hombach et al. 2004). Currently, a detailed search for retinal defects in Cx57-deficient mice is underway.

1.3 Pannexin genes

Innexin genes have been identified as the invertebrate counterparts to connexins in several invertebrate genomes. Innexin proteins do not exhibit any sequence identity to connexin proteins, but probably evolve by convergent evolution. They show the same general tetra-span membrane topology as connexins and can be assembled to functional intercellular channels (cf. Phelan 2004). Based on a computer-aided search for innexin-like sequences in the genome of higher vertebrates, Panchin et al. (2000) first identified pannexin genes in the mouse and human genome. More recently, the structure and expression pattern of three mouse (Bruzzone et al. 2003; Baranova et al. 2004) and three orthologous human pannexin genes (Baranova et al. 2004) have been described. In contrast to connexin genes, where in most cases the coding region is located on a single exon (with the exception of mouse Cx36, Cx39 and Cx57 genes), the coding regions of mouse and human pannexin genes are distributed on 5 (pannexin1), 3 (pannexin2), and 4 exons (pannexin3). The mouse and human pannexin1 and -2 genes are located on syntenic chromosomes in the mouse and human genome (see Table 1.2). Mouse and human connexin genes show between 93 and 94% amino acid sequence identity. The expression pattern of rat and human pannexin genes is summarized in Table 1.3, based on Northern blot hybridization and RT-PCR results. Rat pannexin1 and 2 transcripts were found in several tissues, but seem to be particularly prominent in the brain and spinal cord (Bruzzone et al. 2003). The expression of human pannexin1 and 2 transcripts appears to show similarities to the corresponding rat pannexin mRNAs, although Baranova et al. (2004), did not find human pannexin2 mRNA in the spinal cord. Rat pannexin3 transcripts were found in skin which was devoid of rat pannexin1 and 2 mRNAs (Bruzzone et al. 2003). Since no pannexin antibodies have so far been described, in situ hybridization was used to analyze the expression of pannexin1 and 2 mRNAs in rat (Bruzzone et al. 2003) and mouse brain (Baranova et al. 2004). Some brain regions yielded positive results (cortex, striatum, olfactory bulb, hippocampus, thalamus and cerebellum). In hippocampus, pannexin1 and 2 transcripts were found in the pyramidal cell layer and in GABAergic individual neurons in the stratum oriens and stratum radiatum (Bruzzone et al. 2003). Strong labeling of pannexin2 transcripts was found in Purkinje cells of the cerebellum. Similar results were reported in the mouse brain by Baranova et al. 2004.

Bruzzone et al. (2003), expressed rat pannexin RNAs in paired Xenopus oocytes and found that pannexin1 alone, and in combination with pannexin2, formed intercellular channels, whereas pannexin2 alone did not. Hemichannels were detected in *Xenopus* oocytes after their injection of pannexin1 RNA, but not of pannexin2 RNA (Bruzzone et al. 2003). Obviously, much more work needs to be done in order to understand the functions of gap junction channels comprised of pannexin protein subunits.

The gene symbols used for designation of the mouse and human pannexin genes are not uniform. Bruzzone et al. (2003) used the abbreviation Px1–3, whereas the Panchin group wrote Panx1–3. In Tables 1.2 and 1.3, we followed the designations of Baranova et al. (2004).

1.4 Outlook

Although it is likely that all mouse and human connexin genes (cf. Table 1.1) have now been identified in the genomic data bases, the function of these genes in the living organism is only partially understood. So far, 14 out of 20 mouse connexin genes have been deleted by homologous recombination and probably the rest of them will follow in the next few years. In all these cases, however, the coding region or at least a very large part of it was deleted. Thus, these mutants lack the corresponding connexin protein. In contrast, connexin mutations which cause hereditary human diseases are mostly missense mutations leading to a codon alteration. In these cases, dominant effects (on the same connexin) or transdominant effects (on another connexin expressed in the same cell type) can occur. Thus, the exact analysis of human connexin missense mutations (also after expression in transgenic mice) should be very informative regarding interactions of different connexins in the same cell type. So far, we do not know why more than one connexin protein is expressed in the same cell type. Furthermore, the molecular mechanisms of phenotypic abnormalities found in connexin-deficient mice or in patients should be experimentally clarified. Is there a qualitative difference of gap junctions between different cell types or the same cell type? The biological role(s), if any, of connexin hemichannels should be dissected, and the network of interactions of connexins (and pannexins) with cytoskeletal and other binding proteins needs to be characterized. Is there any cross talk with other cell contact mediated signal transduction systems? Finally, many of the questions raised on the 20 or so connexin genes can also be asked regarding the three pannexin genes which are far less characterized compared to connexins. Do gap junction channels with connexin protein subunits form a network independent of gap junction channels with pannexin subunit proteins? Are there shared functions among pannexin proteins? How do chemical synapses interact with electrical synapses consisting of connexin or pannexin protein subunits? It seems that gap junction researchers are to be challenged with tough problems during the next few years.

Acknowledgements. Work in our laboratory was supported by several grants of the German Research Association to K.W.

References

Altevogt BM, Kleopa KA, Postma FR, Scherer SS, Paul DL (2002) Connexin29 is uniquely distributed within myelinating glial cells of the central and peripheral nervous systems. J Neurosci 22:6458–6470

Altevogt BM, Paul DL (2004) Four classes of intercellular channels between glial cells in the CNS. J Neurosci 24:4313–4323

Baranova A, Ivanov D, Petrash N, Pestova A, Skoblov M, Kelmanson I, Shagin D, Nazarenko S, Geraymovych E, Litvin O, Tiunova A, Born TL, Usman N, Staroverov D, Lukyanov S, Panchin Y (2004) The mammalian pannexin family is homologous to the invertebrate innexin gap junction proteins. Genomics 83:706–716

Bruzzone R, Hormuzdi SG, Barbe MT, Herb A, Monyer H (2003) Pannexins, a family of gap junction proteins expressed in brain. Proc Natl Acad Sci USA 100:13644–13649

Fishman GI, Spray DC, Leinwand LA (1990) Molecular characterization and functional expression of the human cardiac gap junction channel. J Cell Biol 111:589–598

Güldenagel M, Ammermüller J, Feigenspan A, Teubner B, Degen J, Söhl G, Willecke K, Weiler R (2001) Visual transmission deficits in mice with targeted disruption of the gap junction gene connexin36. J Neurosci 21:6036–6044

Hombach S, Janssen-Bienhold U, Söhl G, Schubert T, Bussow H, Ott T, Weiler R, Willecke K (2004) Functional expression of connexin57 in horizontal cells of the mouse retina. Eur J Neurosci 19:2633–2640

Jan AY, Amin S, Ratajczak P, Richard G, Sybert VP (2004) Genetic heterogeneity of KID syndrome: identification of a Cx30 gene (GJB6) mutation in a patient with KID syndrome and congenital atrichia. J Invest Dermatol 122:1108–1113

Krüger O, Plum A, Kim JS, Winterhager E, Maxeiner S, Hallas G, Kirchhoff S, Traub O, Lamers WH, Willecke K (2000) Defective vascular development in connexin 45-deficient mice. Development 127:4179–4193

Kumai M, Nishii K, Nakamura K, Takeda N, Suzuki M, Shibata Y (2000) Loss of connexin45 causes a cushion defect in early cardiogenesis. Development 127:3501–3512

Manthey D, Bukauskas F, Lee CG, Kozak CA, Willecke K (1999) Molecular cloning and functional expression of the mouse gap junction gene connexin-57 in human HeLa cells. J Biol Chem 274:14716–14723

Maxeiner S, Dedek K, Janssen-Bienhold U, Ammermüller J, Brune H, Kirsch T, Pieper M, Degen J, Krüger O, Willecke K, Weiler R (2005) Deletion of connexin45 in mouse retinal neurons disrupts rod/cone signalling pathway between AII amacrine and ON cone bipolar cells and leads to impaired visual transmission. J Neurosc 25:566–576

Menichella DM, Goodenough DA, Sirkowski E, Scherer SS, Paul DL (2003) Connexins are critical for normal myelination in the CNS. J Neurosci 23:5963–5973

Nagy JI, Ionescu AV, Lynn BD, Rash JE (2003) Connexin29 and connexin32 at oligodendrocyte and astrocyte gap junctions and in myelin of the mouse central nervous system. J Comp Neurol 464:356–570

Nelles E, Butzler C, Jung D, Temme A, Gabriel HD, Dahl U, Traub O, Stumpel F, Jungermann K, Zielasek J, Toyka KV, Dermietzel R, Willecke K (1996) Defective propagation of signals generated by sympathetic nerve stimulation in the liver of connexin32-deficient mice. Proc Natl Acad Sci USA 93:9565–9570

Nielsen PA, Kumar NM (2003) Differences in expression patterns between mouse connexin-30.2 (Cx30.2) and its putative human orthologue, connexin-31.9. FEBS Lett 540:151–156

Nishii K, Kumai M, Egashira K, Miwa T, Hashizume K, Miyano Y, Shibata Y (2003) Mice lacking connexin45 conditionally in cardiac myocytes display embryonic lethality similar to that of germline knockout mice without endocardial cushion defect. Cell Commun Adhes 10:365–369

Odermatt B, Wellershaus K, Wallraff A, Seifert G, Degen J, Euwens C, Fuss B, Bussow H, Schilling K, Steinhauser C, Willecke K (2003) Connexin 47 (Cx47)-deficient mice with enhanced green fluorescent protein reporter gene reveal predominant oligodendrocytic expression of Cx47 and display vacuolized myelin in the CNS. J Neurosci 23:4549–4559

Panchin Y, Kelmanson I, Matz M, Lukyanov K, Usman N, Lukyanov S (2000) A ubiquitous family of putative gap junction molecules. Curr Biol 10:R473–R474

Phelan P (2004) Innexins: members of a exolutionary caused family of gap junction proteins. Biochem Biophys Act doi 10.1016/jbbamem.2004.10.004

Söhl G, Willecke K (2003) An update on connexin genes and their nomenclature in mouse and man. Cell Commun Adhes 10:173–180

Söhl G, Willecke K (2004) Gap junctions and the connexin protein family. Cardiovasc Res 62:228–232

Uhlenberg B, Schuelke M, Ruschendorf F, Ruf N, Kaindl AM, Henneke M, Thiele H, Stoltenburg-Didinger G, Aksu F, Topaloglu H, Nürnberg P, Hübner C, Weschke B, Gartner J (2004) Mutations in the gene encoding gap junction protein alpha 12 (connexin 46.6) cause Pelizaeus-Merzbacher-like disease. Am J Hum Genet 75:251–260

van Steensel MA (2004) Gap junction diseases of the skin. Am J Med Genet 131C:12–19

von Maltzahn J, Euwens C, Willecke K, Söhl G (2004) The novel mouse connexin39 gene is expressed in developing striated muscle fibers. J Cell Sci 117:5381–5392

2 Essential Role of Gap Junctions During Development and Regeneration of Skeletal Muscle

Julia von Maltzahn, Klaus Willecke[1]

2.1 Introduction

Intercellular communication via gap junction channels has been studied during embryonic development of various tissues and involves direct exchange of metabolites, ions or second messengers. Gap junctions are built up of connexin protein subunits that are coded by a family of 20 genes in the mouse genome and are cell type specifically expressed, often more than one isoform per cell type with overlapping specificity (cf. Willecke et al. 2002). Six connexin proteins assemble to a hemichannel and two hemichannels in the plasma membranes of contacting cells can dock to each other to form intercellular protein conduits. Gap junctions are absent from adult skeletal muscles, but are expressed during muscle development and regeneration. Several experiments have suggested an important role of gap junctional intercellular communication with the formation of skeletal muscles, since myogenic precursor cells and developing muscle fibers express functional gap junctions (cf. Constantin and Cronier 2000). During regeneration of adult skeletal muscles, the expression of connexins appears to be recapitulated (Araya et al. 2005; von Maltzahn et al. 2004) similar to embryonic myogenesis. In this chapter we review the current knowledge on gap junctional intercellular communication and connexin expression during formation of skeletal muscle fibers.

2.2 Development of Skeletal Muscles

2.2.1 Embryonic Origin of Myoblasts

The development of skeletal muscles is characterized by a sequence of cellular events that are linked to the myogenic lineage of multipotential pre-

[1]Institut für Genetik, Abt. Molekulargenetik, Römerstr. 164, 53117 Bonn, Germany, E-mail: genetik@uni-bonn.de

Gap Junctions in Development and Disease
(ed. by E. Winterhager)

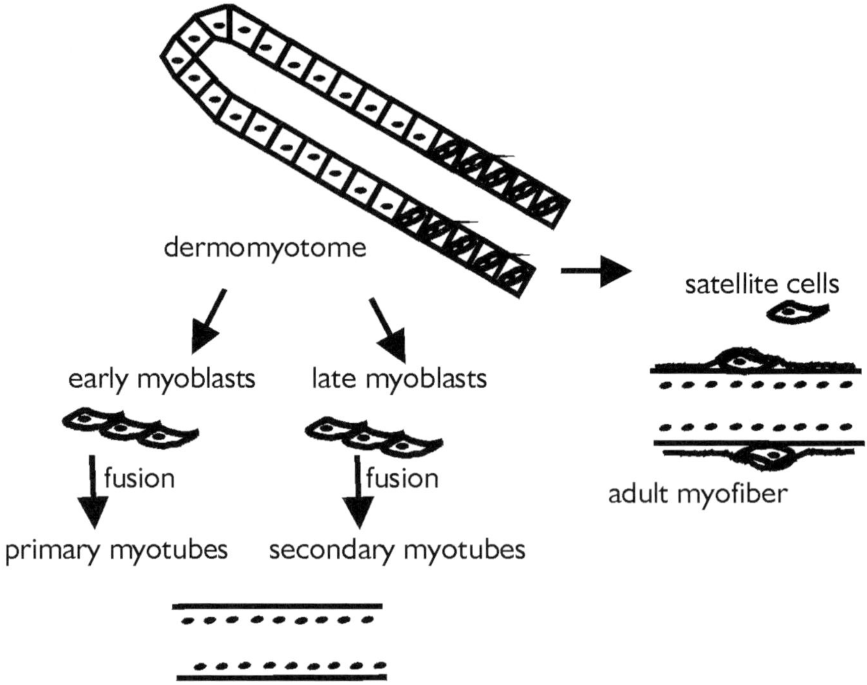

Fig. 2.1. Scheme of the skeletal muscle development in vertebrates. During embryonic development, the dermomyotome gives rise to early and late myoblasts as well as satellite cells residing at the basal lamina and surrounding individual myofibers. Early myoblasts fuse to form primary myotubes, late myoblasts fuse to generate secondary myotubes

cursor cells, derived from the somites of the paraxial mesoderm (Pourquie et al. 2001). These somitic cells differentiate along the dorsal-ventral axis to form the dorsally located dermomyotome or sclerotome (Fig. 2.1). The dermomyotome gives rise to the dermis and the skeletal muscles of the limbs and the trunk, whereas the sclerotome forms the cartilage and the bone of the vertebrae as well as the ribs (cf. Parker et al. 2003). The myogenic precursor cells of the dermomyotome are characterized by expression of the transcription factors Pax3, Pax7 and a low level of Myf5 (cf. Parker et al. 2003).

Determination of the precursor cells results in the development of replicative myoblasts. The terminal differentiation of these replicative myoblasts into postmitotic myoblasts is marked by the expression of the transcription factors MyoD and Myf5 as well as by cell fusion which gives rise to multinucleated myotubes (Ordahl et al. 2000). Both transcription factors belong to the family of primary myogenic basic helix-loop-helix proteins (Parker et al. 2003) which interact with each other and positively regulate transcription of downstream genes. MyoD and Myf5 have been proposed to be involved in the determination of myoblasts, whereas the subsequently expressed myogenin and MRF4 activate terminal differentiation accompanied by exit from the cell cycle (Andrés and Walsh 1996, cf. Hawke and Garry 2001). In addition, external factors like growth factors released by neighboring tissues control cytodifferentiation, either positively or negatively (cf. Hawke and Garry 2001).

A few hours prior to fusion, human and murine myoblasts express functional voltage-operated sodium channels and acetylcholine-gated nicotinic receptor channels, both essential for generation of action potentials in response to a presynaptic release of acetylcholine. Myoblasts are devoid of voltage-dependent systems for Ca^{2+} signaling and of Ca^{2+} channels at the internal sites of the sarcomeric reticulum, necessary for activation of contraction. However, other pathways permitting the generation of Ca^{2+} signals and regulation of intercellular Ca^{2+} homeostasis have been found in cell culture (cf. Constantin and Cronier 2000).

Cell aggregation is required for commitment and differentiation of skeletal precursor cells, known as the "community effect" (George-Weinstein et al. 1998). In several experiments, it was shown that close contacts between cells are important for the expression of muscle-specific differentiation markers such as desmin or myogenin, supporting the previous notion (cf. Constantin and Cronier 2000).

2.2.2 Myotube Formation by Myoblast Fusion

Two phases of myotube formation have been distinguished. Early myotubes or primary myotubes, which are formed first, coexist with a great number of myoblasts that form the secondary myotubes by fusion to primary myotubes. Intercellular communication via gap junctions might allow the transfer of regulatory factors between myoblasts or between myoblasts and myotubes which are essential for myotube formation.

Ca^{2+}-dependent proteases such as calpain (Dourdin et al. 1999; Dedieu et al. 2002) have been implicated to be active during the fusion process and may control the cleavage of proteins of the extracellular matrix or the cytoskeleton such as fibronectin or desmin (Dourdin et al. 1999), thus leading to a reorganization of the cytoskeleton.

Prior to fusion of confluent myoblasts, an increase in cytosolic Ca^{2+} concentration occurs which remains elevated in the newly formed myotubes (Constantin et al. 1996). In many processes accompanying myoblast fusion, such as cell surface interactions via M-cadherins or protease activities, Ca^{2+} ions appear to be involved (cf. Constantin and Cronier 2000). It was proposed that Ca^{2+} influx into myoblasts is necessary to promote myotube formation (Constantin et al. 1996) leading to the question which pathways might mediate the Ca^{2+} shifts in myoblasts. Fusion-competent myoblasts exhibit a high resting membrane potential, whose depolarization is initiated by ion channels triggered by prostaglandin and acetylcholine. Both appear to control Ca^{2+} permeability through voltage-dependent calcium channels, thus leading to myoblast fusion.

In primary cultures, however, neither dihydropyridine receptors nor voltage-dependant Ca^{2+} currents could be demonstrated even during the onset of fusion (Constantin et al. 1996). Thus, the elevation of internal Ca^{2+} concentration may result from other mechanisms such as nicotinic receptor channels operated by cationic messengers. External Ca^{2+}-dependent Ca^{2+} transients were obtained in response to acetylcholine (Constantin et al. 1996) suggesting the possible influx of Ca^{2+} ions through nicotinic receptor channels. In addition, myoblast fusion depends on the external potassium concentration and the involvement of the Na^+/K^+ pump (cf. Constantin and Cronier 2000).

2.2.3 Characteristics of Primary and Secondary Myotubes

Mammalian muscles develop from two populations of myotubes, the primary and the secondary myotubes. The primary myotubes form the initial population of embryonic muscle fibrils which act as a scaffold for the later developing secondary myotubes. The primary myotubes evolve by fusion of myoblasts and later coexist with a large number of myoblasts, whereas the secondary myotubes emerge from fusion of myoblasts to the earlier generated myotubes on the surface beneath the basal lamina (Maier and Mayne 1995). Both primary and secondary myotubes form a syncytium with nuclei located at the borders of the cells and characterized by the expression of myogenin and MRF4 (Chen and Goldhamer 2003). Later in development, muscle fibers are formed by the maturation of myotubes (Parker et al. 2003), characterized by hypertrophy and subsequent specialization of myofibers expressing different isoforms of muscle-specific genes.

2.2.4 Gap Junctional Intercellular Coupling During Myogenesis

Intercellular communication via gap junctions was already described 30 years ago in the myotome of amphibians (cf. Constantin and Cronier 2000). In other vertebrates such as mouse and rat as well as in invertebrates like Drosophila, gap junctional intercellular coupling was also shown to be

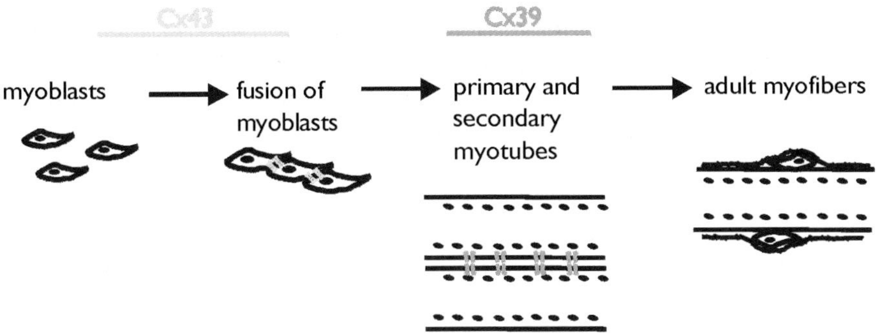

Fig. 2.2. Expression of Cx39 and Cx43 during myogenesis in the mouse. Cx43 is expressed in myoblasts prior to fusion, whereas Cx39 expression was demonstrated in primary and secondary myotubes

involved in myogenesis (Constantin et al. 1997; Todman et al. 1999; Araya et al. 2003).

Gap junctions were observed both between interacting myoblasts and between myoblasts and myotubes as well as between myotubes themselves in developing chick and rat muscles by electron microscopy. These gap junctional structures were not only found in the time interval when fusion activity was predominant, but also later on, leading to the conclusion that gap junctions are present during all stages of myogenesis in vertebrates (Fig. 2.2). During the last few years, the corresponding connexins were identified: Cx40, Cx43 and Cx39 were found to be expressed in gap junctional structures during myogenesis (Dahl et al. 1995; Constantin and Cronier 2000; Araya et al. 2003; von Maltzahn et al. 2004).

Dahl et al. (1995) identified expression of connexin40 (Cx40) in developing myoblasts and myotubes of embryonic mice by in situ hybridization. The highest amount of Cx40 was found in back muscles and a lower amount in the limbs of mouse embryos. Cx40 mRNA was predominantly detected at the outermost edges of muscle bundles where myoblasts fuse to form multinucleated myotubes. From embryonic day 14.5 to 16.5, the level of Cx40 transcripts decreased in back muscles, concomitantly with the differentiation of myoblasts. Until now this is the only evidence that Cx40 is expressed during myogenesis.

2.2.5 Connexin Expression During Fusion of Myoblasts

To verify whether the expression of gap junctions might be related to the fusion of myoblasts, a Northern blot analysis using a cDNA probe for Cx43 and immunofluorescence analysis with antibodies directed against Cx43 were performed. It was shown that Cx43 was developmentally regulated during the early stages of myogenesis in L6 myoblasts and the presence of the Cx43 was correlated to the fusion of myoblasts. The Cx43 transcript was synthesized at high levels in L6 myoblasts, cultured for 12–24 h and decreased 70–140 h after plating when myoblast fusion and myotube differentiation occurred. This suggests that Cx43 mRNA is predominantly expressed in myoblasts prior to the onset of terminal differentiation. By using another mouse myoblastic cell line (C2C12) and primary cultures from newborn mice and rats (Constantin and Cronier 2000; Araya et al. 2005), the localization of Cx43 protein in the plasma membrane of myoblasts was confirmed by indirect immunofluorescence analysis. Heterogeneous expression of Cx43 was demonstrated in cultured C2C12 as well as in primary cultures (Araya et al. 2005). Some of the myoblasts displayed no Cx43 in their plasma membranes that contact neighboring cells, whereas others exhibited Cx43 staining with some of the adjacent cells. This corresponds well to the observation that Cx43 is transiently expressed for a short time prior to myoblast fusion.

In previous studies of electrical and metabolic coupling between interacting myoblasts in primary chick cultures, half of the cell pairs examined were ionically coupled 36 h after plating. Cultured myoblasts are able to develop sodium-dependent action potentials and can be excited when stimulated by injection of an anionic current (Constantin et al. 1996). Myoblasts possess voltage-operated sodium channels and acetylcholine-gated nicotinic receptor channels, which are specialized in membrane excitability. The action potential might be propagated to the adjacent cells through the low resistance gap junctional pathway. It is not yet known whether the excitability of myoblast membranes is involved in the differentiation or fusion process of myoblasts. Since fusion of myoblasts is dependent on the external potassium concentration and the control of the resting membrane potential (cf. Constantin and Cronier 2000), voltage-dependent pathways might be important for the regulation of myoblast fusion. Changes in membrane potential between contacting myoblasts during achievement of fusion competence might be coordinated via electrical coupling through gap junctions.

Metabolic coupling was investigated by intercellular transfer of [^{3}H]thymidine between interacting chick myoblasts. Since coupling was not detected in all interacting donor-recipient cells, intercellular communication between myoblasts was interpreted to be a transient phenomenon. Dye coupling between L6 myoblasts was shown by injection of fluorescent carboxyfluorescein (molecular mass: 520 Da; cf. Constantin and Cronier 2000); progressive dye transfer was noticed between adjacent myoblasts of the first order and some cells of the second order. The pattern of dye spreading was consistent with gap junctional coupling and decreased proportionally to the expression of Cx43. It was suggested that coupling via gap junctions occurred transiently and disappeared during terminal differentiation. The notion that L6 myoblasts are coupled via gap junctions was supported by experiments using the gap junction blocker heptanol, reversibly blocking Lucifer Yellow (443 Da) transfer. In primary cultures of confluent rat myoblasts, gap junctional communication was also observed, and reversibly blocked by heptanol (Constantin et al. 1997). The extent of intercellular diffusion of 6-carboxyfluorescein into cultured myoblasts was similar to the one between human fibroblast or nonstimulated trophoblasts, suggesting moderate coupling. Dye coupling between L6 myoblasts and newly formed myotubes was occasionally observed, consistent with the immunofluorescence staining pattern for Cx43 in cultured L6 cells or in rat or mouse primary cultures (cf. Constantin and Cronier 2000; Araya et al. 2003).

L6 myoblasts treated with the gap junction blockers octanol or 18 beta-glycyrrhetinic acid exit from the cell cycle but do not fuse to myotubes. Myogenin and MRF4, two muscle regulatory factors (MRFs) which are activated at the onset of differentiation, were not turned on in the blocked cells. Removing the gap junction blockers from the culture medium resulted in activation of both myogenin and MRF4, thus ruling out reversible inhibition (Proulx et al. 1997a). These results were confirmed by a recent study of Araya et al. (2003) using 18 beta-glycyrrhetinic acid in C2C12 cells. A reversible block of myotube formation and an inhibition of the creatine kinase activity was observed. In primary cell cultures of myoblasts derived from Cx43fl/fl:Mx-cre newborn mice after induction of Cx43 depletion, a decrease in myogenin expression (Araya et al. 2005) was shown. These results are consistent with the notion that gap junctional intercellular coupling plays an important role in terminal differentiation of normal skeletal muscles.

These results suggest that gap junctions are transiently involved in the exchange of small molecules as well as in electrical coupling during the lag period in myoblasts preceding the first wave of fusion which results in the formation of multinucleated myotubes. Overexpression of Cx43 in communication-deficient RDL6 rhabdomyosarcoma cells led to myogenic differentiation (Proulx et al. 1997b), supporting the notion that gap junctional coupling is involved in myogenic differentiation.

2.2.6 Connexin Expression in Myotubes

Gap junctional structures are found both between myoblasts, myoblasts and primary myotubes and between myotubes. An electron microscopy study by Ling et al. (1992) demonstrated gap junctions between primary and secondary myotubes in the embryonic rat diaphragm. Cx39 immunosignals were found between primary myotubes as well as between primary and secondary myotubes in developing mouse diaphragm and in other skeletal muscles (von Maltzahn et al. 2004). Cx39 transcript and protein were detected in mouse embryos from embryonic day 13.5 up to several days after birth in diaphragm, intercostal muscle and hindlimb musculature. Coupling with Alexa488 dye (570 kDa), observed between myotubes of neonatal diaphragm, suggested expression of functional gap junctions (von Maltzahn et al. 2004).

A recent study of Reinecke et al. (2004) has described the overexpression of Cx43 in cultured mouse myotubes. A viral constitutively active promoter was chosen for expression of Cx43 in MM14 mouse myoblasts. Transfection of these myoblasts led to significant cell death upon differentiation. When differentiated myotubes were transfected with Cx43, no toxic effects were observed, assuming a "window of vulnerability" during differentiation. To examine this hypothesis further, Cx43 was expressed under control of the muscle creatine kinase promoter specifically after myocyte differentiation. These experiments resulted in a strong expression of Cx43 in differentiated myotubes, but did not cause cell death. In contrast, Suzuki et al. (2001) had previously observed more rapid fusion and differentiation into multinucleated myotubes after transfection of L6 rat myoblasts with Cx43 cDNA.

During skeletal muscle development, gap junctions are expressed and the transfer of small molecules as well as electrical coupling can be observed. Differential expression of several connexins (Cx39, Cx40, Cx43) may be important due to the specific properties of permeability and homotypic or heterotypic channel assembly (Saéz et al. 2003). A switch in the expression of connexins during myogenesis could be involved in the recruitment of different homotypic or heterotypic gap junctional channels, with each being responsible for different forms of gap junctional intercellular coupling. Gap junctional channels might permit electrical coupling between myogenic cells, thus permitting the subsequent metabolic coupling. Electrically coupled myoblasts could coordinate their transmembrane potential and may afterwards exchange small molecules like second messengers regulating the time course of the fusion process and the terminal differentiation of interacting myoblasts. So far, only little is known about the nature of these intracellular signals which are directly exchanged. Probably Ca^{2+}, IP_3 and cAMP diffuse through gap junctions during the prefusion lag period, since it was demonstrated that Ca^{2+} and cAMP are involved in the acquirement of fusion competence. In many other cell types, Ca^{2+} waves were observed to spread from cell to cell via gap junctions. In these cellular models, the second messengers diffuse through gap junctions and release Ca^{2+} from intracellular stores of the neighboring cells by binding to their receptors. Gap junctions might synchronize a population of adjacent myoblasts, generating fusion competence and leading to final commitment of terminal differentiation. Myotubes might cooperate in the regulation of the membrane potential and exchange of second messengers triggering intracellular cascades of regulating molecules and gene transcription necessary for myofiber formation. Not only in myogenesis was communication through gap junctions observed, other tissues such as trophoblasts also communicate via gap junctions. The function of gap junctional intercellular coupling might be necessary for the development of tissues that end up in the formation of a syncytium.

2.3 Regeneration of Adult Skeletal Muscles

The degenerative phase of muscle regeneration starts with necrosis of the injured muscle fibers. Disruption of the myofiber sarcolemma results in an increased myofiber permeability allowing an increased Ca^{2+} influx, thereby disrupting Ca^{2+} homeostasis. One hypothesis claims that increased Ca^{2+}-dependent proteolysis, e.g., via calpains, drives tissue degeneration (Belcastro et al. 1998; Alderton and Steinhardt 2000).

The process of muscle degeneration is followed by the muscle repair process. The increased number of myogenic cells provide a sufficient reservoir for muscle repair (Grounds et al. 2002). These myogenic cells proliferate and fuse after differentiation to the existing damaged fibers or to one another, thus forming new myofibers (Hawke and Garry 2001). The newly formed myofibers display morphological differences from undamaged fibers, such as a smaller diameter and centrally located myonuclei (cf. Chargé and Rudnicki 2004). They are often neutrophilic reflecting high protein synthesis and express embryonic forms of myosin heavy chain (MHC) protein demonstrating de novo fiber formation (cf. Chargé and Rudnicki 2004). Central nuclei are located in discrete portions of regenerating muscle fibers or along the whole new fiber. Thus, cell fusion does not occur diffusely during regeneration, but rather focally at the site of injury (Blaveri et al. 1999). After the fusion of myogenic cells is completed, the newly formed myofibers increase in size and the nuclei move to the periphery of the muscle fiber until the regenerated muscle is morphologically and functionally indistinguishable from the undamaged one.

2.3.1 Satellite Cells

Satellite cells comprise a population of undifferentiated mononuclear myogenic cells discovered in many vertebrates including mammalian and avian skeletal muscles. Muscle satellite cells represent a distinct population of myoblasts at about embryonic day 17.5 in mice, residing at the basal lamina and surrounding individual myofibers adjacent to the plasma membrane of the muscle fiber and the basement membrane (cf. Hawke and Garry 2001). Other morphological characteristics are an increased nuclear-to-cytoplasmic ratio, a reduced organelle content and a smaller size of the nucleus with a large amount of heterochromatin consistent with their quiescent state (cf. Chargé and Rudnicki 2004). The number of satellite cells decreases postnatally. Around birth, satellite cells account for 30% of sublaminar muscle nuclei in mice followed by a decrease to less than 5% in 2-month-old animals (cf. Hawke and Garry 2001).

Two hypotheses can be discussed concerning the origin of satellite cells: a somitic versus an endothelial origin (cf. De Angelis et al. 1999; Chargé and Rudnicki 2004). Although the embryonic origin of satellite cells has not yet been proven, one gene responsible for specification of progenitor cells to the satellite cell lineage has been identified (Seale et al. 2000). Pax7 is expressed in proliferating satellite-derived myoblasts and rapidly downregulated upon myogenic differentiation. Pax7-deficient mice are devoid of satellite cells and die within 2 weeks after birth (Mansouri et al. 1996; Seale et al. 2000), suggesting a key role of Pax7 in specification of myogenic precursor cells to the satellite lineage.

Upon muscle injury, satellite cells are activated (Grounds et al. 2002). During regeneration of adult skeletal muscle (Fig. 2.3), satellite cells leave their quiescent state and start to proliferate before they migrate and attach to the injured myofibers guided by chemotactic signaling (Hawke and Garry 2001). Two mechanisms of generating a new fiber are possible. In hypertrophy, the satellite cells fuse to the damaged myofiber, whereas during hyperplasia the satellite cells fuse with each other to form new myofibers. In both cases, the newly formed myofibers are characterized by central nuclei. Later, resting myofibers arise from the regenerated myofibers (Hawke and Garry 2001).

The activation of satellite cells is not restricted to the damaged site in the muscle. Damage at one end of a muscle fiber will activate satellite cells along the fiber, leading to proliferation and migration of the satellite cells. Recruitment of satellite cells from neighboring muscles was rarely observed and occurred only when the connective tissue between these muscles was damaged (cf. Chargé and Rudnicki 2004). The process of satellite cell activation parallels embryonic muscle development. In both cases, MRFs (MyoD,

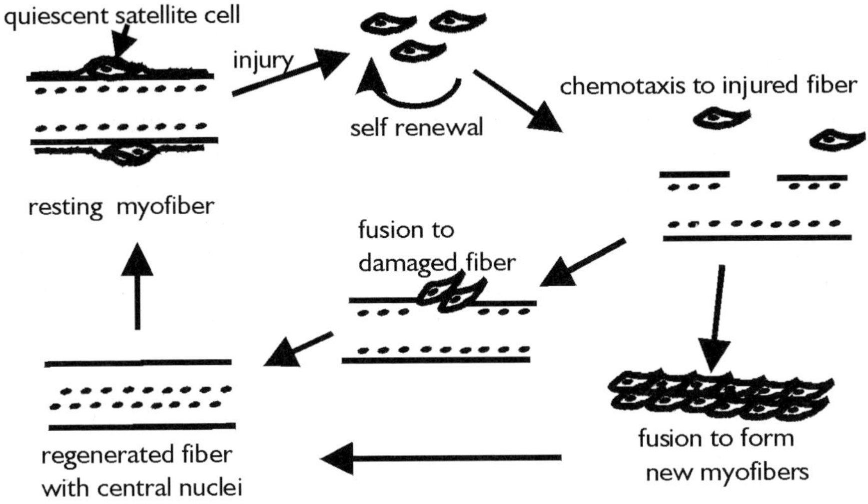

Fig. 2.3. Response of satellite cells to myotrauma. In response to a myotrauma, satellite cells at the basal lamina of myofibers are activated and start to proliferate. Through the process of self-renewal, the pool of quiescent satellite cells is replenished. Other satellite cells migrate to the injured fiber stimulated by chemotaxis and either fuse to the damaged fiber or fuse to form new myofibers

myogenin, Myf5, MRF4) play an important role (Chargé and Rudnicki 2004). Activated and proliferating satellite cells are also called myogenic precursor cells or adult myoblasts. The activation of satellite cells is characterized by the upregulation of Myf5 and MyoD (Cornelison and Wold 1997; Cooper et al. 1999). Myf5 promotes the self-renewal of satellite cells, whereas MyoD promotes satellite cell progression to terminal differentiation (cf. Chargé and Rudnicki 2004). Subsequent to the proliferation phase, the expression of myogenin and MRF4 is upregulated in cells which are committed to terminal differentiation (Cornelison and Wold 1997). This upregulation is followed by activation of p21 protein, which promotes an arrest and a permanent exit from the cell cycle. Since MRF4 is expressed in newly regenerated myofibers or young myotubes after fusion, a distinct role for Myf5, MyoD and myogenin in myofiber during the process of maturation was suggested (Zhou and Bornemann 2001). M-cadherin has been proposed to be an important molecule in the fusion process of myoblasts during embryonal development as well as during muscle regeneration (Kaufmann et al. 1999, cf. Chargé and Rudnicki 2004), but other cadherins might compensate for the loss of M-cadherin shown in M-cadherin knockout mice (Hollnagel et al. 2002).

2.3.2 Connexin Expression During Muscle Regeneration

During the regeneration process, different connexins are expressed (Cx39, Cx43, Cx45). Cx43 is known to be expressed in myoblasts during muscle development and upregulated during regeneration of adult skeletal muscles (Araya et al. 2005). The tibialis anterior muscle of adult wild-type mice and induced Cx43fl/fl:Mx-cre mice was wounded through injection of $BaCl_2$. The highest amount of Cx43 in wild-type animals was detected between day 5 and 7 after wounding. Expression of Cx43 in endothelial cells and satellite cells was shown by colocalization with van Willebrand factor (VWF), M-Cadherin and desmin, i.e., markers for satellite cells (Hawke and Garry 2001; Chargé and Rudnicki 2004). Seven days after injection of $BaCl_2$, Cx43 protein was detected between regenerated myofibers and newly formed myofibers.

To further investigate the role of Cx43 during muscle regeneration in mice, the inducible Mx-cre transgene (Kühn et al. 1995) was bred into Cx43 fl/fl animals, since Cx43-deficient mice died right after birth (Reaume et al. 1995). The regeneration process in induced Cx43 fl/fl:Mx-cre animals was dramatically slowed down compared to uninduced littermates, taking nearly twice as long when examined by measurement of myogenin expression (Araya et al. 2005). Thus, Cx43 appears to be important for the normal time course of regeneration.

Besides Cx43, Cx45 and Cx39 are expressed during regeneration of wild-type adult skeletal muscle. The highest expression level of Cx45 in regenerating tibialis anterior muscle was observed 3 days after injection of $BaCl_2$ (Araya et al. 2005). Using cell type specific markers on uninjured tibialis anterior muscle, Cx45 expression was found to be restricted to microvessels and to endothelial cells identified by van Willebrand factor. After destruction of the muscle, Cx45 expression was colocalized with van Willebrand factor and desmin, a marker for activated and proliferating satellite cells (cf. Chargé and Rudnicki 2004). Five days postinjection, Cx45 was detected in numerous myofibers. Expression of Cx39 was also found in regenerating adult skeletal muscle (tibialis anterior) after $BaCl_2$ treatment (von Maltzahn et al. 2004). In further investigations the time course of expression will be investigated.

Until now three connexin isoforms are known to be expressed during regeneration of adult skeletal muscles in mice. The functional significance of the expression of three different connexins may be important, since each connexin displays specific properties of permeability and regulation of homotypic gap junctional channels (Saèz et al. 2003). Functional gap junctions might be responsible for the normal time course and coordinated regeneration by intercellular transfer of signals like second messengers. Synchronization of satellite cells in the regeneration process might be achieved in a similar way as during myogenesis.

The similarities between embryonic myogenesis and regeneration of adult skeletal muscles are obvious. Reactivation of transcription factors like MyoD or myogenin resembles the expression of the different connexins which are expressed subsequently. These similarities were also observed in other tissues like liver regeneration or folliculogenesis (Fladmark et al. 1997; Ackert et al. 2001). Since the expression patterns in development and regeneration resemble each other, one can speculate that similar signaling molecules may diffuse through gap junction channels during myogenesis and regeneration. It is challenging to identify these signaling molecules for intercellular communication through gap junctional channels during development and regeneration of skeletal muscle.

Acknowledgements. Work in our laboratory was supported by a grant of the German Research Association Wi 270–22 to K.W.

References

Ackert CL, Gittens JE, O'Brien MJ, Eppig JJ, Kidder GM (2001) Intercellular communication via connexin43 gap junctions is required for ovarian folliculogenesis in the mouse. Dev Biol 233:258–270

Alderton JM, Steinhardt RA (2000) Calcium influx through calcium leak channels is responsible for the elevated levels of calcium-dependent proteolysis in dystrophic myotubes. J Biol Chem 275:9452–9460

Andrés V, Walsh K (1996) Myogenin expression, cell cycle withdrawal, and phenotypic differentiation are temporally separable events that precede cell fusion upon myogenesis J Cell Biol 132:657–666

Araya R, Eckardt D, Riquelme MA, Willecke K, Saez JC (2003) Presence and importance of connexin43 during myogenesis. Cell Commun Adhes 10(4–6):451–456

Araya R, Eckardt D, Maxeiner S, Krüger O, Theis M, Willecke K, Sáez JC (2005) Expression of connexins during differentiation and regeneration of skeletal muscle: functional relevance of connexin 43. J Cell Sci 118:27–31

Belcastro AN, Shewchuk LD, Raj DA (1998) Exercise-induced muscle injury: a calpain hypothesis. Mol Cell Biochem 179:135–145

Blaveri K, Heslop L, Yu DS, Rosenblatt JD, Gross JG, Partridge TA, Morgan JE (1999) Patterns of repair of dystrophic mouse muscle: studies on isolated fibers. Dev Dyn 216:244–256

Chargé SB, Rudnicki MA (2004) Cellular and molecular regulation of muscle regeneration. Physiol Rev 84:209–238

Chen JC, Goldhamer DJ (2003) Skeletal muscle stem cells. Reprod Biol Endocrinol 1(1):101

Constantin B, Cognard C, Raymond G (1996) Myoblast fusion requires cytosolic calcium elevation but not activation of voltage-dependent calcium channels. Cell Calcium 19:365–374

Constantin B, Cronier L, Raymond G (1997) Transient involvement of gap junctional communication before fusion of newborn rat myoblasts. C R Acad Sci III 320(1):35–40

Constantin B, Cronier L (2000) Involvement of gap junctional communication in myogenesis. Int Rev Cytol 196:1–65

Cooper RN, Tajbakhsh S, Mouly V, Cossu G, Buckingham M, Butler-Browne GS. (1999) In vivo satellite cell activation via Myf5 and MyoD in regenerating mouse skeletal muscle. J Cell Sci 112:2895–2901
Cornelison DD, Wold BJ (1997) Single-cell analysis of regulatory gene expression in quiescent and activated mouse skeletal muscle satellite cells. Dev Biol 191:270–283
Dahl E, Winterhager E, Traub O, Willecke K (1995) Expression of gap junction genes, connexin40 and connexin43, during fetal mouse development. Anat Embryol (Berl) 191:267–278
De Angelis L, Berghella L, Coletta M, Lattanzi L, Zanchi M, Cusella-De Angelis MG, Ponzetto C, Cossu G. (1999) Skeletal myogenic progenitors originating from embryonic dorsal aorta coexpress endothelial and myogenic markers and contribute to postnatal muscle growth and regeneration. J Cell Biol 147:869–878
Dourdin N, Balcerzak D, Brustis JJ, Poussard S, Cottin P, Ducastaing A (1999) Potential m-calpain substrates during myoblast fusion. Exp Cell Res 246:433–442
Fladmark KE, Gjertsen BT, Molven A, Mellgren G, Vintermyr OK, Doskeland SO (1997) Gap junctions and growth control in liver regeneration and in isolated rat hepatocytes. Hepatology 25:847–855
George-Weinstein M, Gerhart J, Mattiacci-Paessler M, Simak E, Blitz J, Reed R, Knudsen K (1998) The role of stably committed and uncommitted cells in establishing tissues of the somite. Ann N Y Acad Sci 842:16–27
Grounds MD, White JD, Rosenthal N, Bogoyevitch MA (2002) The role of stem cells in skeletal and cardiac muscle repair. Histochem Cytochem 50:589–610
Hawke TJ, Garry DJ (1997) Myogenic satellite cells: physiology to molecular biology. J Appl Physiol 91:534–551
Hollnagel A, Grund C, Franke WW, Arnold HH (2002) The cell adhesion molecule M-cadherin is not essential for muscle development and regeneration. Mol Cell Biol 22:4760–4770
Kaufmann U, Kirsch J, Irintchev A, Wernig A, Starzinski-Powitz A (1999) The M-cadherin catenin complex interacts with microtubules in skeletal muscle cells: implications for the fusion of myoblasts. J Cell Sci 112:55–68
Kühn R, Schwenk F, Aguet M, Rajewsky K (1995) Inducible gene targeting in mice. Science 269:1427–1429
Ling Y, Appelt D, Kelly AM, Franzini-Armstrong C (1992) Differences in the histogenesis of EDL and diaphragm in rat. Dev Dyn 193:359–369
Maier A, Mayne R (1995) Basal lamina development in chicken muscle spindles. Dev Dyn 202:284–293
Mansouri A, Hallonet M, Gruss P (1996) Pax genes and their roles in cell differentiation and development. Curr Opin Cell Biol 8:851–857
Parker MH, Seale P, Rudnicki MA. (2003) Looking back to the embryo: defining transcriptional networks in adult myogenesis. Nat Rev Genet 4:497–507
Pourquie O (2001) Vertebrate somitogenesis. Annu Rev Cell Dev Biol 17:311–350
Proulx A, Merrifield PA, Naus CC (1997a) Blocking gap junctional intercellular communication in myoblasts inhibits myogenin and MRF4 expression. Dev Genet 20:133–144
Proulx AA, Lin ZX, Naus CC (1997b) Transfection of rhabdomyosarcoma cells with connexin43 induces myogenic differentiation. Cell Growth Differ 8:533–540
Reaume AG, de Sousa PA, Kulkarni S, Langille BL, Zhu D, Davies TC, Juneja SC, Kidder GM, Rossant J (1995) Cardiac malformation in neonatal mice lacking connexin43. Science 267:1831–1834
Reinecke H, Minami E, Virag JI, Murry CE (2004) Gene transfer of connexin43 into skeletal muscle. Hum Gene Ther 15:627–636

Reinecke H, MacDonald GH, Hauschka SD, Murry CE (2004) Expression of connexins during differentiation and regeneration of skeletal muscle: Functional relevance of connexin43. J Cell Sci 149(3):731–740

Seale P, Sabourin LA, Girgis-Gabardo A, Mansouri A, Gruss P, Rudnicki MA (2000) Pax7 is required for the specification of myogenic satellite cells. Cell 102(6):777–786

Suzuki K, Brand NJ, Allen S, Khan MA, Farrell AO, Murtuza B, Oakley RE, Yacoub MH (2001) Overexpression of connexin 43 in skeletal myoblasts: Relevance to cell transplantation to the heart. J Thorac Cardiovasc Surg 122(4):759–766

Todman MG, Baines RA, Stebbings LA, Davies JA, Bacon JP (1999) Gap-Junctional communication between developing Drosophila muscles is essential for their normal development. Dev Genet 24(1–2):57–68

von Maltzahn J, Euwens C, Willecke K, Sohl G (2004) The novel mouse connexin39 gene is expressed in developing striated muscle fibers. J Cell Sci 117:5381–5392

Willecke K, Eiberger J, Degen J, Eckardt D, Romualdi A, Guldenagel M, Deutsch U, Sohl G (2002) Structural and functional diversity of connexin genes in the mouse and human genome. Biol Chem 383(5):725–737

Zhou Z, Bornemann A (2001) MRF4 protein expression in regenerating rat muscle. J Muscle Res Cell Motil 22(4):311–316

3 Connexins in Cardiac Development: Expression, Role, and Transcriptional Control

Daniel B. Gros[1], Sébastien Alcoléa[1], Laurent Dupays[1], Sonia Meysen[1], Magali Théveniau-Ruissy[1], Birgit E.J. Teunissen[2], Marti F.A. Bierhuizen[2]

3.1 Introduction

Gap junction channels provide a direct pathway for the movement of signaling molecules and ionic current between adjacent cells. In diploblast organisms and protostomes, these intercellular channels are composed of innexins (Phelan et al. 1998; Alexopoulos et al. 2004). In deuterostomes, they are composed of pannexins (Pxs; Panchin et al. 2000), innexin orthologues or of connexins (Cxs), a protein family unrelated to innexins. Three *Pxs* genes have been identified in the genomes of mammals (Bruzzone et al. 2003; Baranova et al. 2004). The *Px2* gene is expressed in the rat heart (Bruzzone et al. 2003), but the distribution and the role of Px2 are still unknown, investigations on Pxs still being in the rudimentary stages. In the mouse genome, 20 Cxs have been identified, 21 in the human genome, and 36 in the zebrafish (Eastman and Iovine 2004; Söhl and Willecke 2004). *Cx* genes have also been identified in a variety of other vertebrates whose genomes have not been fully sequenced yet, including *Xenopus*, chicken, pig, dog, bovine, etc. The structure of Cxs, their association into connexons (or hemichannels), the incorporation of the latter into intercellular gap junction channels, and the biophysical properties of these channels have been the subject of excellent reviews (Harris 2001; Evans and Martin 2002; Sáez et al. 2003; Martin and Evans 2004; Moreno 2004). Recent investigations have indicated that the unpaired Cx hemichannels were also active in single plasma membranes (John et al. 2003; Stout et al. 2004); others have suggested that Cx proteins may mediate cell signaling independently of their channel functions (Wei et al. 2004).

In the heart, Cxs have been shown to be involved in the propagation of electrical activity, the developmental processes, and the volume regulation of cardiomyocytes (Gros et al. 2004; Röhr 2004; Schultz and Heusch

[1]Laboratoire de Génétique et Physiologie du Développement (UMR CNRS 6545), Institut de Biologie du Développement de Marseille (IBDM, IFR 15), Université de la Méditerranée, 13288 Marseille, Cedex 09, France, E-mail: gros@ibdm.univ-mrs.fr
[2]Department of Medical Physiology, University Medical Center Utrecht, Yalelaan 50, 3584 CM Utrecht, The Netherlands

Gap Junctions in Development and Disease
(ed. by E. Winterhager)

2004). In this chapter, the expression of Cxs in the mammalian heart, their transcriptional control and their roles during cardiac development are analyzed. The reader may refer to various reviews on the same themes that have recently been published (Gros et al. 2004; Teunissen and Bierhuizen 2004; Wei et al. 2004).

3.2 Multiple Connexin Genes Are Expressed in the Heart

3.2.1 The Cardiac Connexins

Since the identification of Cx43 (Beyer et al. 1987), the major Cx of the mammalian heart, the products of several other Cx genes have been detected in varying degrees of abundance in this organ. Human and mouse are the species in which the expression of cardiac Cxs is the most thoroughly documented. Eight Cx genes have thus been found to be expressed in the human heart. These are: *hCx30.2* (The letters h and m, associated with the abbreviation Cx, indicate human and mouse, respectively.), *-31.9*, *-37*, *-40.1*, *-40*, *-43*, *-45*, and *-62* (Gros and Jongsma 1996 and cited references; Nielsen et al. 2002; Söhl et al. 2003). In the mouse heart, the expression of five Cx genes has been demonstrated so far. These are: *mCx30.2*, orthologue of the *hCx31.9* gene, and *mCx37*, *-40*, *-43*, and *-45*, orthologues of the *hCx37*, *-40*, *-43*, and *-45* genes, respectively (Gros and Jongsma 1996 and cited references; Söhl et al. 2003). The expression of the *mCx57* gene (orthologue of the *hCx62* gene), originally reported by Manthey et al. (1999), has been recently re-assessed (Söhl et al. 2003) and was not found to be significantly expressed. In addition, neither mRNA of *mCx29*, orthologue of *hCx30.2*, nor that of *mCx39*, orthologue of *hCx40.1*, have been detected by the Northern blot technique in the mouse heart (Söhl et al. 2003). The repertoire of human heart Cxs and that of the mouse heart thus do not appear to be quite identical. In addition, they are probably not definitive.

The cardiomyocytes, which occupy 90% of the mass of the heart, represent only 30–40% of the total number of cells (Nag 1980). Various other cell types, such as endothelial cells, smooth muscle cells, macrophages, pericytes, fibroblasts (which contribute to cardiac electrophysiological properties), are also present in the heart, and one of the first questions to address after the identification of a cardiac Cx gene is to know in which cell type the gene products are expressed. To date, the cell types which express cardiac Cxs have been identified for only a few of them. Cx37 is synthesized by the endothelial cells of both coronary vessels and endocardium (Haefliger et al. 2004); Cx40, -43 and -45, by cardiomyocytes. The latter Cxs that we shall qualify by oversimplification as "cardiomyocyte-related Cxs", are, however, also expressed in other cardiac cell types. Thus, Cx43 was found to be expressed in the endothelial cells of the cardiac valves in the rat heart (Inai et al. 2004); Cx40 is synthesized in the endothelial cells of coronary vessels, along with Cx37, and, at least in the rabbit heart, in the fibroblasts of the sinoatrial (SA) node, along with Cx45 (Camelliti et al. 2005). Cx45 is also expressed in the smooth muscle cells of the aorta and coronary vessels (Alcoléa et al. 1999; Krüger et al. 2000), and in the sheep heart, in the early-infiltrating fibroblasts of ischemic regions (Camelliti et al. 2005). Beside the Cxs listed above, two other *Cx* genes are expressed in cardiomyocytes of the mammalian heart. These are Cx46 that has been detected in the rabbit SA node myocytes (Verheule et al. 2001), and the protein encoded by the *mCx30.2* gene that has recently been identified in mouse heart cardiomyocytes (Nielsen and Kumar 2003). Expression of these Cxs in the heart of

Table 3.1. Cx genes expressed in the human heart and the mouse heart, and cardiac cell types expressing Cxs

Human		Cell types	Mouse		Cell types
hCx30.2	+	nd	mCx29	–	–
hCx31.9	+	nd	mCx30.2	+	Cardiomyocytes
hCx37	+	Endothelial cells of coronary vessels and endocardium	mCx37	+	Endothelial cells of coronary vessels and endocardium
hCx40.1	+	nd	mCx39	–	–
hCx40	+	Cardiomyocytes and endothelial cells of coronary vessels	mCx40	+	Cardiomyocytes and endothelial cells of coronary vessels
hCx43	+	Cardiomyocytes	mCx43	+	Cardiomyocytes
hCx45	+	Cardiomyocytes, coronary smooth muscle cells, endocardium (?)	mCx45	+	Cardiomyocytes, coronary smooth muscle cells, endocardium
hCx62	+	nd	mCx57	–	–

The *mCx29*, *mCx30.2* and *mCx57* genes are the orthologues of the *hCx30.2*, *hCx31.9*, and *hCx62* genes, respectively (Söhl et al. 2003; Söhl and Willecke 2004). The signs + and – indicate whether gene expression has been detected or not in the cardiac tissues. nd, unidentified other mammals remains to be confirmed. Finally, it should be noted that the expression and distribution of Cxs in the heart may be remodeled in cardiac diseases. This point is analyzed in another chapter of this volume. The Cx-expressing cell types which have been identified in the human and murine hearts are listed in Table 3.1.

Identification of the cell types responsible for the synthesis of Cxs and the establishment of expression maps are not trivial problems and run up against technical difficulties, including, for human, access to clearly identified specimens of healthy cardiac tissues. This is the case for the heart as well as other complex organs, and finally the most extensively documented expression patterns are those of Cx45, -43 and -40, synthesized by cardiomyocytes, in the mouse heart. These expression patterns are described below.

3.2.2 Expression Patterns of Cardiomyocyte-Related Cx43, -40 and -45 in the Adult Mouse Heart

Cx43 is abundantly expressed in all the cardiac compartments with the exception of the SA and atrioventricular (AV) nodes, the His bundle, and the proximal parts of the bundle branches (Table 3.2, Fig. 3.1). In addition, its expression in the ventricular free walls is not uniform. Cx43 is in fact twice as abundant in midmyocardial and endocardial regions compared with epicardial layers (Yamada et al. 2004). Cx40 is expressed in the atria and the conduction system. Consequently, specific expression of the EGFP (enhanced-green fluorescent protein) reporter protein under the control of the *Cx40* gene has recently made it possible to clearly demonstrate the architectural and functional asymmetry of the right and left parts of the ventricular conduction system in the mouse heart (Miquerol et al. 2004). However, it should be noted that Cx40 has not been detected in the cardiomyocytes of the SA node, and that in the AV node it is expressed in the central region only (Table 3.2, Fig. 3.1). Cx45 is synthesized in low abundance in the mouse conduction system, including the cardiomyocytes of the SA and AV nodes, and in the most peripheral regions of the interventricular septum (IVS), i.e., in the vicinity of conduction myocytes (Coppen et al. 1998, 1999; Verheijck et al. 2001; Table 3.2, Fig. 3.1). Its expression in the ventricular working cardiomyocytes is open to controversy. Using immunofluorescence techniques, some authors have shown that Cx45 was uniformly expressed in the cardiomyocytes across the ventricular free walls, and colocalized with Cx43, whereas others have not detected it (Table 3.2). This discrepancy could be explained by the greater or lesser affinity of the antibodies for Cx45 which come from different sources, but in some cases the same antibodies were used and gave different results (van Veen et al.

Table 3.2. Distribution of cardiomyocytes-related Cx43, -40 and -45 in the adult mouse heart

Cardiac compartments	Cx43	Cx40	Cx45 (Reference)[c]
Atria	+ (ref 2)	+ (ref 2)	− (3, 4, 5)[c]
Sinoatrial node	− (ref 1)	− (ref 1)	+ (1)[c]
Atrioventricular node	− (ref 6, 7, 8[a])	+ (ref 8)	+ (8, 9)[c]
Ventricles	+ (ref 2)	− (ref 2)	− (3, 4, 5)[c]; ± (9)[c]; + (10, 11[b])[c]
Interventricular septum	+ (ref 2)	− (ref 2)	+ (8)[c]
His bundle	− (ref 2, 7)	+ (ref 2, 5, 8)	+ (4, 5, 8, 9)[c]
Bundle branches			
Proximal regions	− (ref 2, 5)	+ (ref 2, 5)	+ (3, 5, 9)[c]
Distal regions	+ (ref 2, 5)	+ (ref 2, 5)	+ (3, 5, 9)[c]
Purkinje fibers	+ (2, 3, 5)	+ (ref 2, 5)	+ (3, 4, 5)[c]

This table indicates the expression of *Cx43*, *-40* and *-45* gene products in the cardiomyocytes of the various compartments of the adult mouse heart. The cardiac conduction system includes the sinoatrial node (impulse generation system), the His bundle, its two branches (bundle branches), and the Purkinje fibers. The signs + and − indicate whether gene expression has been detected or not in the cardiac tissues

[a]Detected in trace amounts

[b]See text for detailed explanations on Cx45 expression in the ventricles

[c] 1 Verheijck et al. 2001; 2 Gros and Jongsma 1996, and cited references; 3 Coppen et al. 1998; 4 Alcoléa et al. 1999; 5 van Veen et al. 2001; 6 Coppen et al. 2003; 7 Franco and Icaro 2001; 8 Coppen et al. 1999; 9 Krüger et al. 2000; 10 Johnson et al. 2002; 11 Yamada et al. 2004.

2001; Johnson et al. 2002). The $Cx45^{+/-}$ mice (Krüger et al. 2000), in which one allele of the *Cx45* gene was replaced by the *lacZ* reporter gene, may offer an elegant means to settle this question (though the half-lives of LacZ and Cx45 are probably not similar). In these mice, some cells in the working myocardium, described as cardiomyocytes, show faint lacZ staining, and are therefore Cx45 positive. However, the identity of these cells remains to be determined. Double-labeling experiments should make it possible to rule out the possibility that these cells may be fibroblasts.

The expression patterns described above for the adult mouse heart are conserved, with a very few variations, in the heart of other mammals (Van Kempen et al. 1995; Gros and Jongsma 1996; Miquerol et al. 2003). In regions where the cardiomyocytes express two or three Cxs, there is a possibility that a wide range of molecular arrangements including heteromeric connexons and heterotypic channels may exist. The diversity of these channels suggests that the electrical coupling between cardiomyocytes will have a particularly complex set of properties and regulatory modes. A higher level of complexity is reached when the coupling between cardiomyocytes is achieved by means of fibroblasts (Camilliti et al. 2005).

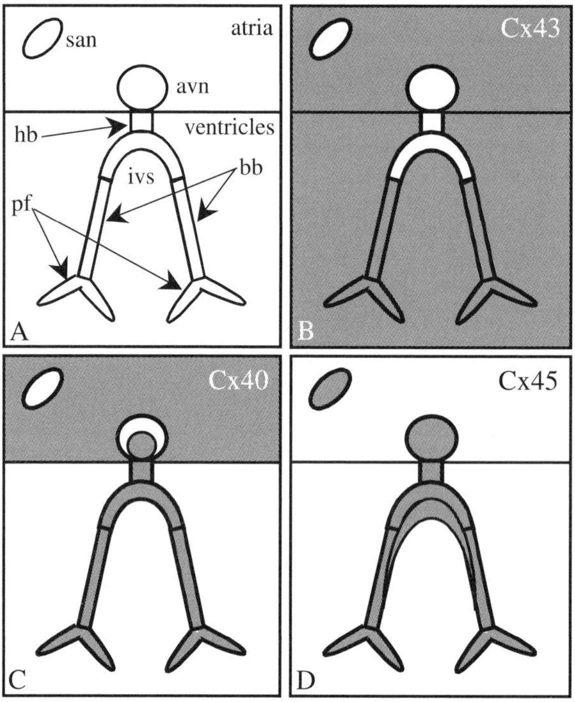

Fig. 3.1A–D. Expression patterns of cardiomyocyte-related Cxs in the adult mouse heart. **A** Diagram of the heart. The conduction system includes the sinoatrial node (*san*), in which the cardiac electrical activity is spontaneously generated, the atrioventricular (*avn*) node, the His bundle (*hb*), the bundle branches (*bb*) and the Purkinje fibers (*pf*). *ivs* Interventricular septum. **B, C, D** Expression (*dark gray*) of Cx43, -40 and -45, respectively

3.2.3 Spatiotemporal Expression Patterns of Cardiomyocyte-Related Cx43, -40 and -45 in the Developing Mouse Heart

The expression patterns of cardiomyocyte-related Cx43, -40 and -45 in the adult mouse heart result from strictly controlled spatiotemporal regulation which operates during heart development. At the stage of the first cardiac contractions, i.e., at E8.5 (embryonic day 8.5) in the mouse, transcripts of all three genes have been weakly detected in the heart, but only Cx45 protein has been shown to be expressed in detectable amount (Delorme et al. 1997; Alcoléa et al. 1999; Kumai et al. 2000; Fig. 3.2). At further stages of development, the expression of these genes develops independently of each other.

Cx43 protein has been weakly detected at E9.5, and more strongly at E10.5, in the ventricular walls, then in the atria at E12.5 (Fig. 3.2). At

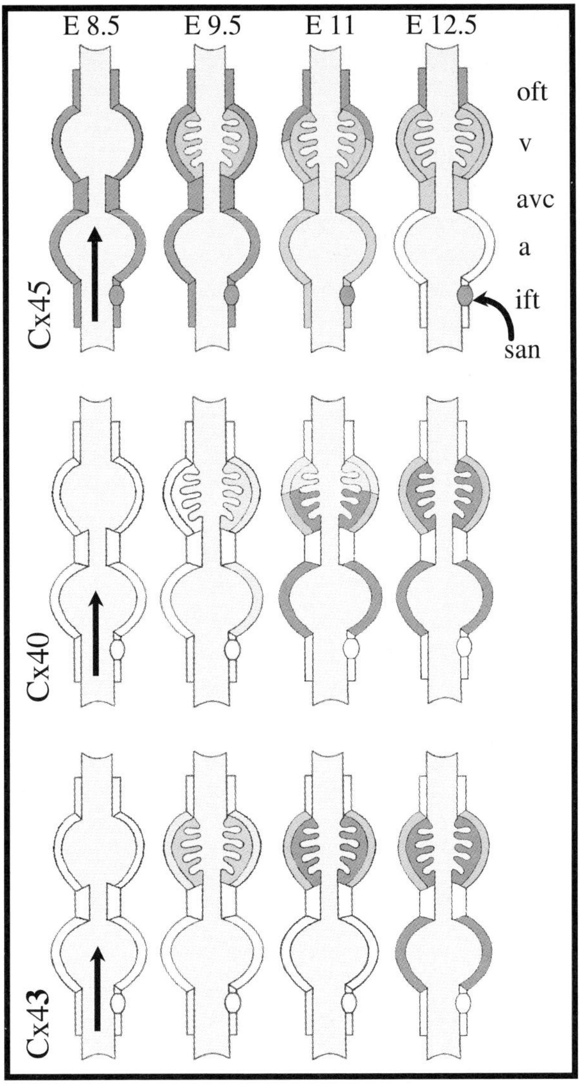

Fig. 3.2. Expression patterns of cardiomyocyte-related Cx45, -40 and -43 during mouse heart development. The heart is schematized as a series of compartments which are, along the caudorostral axis: the inflow tract (*ift*) with the sinoatrial node (*san*), the atrial compartment (*a*), the atrioventricular canal (*avc*), the ventricular compartment (*v*), and the outflow tract (*oft*). In the ventricular compartment, the scalloped region represents the trabeculae which develops from E9. Blood flow direction is indicated by *arrows*. The figures are explained in the text. For the development of the heart, see Moorman and Christofels (2003)

this stage abundance of Cx43 increases in both compartments to reach 2–3 weeks after birth an expression level comparable to that of the adult heart (Fromaget al. 1990; Delorme et al. 1997). At the embryonic stages Cx43 has never been detected in the AV canal, nor later, at the fetal stages, in the proximal parts of the developing His-Purkinje system (Fig. 3.1). Quantitative investigations of *Cx43* gene products in the adult and developing mouse heart have indicated that a change in the post-transcriptional regulation of gene expression occurs 2 weeks after birth (Fromaget et al. 1990), when expression of the transcript declines very sharply, whereas the abundance of the protein is maintained.

At E8.5, EGFP expressed under the control of the *Cx40* gene is clearly visible in the heart, but its synthesis is restricted to the endocardial endothelial cells (Fig. 3.3). In the myocardium Cx40 protein is first weakly detected in the roof of the atrium and the outermost curvature of the primitive ventricle at E9.25–E9.5. Its expression amplifies and occupies the left free wall of the primitive ventricle at E10.5 (Fig. 3.3), then the right wall at E11.5 (Delorme et al. 1997; Coppen et al. 2003). Abundance of the protein peaks at E14 (Delorme et al. 1995), but remains undetectable in the AV canal and the IVS. From E14 expression of the*Cx40* gene is progressively downregulated in the ventricular compartment, and at the adult stage only the expression in the His-Purkinje system persists in this compartment (Delorme et al. 1995; Fig. 3.1). Differentiation of the His bundle and its branches, at the top and on the flanks of the developing IVS, can be followed by either immunodetection of Cx40 (Delorme et al. 1995), or by means of EGFP expressed under the control of the *Cx40* gene (Fig. 3.3). Quantitative investigations of *Cx40* gene products have shown that the expression level of the Cx40 protein reflected that of the transcript during heart development and at the adult stage (Delorme et al. 1995).

Cx45 protein has been detected as early as E8.5 in all cardiac compartments including the AV canal (Alcoléa et al. 1999; Fig. 3.2). Widespread expression of the *Cx45* gene throughout the heart at this stage was confirmed

Fig. 3.3A–E. $Cx40^{EGFP/+}$ transgenic mouse embryos. **A** E8.5 embryo observed with normal light (* indicates the heart). **B** EGFP imaging of the E8.5 embryo shown in **A**. EGFP fluorescence appears as small dots scattered throughout the heart and the aortae (*arrows*). Section (**C**) through the heart indicates that the fluorescent signals are restricted to cells that limit the lumen of the cardiac tube, i.e., the endothelial endocardial cells (*arrows*). **D** Heart isolated from a E10.5 embryo. EGFP signals, seen through the tissues, are visible in the roof of the primitive atrium (*arrowhead*), the left part of the primitive ventricle (*lpv*), and the outflow tract (*oft*) (endothelial cells). **E** Section through the heart isolated from an E14.5 embryo. Note the gradient of EGFP signal in the ventricular walls. *Arrows* indicate the His bundle branches differentiating on the flanks of the interventricular septum. *at* Atria ▶

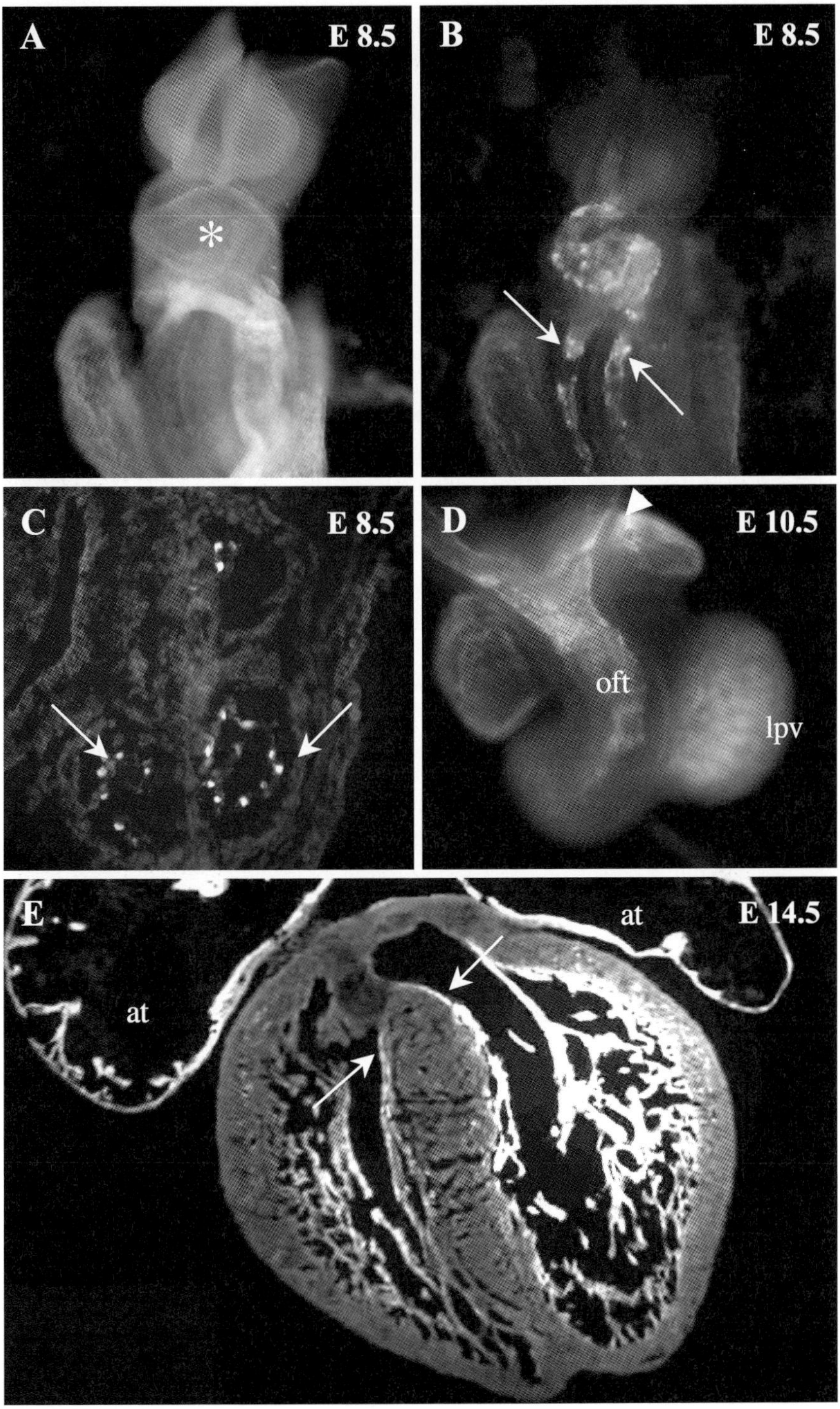
A
E 8.5
*
B
E 8.5
C
E 8.5
D
E 10.5
oft
lpv
E
at
at
E 14.5

by the investigation of $Cx45^{+/lacZ}$ mouse embryos (Kumai et al. 2000). The latter investigation also makes it possible to demonstrate that Cx45 was expressed in the endocardium endothelial cells. Widespread Cx45 expression persists at E9.5, then, as development proceeds, expression declines, but remains prominent in the developing conduction system. Downregulation of the protein occurs earlier or later according to different authors (at E10.5 for Alcoléa et al. 1999, after E12.5 for Coppen et al. 2003), these differences likely reflecting the sensitivity of the immunofluorescence techniques used. In situ hybridization techniques carried out by Alcoléa et al. (1999) have indicated that Cx45 transcript abundance decreased first in the left ventricle at E11, while Kumai et al. (2000) have noted a general downregulation of LacZ staining in $Cx45^{+/lacZ}$ embryos from E10.5. At E12.5 the transcript is weakly expressed in both ventricles, with a more distinct expression in the outflow tract and the anterior part of the right ventricle, but is no longer detected in the atria. At further stages and until E18, the Cx45 transcript is detected in the ventricular walls as a very weak and diffuse signal, the latter being slightly more pronounced in the outflow tract myocardium (Alcoléa et al. 1999). At birth and later, Cx45 immunoreactivity has been detected in the ventricular free walls, but the extent of labelling was insignificant compared with that of the conduction tissue (Coppen et al. 1998, 2001).

Investigations similar to those described above for the mouse heart have also been conducted on the developing rat heart (Van Kempen et al. 1991, 1996; Gourdie et al. 1992; Coppen et al. 1999). These studies have also demonstrated a dynamic regulation of the expression patterns of Cxs, which certainly reflects the adaptation necessary for optimum propagation of the cardiac impulse at the various stages of heart development.

3.3 Role of Cxs in Heart Development

In this section, only the role of Cxs in the heart developmental process will be examined. Their role in cardiac impulse propagation has been recently reviewed (Gros et al. 2004; Röhr 2004).

3.3.1 No Intrinsic Cardiomyocyte-Autonomous Requirement for Cx43 During Heart Development

Germline disruption of both alleles of the Cx43 genes results in the death of the mice by asphyxiation shortly after delivery. The cardiac phenotype of these knockout mice is a pulmonary outflow obstruction, which prevents the blood flow from reaching the lungs with conotruncal malformations, and defects in the patterning of the coronary arteries. These anomalies have been shown to arise from the disturbance of two migratory cell types which play a role in heart formation, and express Cx43 in wild-type mice: the cardiac neural crest cells, and proepicardium-derived cells (reviewed by Wei et al. 2004). Therefore, these results indicate there is no intrinsic cardiomyocyte-autonomous requirement for Cx43 during heart development since the defects observed come from alterations of non-myocyte lineages. Analysis of Cx43-conditional deficient mice and patients with oculodentodigital dysplasia (ODDD) corroborates this conclusion. The Cx43-conditional deficient mice which result from the crossing of $Cx43^{flox/flox}$ mice with either αMHC-Cre or MLC2v-Cre mice (expressing the Cre recombinase under the control of α-myosin heavy chain or myosin light chain v2 promoter, respectively), and in which the *Cx43* gene is thus inactivated in the cardiomyocytes only, have indeed a structurally normal heart (Gutstein et al. 2001). ODDD which is due to mutations in the *Cx43* gene (Paznekas et al. 2003; Kjaer et al. 2004) is a genetic disorder that affects many body structures and systems including the eyes, face, teeth, fingers and toes, and may also cause neurological problems. Heart troubles (seizure) in ODDD have only rarely been observed and it has not been reported that they could be due to an abnormal structure of the heart.

3.3.2
Cx40 is Involved in the Septation Process

Lethality in mice knockout for the *Cx40* gene was reported to be either low (Kirchhoff et al. 1998: Simon et al. 1998), or high (Kirchhoff et al. 2000), suggesting a strong influence of the genetic background on animal survival. The surviving mice are fertile, but a high percentage of these mice have cardiac malformations which have been described in detail in newborn and fetuses (Gu et al. 2003). In the $Cx40^{+/-}$ mice the malformations include bifid atrial appendages, ventricular septal defects, tetralogy of Fallot, and aortic arch abnormalities. In the $Cx40^{-/-}$ mice, the most common abnormalities are double-outlet right ventricle, tetralogy of Fallot, and partial endocardial cushion defects. The incidence of cardiac malformations was 18 and 33% in the heterozygous and homozygous mutant mice, respectively. The incidence of malformations reached 44%, with a predominance of conotruncal malformations (double-outlet right ventricle or tetralogy of Fallot) in the $Cx40^{-/-}$ mice resulting from the crossing of $Cx40^{-/-}$ mice. These results indicate that Cx40 is mainly involved in the septation process. Given that malformations are present only in a fraction of the animals, these data also suggest the existence of genetic modifiers that influence septation, and which either compensate for the absence of Cx40, or are modulated by Cx40 protein. In this context, it is interesting to note that Cx40 has never been detected in the adult or developing interatrial or interventricular septa.

3.3.3 Cx45 is Required for the Normal Progress of Cardiogenesis

Germline disruption of both alleles of the *Cx45* gene is lethal, and all embryos die of heart failure around E10 (Krüger et al. 2000: Kumai et al. 2000). Cardiac contractions are initiated at E8.5 in the mutant embryos, as in the wild types, but AV conduction blocks appearing at E9.5 are associated with the lack of coordination between the contractions of the primitive ventricle and those (weak, and sometimes absent) of the outflow tract (Nishii et al. 2003). The absence of Cx45 in the AV canal of the mutant embryos accounts well for the conduction blocks given the expression pattern of Cxs (Fig. 3.2), but the identification, by electron microscopy, of gap junctions in the cardiac chambers of E9 mutant embryos also indicates that Cx45 is not the only Cx expressed at this stage in the heart. Morphological abnormalities which have been observed in the heart of the mutant embryos include a looping defect, dilation of the chambers, poor or no development of trabeculae and defects in endocardial cushion formation (Kumai et al. 2000). Furthermore, the vascular development is interrupted, the placental function is impaired and there is massive apoptosis in these embryos (Krüger et al. 2000).

Disruption of the *Cx45* gene was also shown to prevent the nuclear translocation of the Ca^{2+}-dependent transcription factor NFATc1 in the endocardium (Kumai et al. 2000). It was then suggested that in the absence of Cx45 gap junction channels between the endocardial cells, a diffusion gradient of Ca^{2+} cannot be established between these cells, resulting in the inactivation of the Ca^{2+}/calcineurin/NFATc1 pathway, and consequently in the impairment of the epithelial-mesenchymal transition that is required for the formation of the endocardial cushions (Nishii et al. 2001; Crabtree and Olson 2002). This hypothesis has been strengthened by the analysis of the Cx45-conditional deficient embryos in which the *Cx45* gene was only inactivated in the myocardium (Nishii et al. 2003). These embryos which result from the crossing of $Cx45^{flox/flox}$ mice with cardiac α actin-Cre mice, still die of heart failure around E10, but show no endocardial cushion defects. However, several arguments have been developed by Armstrong and Bischoff (2004) to suggest that cardiac cushion formation in Cx45 knockout embryos may be delayed, rather than disrupted, and that the absence of Cx45 in the endocardium may well be not the only cause of cushion formation defect. Epigenetic factors such as poor myocardial function or lack of blood flow have been shown to regulate the formation of endocardial cushions (Hove et al. 2003; Bartman et al. 2004). In this respect, cardiac contraction disorder in Cx45 knockout embryos might result in poor cardiac function which in turn would be responsible for the cushion formation defects. This reasoning does not, however, seem convincing because Cx45-conditional deficient embryos which still have poor cardiac function as indicated by their contraction disorder, and loss of cardiac output, do not present cushion defects (Nishii et al. 2003). In addition, it remains to be demonstrated that NFATc1 is active in the endocardium of the Cx45-conditional deficient embryos. So clearly the cushion formation defects associated with inactivation of the *Cx45* gene have not yet been fully elucidated. To circumvent the influence of the hemodynamic factors, Egashira et al. (2004) have investigated cardiomyocytes differentiated from $Cx45^{-/-}$ ES cells. The pulse rate of $Cx45^{-/-}$ cardiomyocytes was irregular and higher than that of wild-type cardiomyocytes. In addition, recording of electrical activity propagation between $Cx45^{-/-}$ cardiomyocytes has revealed uncoupling phenomena between groups of cells. Analysis of the phenotypes of $Cx45^{-/-}$ cardiomyocytes and Cx45-conditional deficient embryos therefore indicates that Cx45 is required for coordinated contraction of cardiomyocytes and normal progress of cardiogenesis, respectively. One question then raised is whether the coordinated contraction of cardiomyocytes is necessary for the normal progress of cardiogenesis.

3.4
Transcriptional Control of Cardiomyocyte-Related Cxs

The spatiotemporal expression patterns of Cx43, -40 and -45, observed in the developing and adult mammalian myocardium, originate from the different transcriptional programs which govern their respective genes. All genes have a regulatory region which, through its unique nucleotide sequence, is able to interact with specific transcription factors. The regulatory region of a gene is believed to be made up of independent modules. Each module which contains a number of transcription factor binding sites unique to that gene, is able to drive part of the total overall transcriptional program (Durocher and Nemer 1998; Habets et al. 2003). Thus, in order to understand the mechanisms responsible for the expression pattern of Cxs, and their modulations, in the developing and adult heart, it is necessary to identify both the regulatory modules of each gene and the transcription factors that interact with them. Given the complexity of the expression pattern of the cardiomyocytes-related Cxs, it is a long and arduous task. Knowledge of the Cx43, -40 and -45 gene structure and the control of their regulatory regions in cardiomyocytes is summarized in this section.

Mouse models in which transcription factors, modulating the expression of cardiomyocyte-related Cxs, have been modified are also discussed.

3.4.1 Structure and Regulation of the Cx43 Gene

Structure of the Gene

The structures of the mouse (Sullivan et al. 1993; Pfeifer et al. 2004), rat (Yu et al. 1994) and human (Fishman et al. 1991; Geimonen et al. 1996) *Cx43* genes, with their splicing patterns are represented in Fig. 3.4A. Initially, the *Cx43* gene was reported to consist of two exons (exon1A and exon2 for mouse and rat, exon1 and exon2 for human; Fig. 3.4A), the transcripts of which were detected in the heart of all three species (Sullivan et al. 1993; De Leon et al. 1994; Yu et al. 1994). Exon2 was shown to contain the complete sequence coding for the protein as well as the 3'-untranslated region (UTR), and a small part of the 5'-UTR, whereas the first exon contains the remainder of the 5'-UTR. Recently, eight new transcripts from the mouse *Cx43* gene have been identified in various tissues, including the heart (Pfeifer et al. 2004). These transcripts differ by their 5'-UTRs, indicating that these genes consist of six rather than two exons, five of them coding exclusively 5'-UTR sequences (Fig. 3.4A). In addition to the transcript made up of exon1A/exon2 (1A/2) sequences, the following transcripts were detected by RT-PCR in the mouse heart: $1A_S/2$, $1A_L/2$, 1A/1E/2, $1B_S/2$, $1B_S/1D/2$, $1B_L/2$, 1C/2 and 1C/1D/2. The subscripts S and L refer to shorter or longer versions of a particular exon, the relative sizes of which are indicated by horizontal lines in Fig. 3.4A. Six transcripts (1A/2, $1A_S/2$, $1A_L/2$, 1B/2, $1C_S/2$ and $1C_L/2$) have also been detected in the rat heart, indicating for this species the presence of four exons at least in the *Cx43* gene of which three contain only 5'-UTR sequences (Fig. 3.4A). No transcript other than that derived from exon1/exon2 has been reported for the human *Cx43* gene yet (Fig. 3.4A), but it is highly likely that the organization of this gene is just as complex as that of the mouse and rat genes. The existence of these different transcripts indicates the presence of alternative splicing events and of three different promoters (P1, P2 and P3 in Fig. 3.4A) in the mouse and rat *Cx43* genes. The translational efficiency of these transcripts was postulated to differ depending on their 5'-UTR sequences, thus suggesting post-transcriptional regulatory mechanisms. The functional activity of the mouse P1, P2 and P3 promoters was confirmed by promoter/reporter analyses in noncardiac cell types, the promoter P1 proving to be the most active (Pfeifer et al. 2004). In addition, P1 was shown to be active throughout the mouse heart, whereas P2 and P3 have more regionalized activities. The role

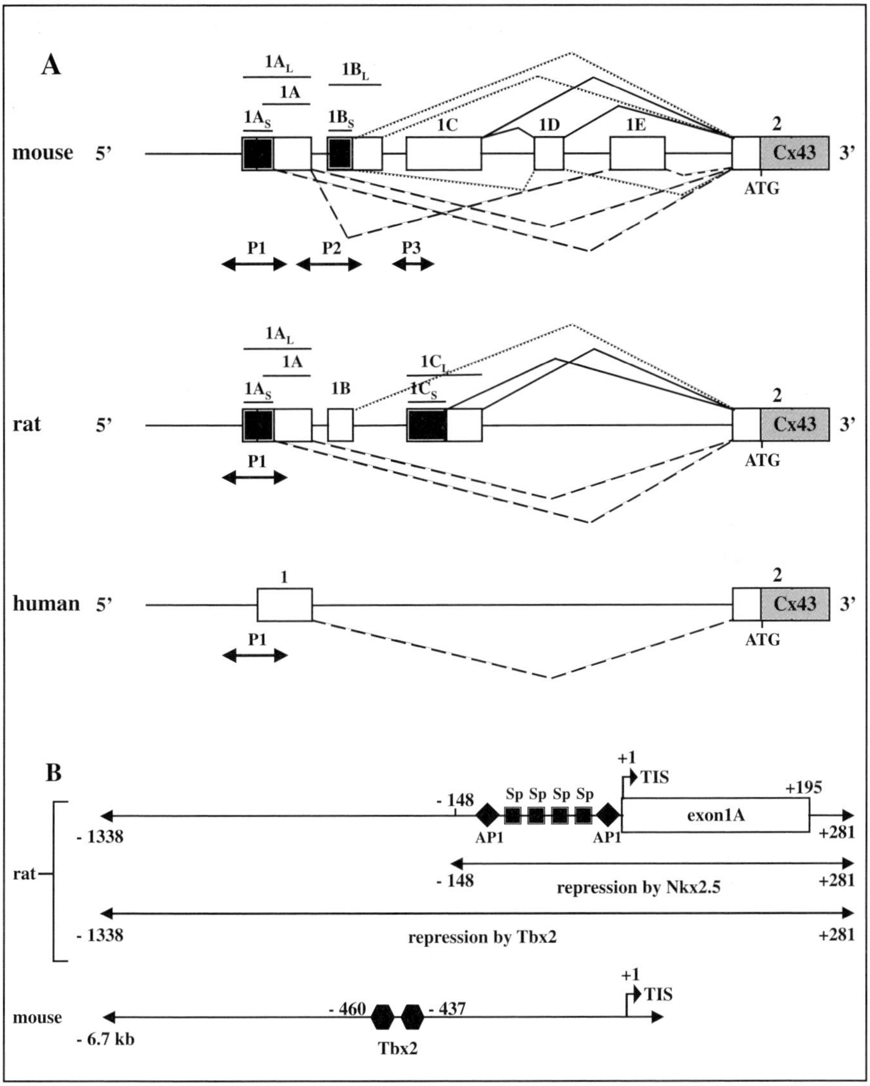

Fig. 3.4. A Organization and splicing patterns of the mouse, rat and human *Cx43* genes. *Boxes* represent exon sequences. Corresponding exons between species are placed above each other. The Cx43-coding sequence (*Cx43*) is shown in *gray* in exon2. The *subscripts L* and *S* refer to the long and short versions, respectively, of one particular exon, whereas *horizontal lines* indicate the sizes of that exon. *Double-headed horizontal arrows* indicate regions with reported promoter activity. Exons and introns are not to scale. Based on current sequence information in the NCBI (http://www.ncbi.nlm.nih.gov) genome databases mouse exon1A, rat exon1A, and human exon1 are localized at 10.1, 9.3, and 11.0 kb from exon2, respectively. The nomenclature of exons of the rat and mouse *Cx43* genes is that of Pfeifer et al. (2004). **B** Regulatory regions and binding elements involved in mouse and rat Cx43 promoter (P1) regulation in cardiac cells or by cardiac transcription factors. The transcription initiation (*TIS*) site reflects the start of exon1A and is defined here as map position +1; regulatory regions are numbered relative to this position

of these different promoters is still far from clearly understood; indeed P1 is to date the only one the regulation of which has been studied in detail in cardiomyocytes and other cell types.

Do Tbx2 and Nkx2–5 Regulate the Expression of the *Cx43* Gene?

De Leon et al. (1994) were the first to map the human Cx43 P1 promoter in rat primary fetal cardiomyocytes and in whole hearts to the (-175, +143) region, relative to the transcription initiation site (TIS, defined as +1; start of exon1 in Fig. 3.4A). Teunissen et al. (2003) extended these findings by mapping the rat Cx43 P1 promoter in rat primary neonatal ventricular myocytes and other cell types to the (−148, +281) region relative to the start of exon1A (Fig. 3.4A). Moreover, evidence was provided that the transcription factors Sp1/Sp3 and AP1 were involved in promoter activation through four Sp and two AP1 binding elements, respectively (Fig. 3.4B). Since these transcription factors are ubiquitously expressed and the promoter activation was similar in both cardiomyocytes and noncardiac cell types, they presumably provide for a basal level of transcription. Other transcription factors are most likely necessary to determine whether or not Cx43 is expressed in particular cell types. With respect to expression of Cx43 in cardiomyocytes, the roles of members of the T-box family and of Nkx2–5, transcription factors that are important for normal cardiogenesis, have been investigated (Bruneau 2002; Moorman and Christoffels 2003).

Both the rat and mouse P1 promoters appear to be responsive to Tbx2, a T-box factor involved in the local repression of chamber-specific gene expression and chamber differentiation in the developing heart. The rat Cx43 (−1338, +281) promoter, for instance, was repressed 2.4-fold in Cos-7 cells by Tbx2 (Christoffels et al. 2004). Similar results were reported in NIH3T3 fibroblasts and ROS 17/2.8 osteoblastic cells for the mouse Cx43 P1 promoter (containing approximately 6.7 kb of upstream sequences, see Fig. 3.4B) upon Tbx2 overexpression (Chen et al. 2004). In the latter study, evidence was provided that two interspecies conserved binding sites located in the (-460,-437) region mediated this repression both in vitro and in vivo. In the osteoblastic cells endogenous Cx43 protein expression was downregulated upon Tbx2 overexpression. Moreover, the spatiotemporal expression pattern of Tbx2 in normal embryonic mouse hearts was found to be complementary to that of Cx43 (Christoffels et al. 2004). Altogether, these results suggest that Cx43 is a transcriptional target of Tbx2.

The effect of modified Nkx2–5 levels on Cx43 expression in the heart has been extensively studied in both mouse models and cultured cardiomyocytes, but a direct role of Nkx2–5 in the in vivo regulation of the *Cx43* gene has not yet been demonstrated. In transgenic mice overexpressing, under control of the β-MHC promoter, a DNA nonbinding mutant of

Nkx2–5 [Nkx2–5(I183P)], expression of Cx43 in the heart was normal at birth, but decreased gradually during the first 3 weeks after birth (Kasahara et al. 2001). In mice expressing the mutant gene, under control of the α-MHC promoter, downregulation of Cx43, assessed in isolated cardiomyocytes, was only observed at 1 year of age. Recombinant adenoviral-mediated overexpression of Nkx2–5(I183P) in adult ventricular myocytes did not affect endogenous Cx43 protein and RNA expression levels (Kasahara et al. 2003). In contrast, transgenic mice expressing, under control of the α-MHC promoter, wild-type Nkx2–5 (Nkx2–5-wild) or a putative transcriptionally active variant of Nkx2–5 (Nkx2–5-ΔC) displayed downregulation of Cx43 expression in the cardiomyocytes at early postnatal developmental stages. Moreover, downregulation of Cx43 protein and RNA was also observed in both rat neonatal and adult ventricular myocytes upon adenoviral-mediated overexpression of Nkx2–5-wild (Kasahara et al. 2003; Teunissen et al. 2003). Under conditions of overexpression of Nkx2–5-wild, rat Cx43 (−148, +281) P1 promoter activity was reduced twofold relative to the controls. Thus, the studies dealing with overexpression of wild-type Nkx2–5 or mutated derivatives of Nkx2–5 in cultured cardiomyocytes or transgenic mice suggest that this factor may function, directly or indirectly, as a transcriptional regulator of the Cx43 promoter. However, analysis of the phenotype of Nkx2–5-deficient mice does not support this hypothesis.

The expression level of the Cx43 protein was found to be similar in the atria and ventricles of both heterozygous Nkx2–5 knockout mice (Nkx2–$5^{+/-}$ mice) and wild-type mice (Jay et al. 2003). In the heart of homozygous Nkx2–5 knockout mouse embryos (9 days postconception), a slight downregulation of the RNA level was detected (Linhares et al. 2004). However, in a ventricular-restricted Nkx2–5 knockout mouse model, no downregulation of Cx43 RNA was observed in the adult ventricular myocardium (Pashmforoush et al. 2004). Clearly, further studies are necessary to determine whether or not Nkx2–5 is really involved in the in vivo regulation of the *Cx43* gene in the heart.

3.4.2
Structure and Regulation of the Cx40 Gene

Structure of the Gene

The schematic structure and the splicing patterns of the mouse (Seul et al. 1997; Anderson et al. 2005), rat (Groenewegen et al. 1998) and human (Dupays et al. 2003) *Cx40* genes are illustrated in Fig. 3.5A. Four, two and three exons have so far been identified in the mouse, rat and human *Cx40* genes, respectively. For the three species the complete sequence coding

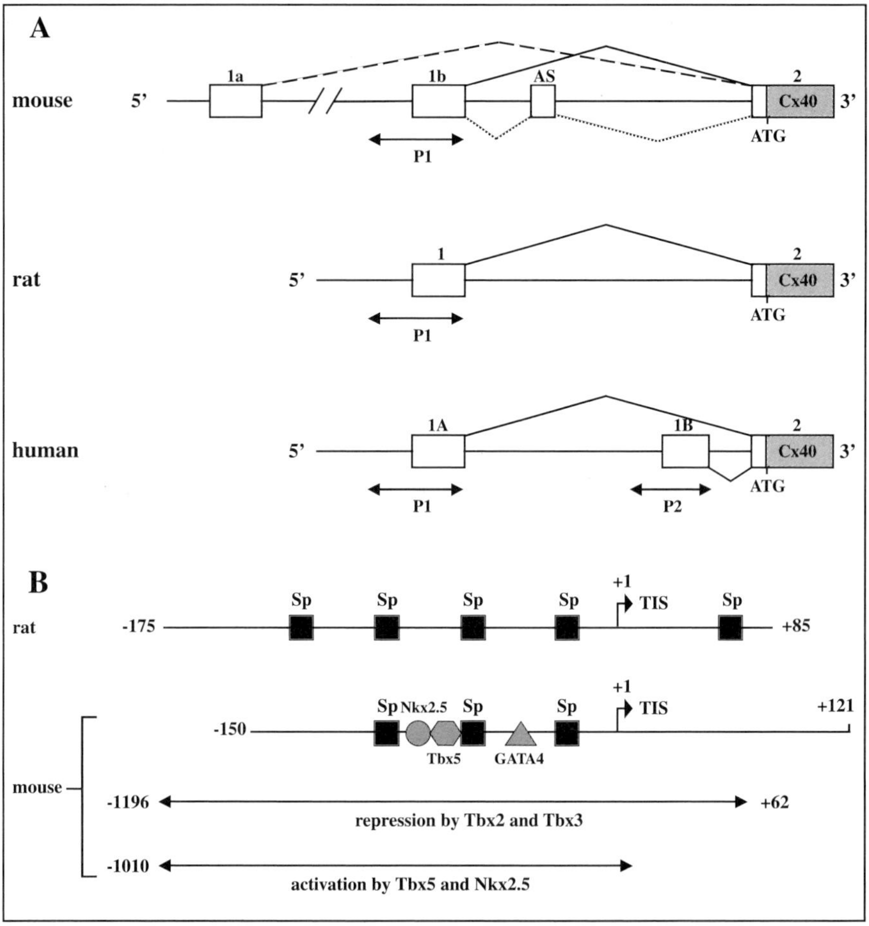

Fig. 3.5. **A** Organization and splicing patterns of the mouse, rat and human *Cx40* gene. The legends are similar to those of Fig. 3.4A. Based on current sequence information in the NCBI genome databases, mouse exon1b, rat exon1, and human exon1A are localized at 18.1, 16.7 and 14.0 kb from exon2, respectively. Mouse exon1a is localized 60.2 kb upstream of exon 1b. The nomenclatures of exons of the mouse and human *Cx40* genes are those of Anderson et al. (2005), and Dupays et al. (2003), respectively. **B** Regulatory regions and binding elements involved in Cx40 promoter (P1) regulation in cardiac cells or by cardiac transcription factors. The transcription initiation (*TIS*) site reflects the start of rat exon1 or mouse exon1b and is defined as map position +1; regulatory regions are numbered relative to this position

for the protein as well as a small part of the 5'- and the complete 3'-UTR are localized in one single exon (exon2). Alternative splicing mechanisms are responsible for the occurrence of multiple transcripts, which differ only by their 5'-UTR. For instance, three different transcripts (1a/2, 1b/2 and 1b/AS/2, see Fig. 3.5A) have been identified in various mouse tissues, including the heart. Two different transcripts (1A/2 and 1B/2) have been identified in the human heart, whereas to date only one (1/2) has been found in the heart and various other tissues of the rat. The organization of the human and mouse *Cx40* genes suggests the presence of more than one promoter, which has indeed been confirmed for the human gene by promoter/reporter analyses (promoters P1 and P2; Dupays et al. 2003). The mouse and rat Cx40 P1 promoters have been mapped and further characterized in cardiomyocytes and other cell types.

Regulation of *Cx40* Gene Expression: In Vitro Investigations

The P1 promoter of the rat *Cx40* gene has been defined to the (−175, +85) region (relative to the start of exon1) in rat primary neonatal ventricular myocytes (Teunissen et al. 2002). Within this region, five Sp binding elements which bind both Sp1 and Sp3 contribute to the promoter activation to provide for presumably a basal level of transcription (Fig. 3.5B). The mouse Cx40 P1 promoter was initially mapped to the (−293, +121) region in the BWEM cell line derived from rat fetal cardiomyocytes (Seul et al. 1997), and recently further defined to the (−150, +121) region in primary rat neonatal cardiomyocytes (Linhares et al. 2004; Fig. 3.5B). As for the rat, the functionality of the conserved Sp binding elements in the murine promoter was investigated by electrophoretic mobility shift assays and confirmed. Moreover, the cardiac transcription factors Nkx2–5 and GATA4 were shown to interact with their respective binding sites in the P1 promoter region (Fig. 3.5B) and to cooperatively activate it in CH3 10T1/2 fibroblasts. The T-box factor Tbx5 was also shown to bind to this region and to be able to repress the activation by both Nkx2–5 and GATA4, but was not able to activate the murine Cx40 (−150, +121) promoter by itself, or in combination with Nkx2–5. This is especially interesting since it has been reported previously that Tbx5 alone or in cooperation with Nkx2–5 could activate a murine Cx40 regulatory fragment containing 1010 base pairs of 5'-upstream sequence (relative to the start of exon1b) in CV-1 cells (Bruneau et al. 2001). Whether this discrepancy is due to the use of promoter fragments of different lengths, or of distinct cell types used for the activation experiments, remains to be determined. Other members of the T-box family have also been implicated in mouse Cx40 promoter regulation. For instance, Tbx20 is able to activate the (−150, +121) promoter region in CH3 10T1/2 fibroblasts by itself or in cooperation with Nkx2–5 and members of the GATA family (GATA4, −5 or −6; Stennard et al. 2003). Binding of Tbx20 to its binding site in this region could not be demonstrated, however. In addition, both Tbx2 and Tbx3 were shown to be able to repress the mouse Cx40 (−1196, +62) promoter approximately fourfold in Cos-7 cells (Christoffels et al. 2004; Hoogaars et al. 2004), but the target binding sites for these factors have not been identified yet. In summary, the mouse P1 promoter appears to be regulated by several transcription factors, including Sp1, Nkx2–5, and several members of the GATA and T-box families. However, it should be kept in mind that these studies have been performed on cell lines, in vitro, with relatively short regulatory fragments. That the above transcription factors are also involved in the in vivo regulation of the *Cx40* gene remains to be demonstrated. Investigation of various mouse models have provided some information on the involvement of these factors in this regulation.

Regulation of *Cx40* Gene Expression: in Vivo Investigations

Tbx2 is expressed in regions of the embryonic heart tube that do not differentiate into the cardiac chambers and its spatiotemporal expression pattern in embryonic mouse hearts was found to be mutually exclusive to that of Cx40. Moreover, transgenic embryos expressing Tbx2, under control of the β-MHC promoter, in prechamber myocardium did not develop cardiac chambers, nor did they express Cx40. A similar complementary expression pattern was also reported for Cx40 and Tbx3, a transcriptional repressor delineating the developing central conduction system (Hoogaars et al. 2004). The latter results, along with those mentioned above on the repressive effect of both Tbx2 and Tbx3 overexpression on mouse Cx40 P1 promoter activity, suggest that these two transcription factors may be involved in the in vivo regulation of *Cx40* gene expression.

Mutations in the *Tbx5* gene have been linked to Holt-Oram syndrome, a human disorder characterized by both skeletal and cardiac congenital defects, among them conduction abnormalities (Packham and Brook 2003). As discussed above, Tbx5 is able to bind to, and regulate in vitro the mouse Cx40 P1 promoter. Expression of the Cx40 transcript in $Tbx5^{+/-}$mice was undetectable in embryonic hearts (atria, left ventricle) and was only 10% of the normal level in the adult atria (Bruneau et al. 2001), suggesting a role for Tbx5 in the in vivo regulation of the *Cx40* gene. However, the morphological and functional defects which affect the conduction system of these mice are not the result of deficient Cx40 expression, but reflect the abnormal development of the conduction system, as has been demonstrated recently (Moskowitz et al. 2004).

Alterations in Cx40 expression have also been reported in mouse models and cardiomyocytes with modified levels of wild-type or mutated Nkx2–5. Neonatal mice overexpressing in the heart Nkx2–5(I183P), under control of the β-MHC promoter, displayed normal Cx40 RNA and protein levels in their atria, but 3 weeks after birth both the RNA and protein levels were markedly downregulated in this tissue (Kasahara et al. 2001).

Adenoviral-mediated overexpression of this mutant in cultured adult rat atrial cardiomyocytes, however, did not alter the Cx40 expression level, whereas overexpression of Nkx2–5-wild abrogated Cx40 expression (Kasahara et al. 2003). In the hearts of homozygous Nkx2–5 knockout mouse embryos (9 days postconception) RT-PCR analysis has revealed a substantial reduction in Cx40 RNA expression (Linhares et al. 2004; Dupays et al. unpublished data). Heterozygous Nkx2–5 knockout mice, which express half as much Nkx2–5 mRNA in their hearts as the corresponding wild-type littermates, have reduced levels of Cx40 protein in the atria but normal Cx40 expression in conductive cardiomyocytes (Jay et al. 2003, 2004). The conduction defects observed in these mice cannot therefore be attributed to abnormalities in Cx40 expression in the conduction system. They are in fact the result of the hypoplastic development of the conduction system (Jay et al. 2004). In ventricular-restricted Nkx2–5 knockout mice Cx40 expression was relatively well-preserved in the heart of neonates, but virtually absent in the AV node and the His bundle of adult mice (Pashmforoush et al. 2004). Again, this lack of Cx40 expression reflects the role of Nkx2–5 in the postnatal maturation and maintenance of the conduction system, and results from the degeneration of conductive cells due to the loss of Nkx2–5. Altogether, these studies on mutant mice suggest that Nkx2–5 may be involved in the regulation of the *Cx40* gene in the atria, but do not provide any clues to its role in the regulation of this gene in the conduction system.

3.4.3 Structure of the Cx45 Gene

To date, only the genomic organization of the mouse *Cx45* gene has been reported (Baldridge et al. 2001; Jacob and Beyer 2001; Anderson et al. 2005). Five exons have been identified (Exons1a, 1b, 1c, 2 and 3) (Fig. 3.6). Exon3

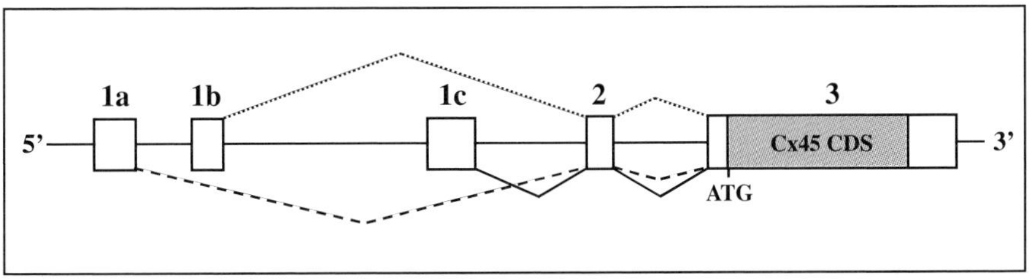

Fig. 3.6. Organization and splicing pattern of the mouse *Cx45* gene. The legends are similar to those of Fig. 3.4A. The Cx45-coding sequence (*Cx45 CDS*) is shown in *grey* in exon3. The distance between the 3'-end of exon1a and the beginning of exon3 is approximately 18.3 kb. The nomenclature of the exons is that of Anderson et al. (2005)

contains a small part of the 5'-UTR, the complete sequence coding for the protein, and the 3'-UTR. RT-PCR analyses have indicated the presence of four different transcripts (1a/2/3, 1b/2/3, 1c/2/3 and 2/3) in several mouse tissues, which only differed in the nature of their 5'-UTR. No reports describing the mapping and the characterization of the Cx45 promoter(s) have been published yet.

3.5 Conclusions

The repertoire of Cxs expressed in the heart in general, and in the cardiomyocytes in particular is not yet complete. Expression of the *Cx43*, *-40* and *-45* genes is determinant for the harmonious development of the heart. The example of Cx43 expressed in the cardiac neural crest cells and proepicardium-derived cells indicates that the cardiomyocyte-related Cxs are not the only Cxs required for the normal development of the heart. Investigations on the transcriptional regulation of *Cx43* and *-40* gene expression in the heart are making slow progress. Tbx5 and Nkx2–5, two transcription factors whose mutations are responsible for cardiac congenital defects, are involved in the in vivo regulation of the *Cx40* gene in the atria. Both factors are in addition involved in the postnatal maturation of the conduction system, thus illustrating the complexity of the relationships between the development of the heart and the regulation of the Cx genes.

Acknowledgements. The research work carried out by the authors was supported by the CNRS, the Université de la Méditerranée, the Ministère de l'Éducation et de la Recherche (ACI "Biologie du Développement et Physiologie Intégrative") (DBG), INSERM (Programme National de recherches sur les maladies cardiovasculaires) (DBG), the Fondation de France (DBG), the Association Française contre les Myopathies (MT-R) and the Netherlands Heart Foundation.

References

Alcoléa S, Théveniau-Ruissy M, Jarry-Guichard T, Marics I, Tzouanacou E, Chauvin JP, Briand JP, Moorman AF, Lamers WH, Gros DB (1999) Downregulation of connexin 45 gene products during mouse heart development. Circ Res 84:1365–1379

Alexopoulos H, Böttger A, Fischer S, Levin A, Wolf A, Fujisawa T, Hayakawa S, Gojobori T, Davies JA, David CN, Bacon JP (2004) Evolution of gap junctions: the missing link. Current Biol 14:R79–R880

Anderson CL, Zundel MA, Werner R (2005) Variable promoter usage and alternative splicing in five mouse connexin genes. Genomics 85:238–244

Armstrong EJ, Bischoff J (2004) Heart valve development. Endothelial cell signaling and differentiation. Circ Res 95:459–470

Baldridge D, Lecanda F, Shin C-S, Stains J, Civitelli R (2001) Sequence and structure of the mouse connexin45 gene. Biosci Rep 21:683–689

Baranova A, Ivanov D, Petrash N, Pestova A, Skoblov M, Kelmanson I, Shagin D, Nazarenko S, Geraymovych E, Litvin O, Tiunova A, Born TL, Usman N, Staroverov D, Lukyanov S, Panchin Y (2004) The mammalian pannexin family is homologous to the invertebrate innexin gap junction proteins. Genomics 83:706–716

Bartman T, Walsh EC, Wen K-K, McKane M, Ren J, Alexander J, Rubenstein PA, Stainier DY (2004) Early myocardial function affects endocardial cushion development in zebrafish. PLoS Biol 2:673–681

Beyer EC, Paul DL, Goodenough DA (1987) Connexin 43, a protein from rat heart homologous to gap junction protein from liver. J Cell Biol 195:2621–2629

Bruneau BG (2002) Transcriptional regulation of vertebrate cardiac morphogenesis. Circ Res 90:509–519

Bruneau BG, Nemer G, Schmitt JP, Charron F, Robitaille L, Caron S, Conner DA, Gessler M, Nemer M, Seidman CE, Seidman JG (2001) A murine model of Holt-Oram syndrome defines roles of the T-box transcription factor Tbx5 in cardiogenesis and disease. Cell 106:709–721

Bruzzone R, Hormuzdi SG, Barbe MT, Herb A, Monyer H (2003) Pannexins, a family of gap junction proteins expressed in brain. Proc Natl Acad Sci USA 100:13644–13649

Camelliti P, Borg TK, Kohl P, (2005) Structural and functional characterization of cardiac fibroblasts. Cardiovasc Res 65:40–51

Chen J-R, Chatterjee B, Meyer R, Yu JC, Borke JL, Isales CM, Kirby ML, Lo CW, Bollag RJ (2004) Tbx2 represses expression of connexin43 in osteoblastic-like cells. Calcif Tissue Int 74:561–573

Christoffels VM, Hoogaars WMH, Tessari A, Clout DEW, Moorman AFM, Campione M (2004) T-box transcription factor Tbx2 represses differentiation and formation of the cardiac chambers. Dev Dyn 229:763–770

Coppen SR, Dupond E, Rothery S, Severs NJ (1998) Connexin45 expression is preferentially associated with the ventricular conduction system in mouse and rat heart. Circ Res 82:232–243

Coppen SR, Severs NJ, Gourdie RG (1999) Connexin45 ($\alpha 6$) expression delineates an extended conduction system in the embryonic and mature rodent heart. Dev Genet 24:82–90

Coppen SR, Gourdie RG, Severs NJ (2001) Connexin 45 is the first connexin to be expressed in the central conduction system of the mouse heart. Exp Clin Cardiol 6:17–23

Coppen SR, Kaba RA, Halliday D, Dupont E, Skepper JN, Elneil S, Severs NJ (2003) Comparison of connexin expression patterns in the developing mouse heart and human foetal heart. Mol Cell Biochem 242:121–127

Crabtree GR, Olson EN (2002) NFAT signaling: choreographing the social lives of the cells. Cell 109 suppl:S67–S79

De Leon JR, Buttrick PM, Fishman GI (1994) Functional analysis of the connexin43 gene promoter in vivo and in vitro. J Mol Cell Cardiol 26:379–389

Delorme B, Dahl E, Jarry-Guichard T, Marics I, Briand JP, Willecke K, Gros D, Théveniau-Ruissy M (1995) Developmental regulation of connexin 40 gene expression in mouse heart correlates with the differentiation of the conduction system. Dev Dynamics 20:358–371

Delorme B, Dahl E, Jarry-Guichard T, Briand JP, Willecke K, Gros D, Théveniau-Ruissy M (1997) Expression pattern of connexin gene products at the early developmental stages of the mouse cardiovascular system. Cir Res 81:423–437

Dupays L, Mazurais D, Rücker-Martin C, Calmels T, Bernot D, Cronier L, Malassiné A, Gros D, Théveniau-Ruissy M (2003) Genomic organization and alternative transcripts of the human connexin40 gene. Gene 305:79–90

Durocher D, Nemer M (1998) Combinatorial interactions regulating cardiac transcription. Dev Genet 22:250–262

Egashira K, Nishii K, Nakamura K-I, Kumaï M, Morimoto S, Shibata Y (2004) Conduction abnormality in gap junction protein connexin45-deficient embryonic stem cell-derived cardiac myocytes. Anat Rec 280A:973–979

Eastman SD, Iovine MK (2004) Identification of 36 connexin genes in Zebrafish. Mol Biol Cell 15 (suppl):p 268a

Evans WH, Martin PE (2002) Gap junctions: structure and function. Mol Membr Biol 19:121–136

Fishman GI, Eddy RL, Shows TB, Rosenthal L, Leinwand LA (1991) The human connexin gene family of gap junction proteins: distinct chromosomal locations but similar structures. Genomics 10:250–256

Franco D, Icardo JM (2001) Molecular characterization of the ventricular conduction system in the developing mouse heart: topographical correlation in normal and congenitally malformed hearts. Cardiovasc Res. 50:610–612

Fromaget C, El Aoumari A, Dupont E, Briand JP, Gros D (1990) Changes in the expression of connexin 43, a cardiac gap junctional protein, during mouse heart development. J Mol Cell Cardiol 22:1245–1258

Geimonen E, Jiang W, Ali M, Fishman GI, Garfield RE, Andersen J (1996) Activation of protein kinase C in human uterine smooth muscle induces *connexin-43* gene transcription through an AP-1 site in the promoter sequence. J Biol Chem 271:23667–23674

Gourdie RG, Green CR, Severs NJ, Thompson RP (1992) Immunolabelling patterns of gap junction connexins in the developing and mature rat heart. Anat Embryo 185:363–378

Groenewegen WA, Van Veen TAB, Van der Velden HMW, Jongsma HJ (1998) Genomic organization of the rat connexin40 gene: identical transcription start sites in heart and lung. Cardiovasc Res 38:463–471

Gros D, Jongsma HJ (1996) Connexins in mammalian heart function. BioEssays 18:719–730

Gros D, Dupays L, Alcoléa S, Meysen S, Miquerol L, Théveniau-Ruissy M (2004) Genetically-modified mice: tools to decode the roles of connexins in the heart, new models for cardiovascular research. Cardiovasc Res 62:299–308

Gu H, Smith FC, Taffet SM, Delmar M. (2003) High incidence of cardiac malformations in connexin40-deficient mice. Circ Res 93:201–206

Gutstein DE, Morley GE, Tamaddon H, Vaidya D, Schneider MD, Chen J, Chien KR, Stuhlmann H, Fishman GI (2001) Conduction slowing and sudden arrhythmic death in mice with cardiac-restricted inactivation of connexin43. Circ Res 88:333–339

Habets PEMH, Moorman AFM, Christoffels VM (2003) Regulatory modules in the developing heart. Cardiovasc Res 58:246–263

Haefliger JA, Nicod P, Meda P (2004) Contribution of connexins to the function of the vascular wall. Cardiovasc Res 62:345–356

Harris AL (2001) Emerging issues of connexin channels: biophysics fills the gap. Quart Rev Biophysics 34:325–472

Hoogaars WMH, Tessari A, Moorman AFM, De Boer PAJ, Hagoort J, Soufan AT, Campione M, Christoffels VM (2004) The transcriptional repressor Tbx3 delineates the developing central conduction system of the heart. Cardiovasc Res 62:489–499

Hove JR, Köster RW, Forouhar AS, Acevedo-Bolton G, Fraser SE, Gharib M (2003) Intracardiac fluid forces are an essential epigenetic factor for embryonic cardiogenesis. Nature 421:172–177

Inai T, Mancuso M, McDonald DM, Kobayashi J, Nakamura K, Shibata Y (2004) Shear stress-induced upregulation of connexin 43 expression in endothelial cells on upstream surfaces of rat cardiac valves. Histochem Cell Biol 122:477–483

Jacob A, Beyer EC (2001) Mouse connexin 45: genomic cloning and exon usage. DNA Cell Biol 20:11–19

Jay PY, Berul CI, Tanaka M, Ishii M, Kurachi Y, Izumo S (2003) Cardiac conduction and arrhythmia: insights from Nkx2–5 mutations in mouse and humans. In: Goode J (ed) Development of the cardiac conduction system, Novartis Foundation, Symposium 250, Wiley, Chichester, pp 227–241

Jay PY, Harris BS, Maguire CT, Buerger A, Wakimoto H, Tanaka M, Kupershmidt S, Roden DM, Schultheiss TM, O'Brien TX, Gourdie RG, Berul CI, Izumo S (2004) *Nkx2–5* mutation causes anatomic hypoplasia of the cardiac conduction system. J Clin Invest 113:1130–1137

John S, Cesario D, Weiss JN (2003) Gap junctional hemichannel in the heart. Acta Physiol Scand 179:23–31

Johnson CM, Kanter EM, Green KG, Laing JG, Betsuyaku T, Beyer EC, Steinberg TH, Saffitz JE, Yamada KA (2002) Redistribution of connexin45 in gap junctions of connexin43-deficient hearts. Cardiovasc Res 53:921-935

Kasahara H, Wakimoto H, Liu M, Maquire CT, Converso KL, Shioi T, Huang W-Y, Manning WJ, Paul D, Lawitts J, Berul CI, Izumo S (2001) Progressive atrioventricular conduction defects and heart failure in mice expressing a mutant Csx/Nkx2–5 homeoprotein. J Clin Invest 108:189–201

Kasahara H, Ueyama T, Wakimoto H, Liu MK, Maguire CT, Converso KL, Kang PM, Manning WJ, Lawitts J, Paul DL, Berul CI, Izumo S (2003) Nkx2–5 homeoprotein regulates expression of gap junction protein connexin43 and sarcomere organization in postnatal cardiomyocytes. J Mol Cell Cardiol 35:243–256

Kirchhoff S, Nelles E, Hagendorff A, Krüger O, Traub O, Willecke K (1998) Reduced cardiac conduction velocity and predisposition to arrhythmias in connexin40-deficient mice. Curr Biol 8:299–302

Kirchhoff S, Kim J-S, Hagendorff A, Thönnissen E, Krüger O, Lamers WH, Willecke K (2000) Abnormal cardiac conduction and morphogenesis in connexin40 and connexin43 double-deficient mice. Circ Res 87:399–405

Kjaer KW, Hansen L, Eiberg H, Leicht P, Opitz JM, Tommerup N (2004) Novel connexin 43 (GJA1) mutation causes oculo-dento-digital dysplasia with curly hair. Am J Med Genet 127A:152–157

Krüger O, Plum A, Kim JS, Winterhager E, Maxeiner S, Hallas G, Kirchhoff S, Traub O, Lamers WH, Willecke K (2000) Defective vascular development in connexin 45-deficient mice. Development 127:4179–4193

Kumai M, Nishii K, Nakamura K-I, Takeda N, Suzuki M, Shibata Y (2000) Loss of connexin45 causes a cushion defect in early cardiogenesis. Development 127:3501–3510

Linhares VLF, Almeida NAS, Menezes DC, Elliott DA, Lai D, Beyer EC, Campos de Carvalho AC, Costa MW (2004) Transcriptional regulation of the murine *connexin40* promoter by cardiac factors Nkx2–5, GATA4 and Tbx5. Cardiovasc Res 64:402–411

Manthey D, Bukauskas F, Lee CG, Kozak CA, Willecke K (1999) Molecular cloning and functional expression of the mouse gap junction gene connexin-57 in human HeLa cells. J Biol Chem 27:14716–14723

Martin PE, Evans WH (2004) Incorporation of connexins into plasma membranes and gap junctions. Cardiovasc Res 62:378–387

Miquerol L, Dupays L, Théveniau-Ruissy M, Alcoléa S, Jarry-Guichard T, Abran P, Gros D (2003) Gap junctional connexins in developing mouse cardiac conduction system. In: Goode J (ed) Development of the cardiac conduction system, Novartis Foundation, Symposium 250, Wiley, Chichester, UK, pp 80–98

Miquerol L, Meysen S, Mangoni M, Bois P, van Rijen HVM, Abran P, Jongsma HJ, Nargeot J, Gros D (2004) Architectural and functional asymmetry of the His-Purkinje system of the murine heart. Cardiovasc Res 63:77–86

Moreno AP (2004) Biophysical properties of homomeric and heteromultimeric channels formed by cardiac connexins. Cardiovasc Res 62:276–286

Moorman AFM, Christoffels VM (2003) Cardiac chamber formation: development, genes, and evolution. Physiol Rev 83:1223–1267

Moskowitz IPG, Pizard A, Patel VV, Bruneau BG, Kim JB, Kupershmidt S, Roden D, Berul CI, Seidman CE, Seidman JG (2004) The T-box transcription factor Tbx5 is required for the patterning and maturation of the murine cardiac conduction system. Development 131:4107–4116

Nag AC (1980) Study of non-muscle cells of the adult mammalian heart: a fine structural analysis and distribution. Cytobios 28:41–61

Nielsen PA, Kumar NM (2003) Differences in expression patterns between mouse connexin-30.2 (Cx30.2) and its putative human orthologue, connexin-31.9. FEBS Lett 540:151–156

Nielsen PA, Beahm DL, Giepmans BN, Baruch A, Hall JE, Kumar NM (2002) Molecular cloning, functional expression, and tissue distribution of a novel human gap junction-forming protein, connexin-31.9. Interaction with zona occludens protein-1. J Biol Chem 277:38272–38283

Nishii K, Kumai M, Shibata Y (2001) Regulation of the epithelial-mesenchymal transformation through gap junction channels in heart development. Trends Cardiovasc Med 11:213–218

Nishii K, Kumai M, Egashira K, Miwa T, Hashizume K, Miyano Y, Shibata Y (2003) Mice lacking connexin45 conditionally in cardiac myocytes display embryonic lethality similar to that of germline knockout mice without endocardial cushion defect. Cell Commun Adhesion 10:365–369

Packham EA, Brook JD (2003) T-box genes in human disorders. Hum Mol Genet 12:R37–R44

Panchin Y, Kelmanson I, Matz M, Lukyanov K, Usman N, Lukyanov S (2000) A ubiquitous family of putative gap junction molecules. Curr Biol 10:R473–R474

Pashmforoush M, Lu JT, Chen H, St Amand T, Kondo R, Pradervand S, Evans SM, Clark B, Feramisco JR, Giles W, Ho SY, Benson DW, Silberbach M, Shou W, Chien KR (2004) Nkx2–5 pathways and congenital heart disease: loss of ventricular myocyte lineage specification leads to progressive cardiomyopathy and complete heart block. Cell 117:373–386

Paznekas WA, Boyadjiev SA, Shapiro RE, Daniels O, Wollnick B, Keegan CE, Innis JW, Dinulos MB, Christian C, Hannibal MC, Jabs EW (2003) Connexin 43 (GJA1) mutations cause the pleiotropic phenotype of oculodentodigital dysplasia. Am J Hum Genet 72:408–418

Pfeifer I, Anderson C, Werner R, Oltra E (2004) Redefining the structure of the mouse connexin43 gene: selective promoter usage and alternative splicing mechanisms yield transcripts with different translational efficiencies. Nucleic Acids Res 32:4550–4562

Phelan P, Bacon JP, Davies JA, Stebbings LA, Todman MG, Avery L, Baines RA, Barnes TM, Ford C, Hekimi S, Lee R, Shaw JE, Starich TA, Curtin KD, Sun YA, Wyman RJ (1998) Innexins: a family of invertebrate gap-junction proteins. Trends Genet 14:348–349

Röhr S (2004) Role of gap junctions in the propagation of the cardiac action potential. Cardiovasc Res 62:309–322

Sáez JC, Berthoud VM, Branes MC, Martinez AD, Beyer EC (2003) Plasma membrane channels formed by connexins: their regulation and functions. Physiol Rev 83:1359–1400

Schultz R, Heusch G (2004) Cx43 and ischemic preconditioning. Cardiovasc Res 62:335–344

Seul KH, Tadros PN, Beyer EC (1997) Mouse connexin40: gene structure and promoter analysis. Genomics 46:120–126

Simon AM, Goodenough DA, Paul DL (1998) Mice lacking connexin40 have cardiac conduction abnormalities characteristic of atrioventricular block and bundle branch block. Curr Biol 8:295–298

Söhl G, Willecke K (2004) Gap junctions and the connexin protein family. Cardiovasc Res 62:228–232

Söhl G, Nielsen PA, Eiberger J, Willecke K (2003) Expression profiles of the novel human connexin genes hCx30.2, hCx40.1, and hCx62 differ from their putative mouse orthologues. Cell Commun Adhes 10:27–36

Stennard FA, Costa MW, Elliott DA, Rankin S, Haast SJP, Lai D, McDonald LPA, Niederreither K, Dolle P, Bruneau BG, Zorn AM, Harvey RP (2003) Cardiac T-box factor Tbx20 directly interacts with Nkx2-5, GATA4, and GATA5 in regulation of gene expression in the developing heart. Dev Biol 262:206–224

Stout C, Goodenough DA, Paul DL (2004) Connexins: functions without junctions. Curr Op Cell Biol 16:507–512

Sullivan R, Ruangvoravat C, Joo D, Morgan J, Wang BL, Wang XK, Lo CW (1993) Structure, sequence and expression of the mouse Cx43 gene encoding connexin 43. Gene 130:191–199

Teunissen BEJ, Bierhuizen MFA (2004) Transcriptional control of myocardial connexins. Cardiovasc Res 62:246–255

Teunissen BEJ, Van Amersfoorth SCM, Opthof T, Jongsma HJ, Bierhuizen MFA (2002) Sp1 and Sp3 activate the rat connexin40 proximal promoter. Biochem Biophys Res Commun 292:71–78

Teunissen BEJ, Jansen AT, Van Amersfoorth SCM, O'Brien TX, Jongsma HJ, Bierhuizen MFA (2003) Analysis of the rat connexin 43 proximal promoter in neonatal cardiomyocytes. Gene 322:123–136

Van Kempen MJA, Fromaget C, Gros D, Moorman AFM, Lamers WH (1991) Spatial distribution of connexin 43, the major gap junction protein, in developing and adult rat heart. Circulation Res 68:1638–1651

Van Kempen MJA, ten Velde I, Wessels A, Oosthoek PW, Gros D, Jongsma HJ, Moorman AFM, Gros D, Lamers WH (1995) Differential connexin distribution accommodates cardiac function in different species. Micros Res Tech 31:420–436

Van Kempen MJA, Vermeulen JLM, Moorman AFM, Gros D, Paul DL, Lamers WH (1996) Developmental changes in the connexin40 and connexin43 mRNA-distribution pattern in the rat heart. Cardiovasc Res 32:886–900

Van Veen AA, van Rijen HV, Opthof T (2001) Cardiac gap junction channels: modulation of expression and channel properties. Cardiovasc Res 51:217–229

Verheule S, van Kempen MJ, Postma S, Rook MB, Jongsma HJ (2001) Gap junctions in the rabbit sinoatrial node. Am J Physiol Heart Circ Physiol 280:H2103–H2115

Verheijck EE, van Kempen MJ, Veereschild M, Lurvink J, Jongsma HJ, Bouman LN (2001) Electrophysiological features of the mouse sinoatrial node in relation to connexin distribution. Cardovasc Res 52:40–50

Wei C-J, Xin X, Lo CW (2004) Connexins and cell signaling in development and disease. Ann Rev Cell Dev Biol 20:811–838

Yamada KA, Kanter EM, Green KG, Saffitz JE (2004) Transmural distribution of connexins in rodents hearts. J Cardiovasc Electrophysiol 15:710–715

Yu W, Dahl G, Werner R (1994) The connexin43 gene is responsive to oestrogen. Proc R Soc Lond B 255:125–132

4 Gap Junction and Connexin Remodeling in Human Heart Disease

Nicholas J. Severs, Emmanuel Dupont, Riyaz Kaba, Neil Thomas[1]

Abstract Gap junctions, assembled from connexins, form the cell-to-cell pathways for propagation of the precisely orchestrated patterns of current flow that govern the normal rhythm of the healthy heart. As in most tissues and organs, multiple connexin types are co-expressed in the heart; connexin43, connexin40 and connexin45 are found in distinctive combinations and relative quantities in different, functionally-specialized subsets of cardiomyocyte. Alterations of gap junction organization and connexin expression are now well established as a consistent feature of human heart disease in which there is an arrhythmic tendency. These alterations may take the form of structural remodeling, involving disturbances in the distribution of gap junctions, and/or remodeling of connexin expression, involving alteration of the amount or type of connexin(s) expressed. In the diseased ventricle, the most consistent quantitative alteration involves heterogeneous reduction in connexin43 expression. In the atria, features of gap junction organization and connexin expression may contribute to both the initiation and persistence of atrial fibrillation. By correlating data from studies on the human patient with those from animal and cell models, alterations in gap junctions and connexins have emerged as important factors to be considered in understanding the pro-arrhythmic substrate found in a variety of forms of heart disease. Our knowledge of the functional correlates of the specific patterns of connexin co-expression in different subsets of myocyte in the healthy and diseased heart still remains limited, however, and the development of new experimental cell models heralds future advances in this field.

4.1 Introduction

Cardiovascular disease is the leading cause of death and disability in most industrialized countries of the developed and developing worlds. Arrhythmias are a common, serious and often fatal complication of many forms of heart disease. As gap junctions mediate the cell-to-cell propagation of the patterns of impulse flow that govern orderly contraction of the healthy heart, considerable attention has been directed to the possible role of these junctions and their component connexins in arrhythmic heart disease. The literature on gap junctions in the healthy and diseased heart is now considerable, and a range of perspectives on this topic can be found in recent

[1]National Heart and Lund Institute, Faculty of Medicine, Imperial College London, Guy Scadding Building, Dovehouse Street, London SW3 6LY, UK,
E-mail: n.severs@imperial.ac.uk

Gap Junctions in Development and Disease
(ed. by E. Winterhager)

reviews (e.g. Jongsma and Wilders 2000, Kanno and Saffitz 2001, Lerner et al. 2001, Severs 2001, Severs et al. 2001, 2004a, 2004b, van der Velden and Jongsma 2002a; Saffitz and Kléber 2004). The focus of the present chapter is to review the nature and possible significance of the alterations in gap junction organization and connexin expression that are now established to occur in adult human heart disease. To provide the background from which disease-related alterations can be recognized, we start with a brief summary of the principal features of gap junction organization and connexin expression in the normal heart.

4.2 Gap Junctions and Connexins in Cardiomyocytes of the Normal Heart

Three principal connexins are expressed by cardiac myocytes, connexin43, connexin40 and connexin45. These three connexins are expressed in characteristic combinations and relative quantities in a chamber-related, myocyte-type-specific and developmentally regulated manner (Gourdie et al. 1993; Van Kempen et al. 1996; Beyer et al. 1997; Coppen et al. 1998, 2001; Alcolea et al. 1999; Vozzi et al. 1999; Severs et al. 2001; Severs 2001, 2002, 2003) Globally, although connexin43 predominates throughout the adult ventricular and atrial contractile myocardium (Beyer et al. 1997; Vozzi et al. 1999; Severs 2001), connexins 40 and/or 45 are expressed locally at high levels.

The working cardiomyocytes of the ventricle are extensively interconnected by clusters of connexin43-containing gap junctions located at the intercalated disks (Fig. 4.1). The intercalated disks of ventricular myocardium generally have a step-like configuration in which the myofibrils link to segments of the interacting membranes that lie at right angles to the long axis of the cell, with the gap junctions situated in the intervening longitudinal membrane segments (Severs 1985). Distinctly larger gap junctions are often seen circumscribing the disk periphery (Hoyt et al. 1989; Gourdie et al. 1991). Preferential propagation of the impulse in the longitudinal axis and hence the normal pattern of anisotropic spread of the impulse of healthy ventricular myocardium depends in part on features of gap junction organization together with aspects of tissue architecture such as the size and shape of the cells.

Atrial cardiomyocytes are slender cells compared with their ventricular counterparts, with shorter, less elaborate intercalated disks. The gap junctions of atrial myocytes of most mammalian species, including humans, contain abundant connexin40 (Vozzi et al. 1999; Dupont et al. 2001a), colocalized with connexin43 within the same individual gap-junctional plaques

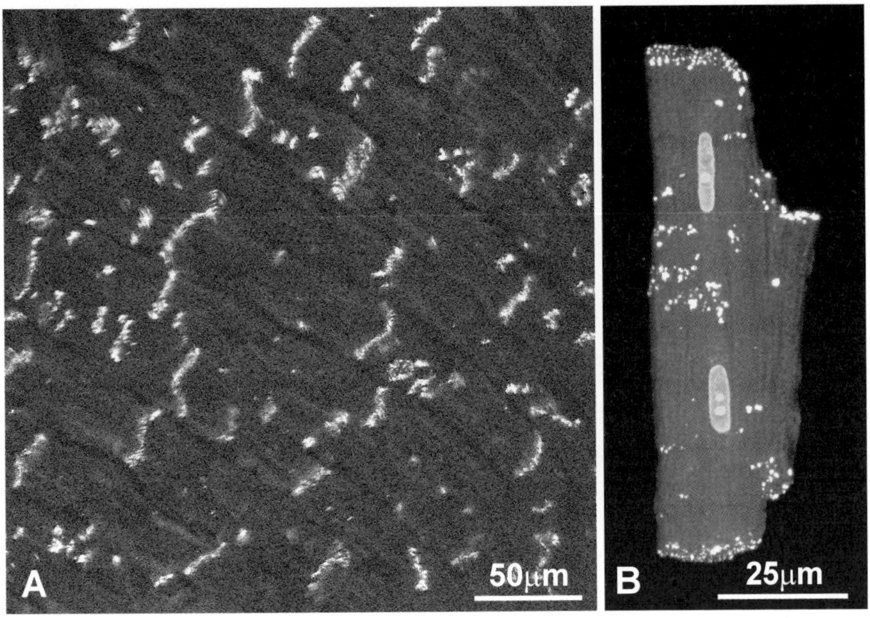

Fig. 4.1A, B. Immunolocalization of connexin43 in left ventricular myocardial tissue (**A**) and in an isolated ventricular myocyte (**B**). Note abundant label for connexin43 marking gap junctions in clusters at the intercalated disks. Confocal reconstruction from serial optical sections. Nuclei in **B** are counterstained with propidium iodide

(Severs et al. 2001). Working ventricular myocytes, by contrast, normally lack detectable connexin40 (Fig. 4.2). In both ventricular and atrial human working myocardium, connexin45 is present in very low quantities, with slightly higher levels in the atria than the ventricles (Coppen et al. 1998; Vozzi et al. 1999; Dupont et al. 2001a).

The cardiomyocytes of the impulse generation and conduction system are quite distinct from those of the contractile ventricular and atrial cells both morphologically (Severs 1989), and with respect to their connexin expression profiles. In experimental animals, the myocytes of the sinoatrial node, the site of impulse generation, and those of the atrioventricular node, the site at which the impulse is slowed before being directed to the ventricles, are equipped with small, dispersed gap junctions rich in connexin45 (Coppen et al. 1999a,b; Honjo et al. 2002), a connexin that is known from in vitro experimental studies to form low conductance channels (Veenstra et al. 1994; Moreno et al. 1995; van Veen et al. 2000). These gap junction features of nodal myocytes suggest relatively poor coupling, a property essential, for example, to the role of the atrioventricular node in slowing of conduction to ensure sequential contraction of the atria and ventricles. In the rodent, the spatial pattern of expression of

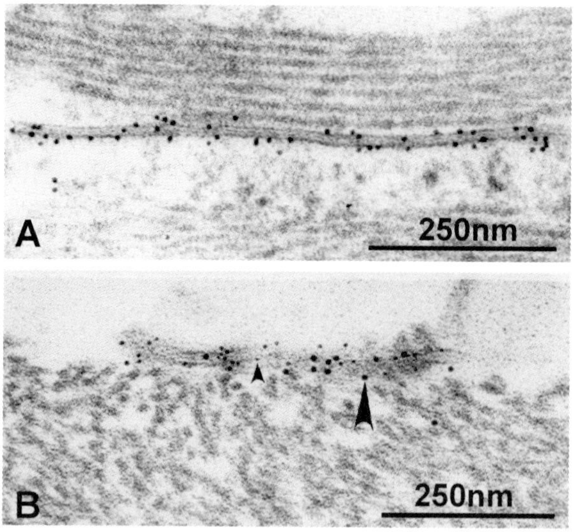

Fig. 4.2A, B. Double immunogold localization of connexins 43 and 40 in thin-section electron micrographs of gap junctions of a ventricular myocyte (**A**) and an atrial myocyte (**B**). Only connexin43 is detected in ventricular gap junctions (10 nm gold label), whereas both connexin 43 (5 nm gold) and connexin40 (10 nm gold) are detected in atrial gap junctions

connexin45 reveals that the atrioventricular node and His bundle form part of an elaborately extended central conduction system circumscribing the atrioventricular and outflow junctional regions (Coppen et al. 1999b).

In the rabbit sinoatrial node, a restricted zone of connexin45/connexin43 co-expression is present at the nodal/crista terminalis border, tentatively identified as the exit route for the impulse into the atrial tissue (Coppen et al. 1999a). In the atrioventricular conduction axis of this species, three-dimensional reconstructions have revealed distinctive compartmentalized connexin expression patterns; one compartment, comprising the His Bundle, lower nodal cells and posterior nodal extension co-expresses connexins 43 and 45, while a second compartment comprising the compact node and transitional cells predominantly expresses connexin45 (Ko et al. 2004). The transitional cells, which are located between atrial muscle and the compact node, connect both to lower nodal cells and the posterior nodal extension.

Downstream from the His Bundle, the conduction system myocytes of most mammals, including man, prominently express connexin40, a connexin which in in vitro experimental studies is associated with high conductance channels (Bastide et al. 1993; Gourdie et al. 1993; Gros et al. 1994; Bukauskas et al. 1995; Davis et al. 1995; Coppen et al. 1998, 2001). Extensive immunolabeling for this connexin, in the form of large, abundant gap junctions, correlates with the fast conduction properties of the bundle branches and Purkinje fiber system which facilitate rapid distribution of the impulse throughout the working ventricular myocardium. In rodents, connexin45 is co-expressed with connexin40 in a central zone of the bundle branches and Purkinje fibers, enveloped by an outer zone in which only connexin45 is found (Coppen et al. 1999b).

As illustrated by the foregoing account, key features of connexin expression are known to be common to many mammalian species. At the same time, however, some features are yet to be demonstrated in the human, and some striking examples of species differences have been identified, e.g. lack of connexin40 expression in rat atrial muscle and in the guinea pig conduction system (Gros et al. 1994, Van Kempen et al. 1995). Species variation involving co-expression within the atrioventricular node has also been suggested, especially in larger mammals which have less need for atrioventricular nodal impulse delay (Coppen and Severs 2002). A more detailed knowledge of such species-specific patterns is important in the context of the ever wider use of transgenic animals for investigating the role of connexins in cardiac function. Extrapolation of data on transgenic mice to the human, for example, depends critically on a sound understanding of the similarities and differences of the connexin expression of these two species (Kaba et al. 2001; Coppen et al. 2003). One area requiring further work is the extent to which the connexin expression patterns of the impulse generation and conduction system of humans resemble those identified in experimental animals.

4.3 Alterations in Gap Junctions and Connexin Expression in Heart Disease

Despite some remaining gaps in our knowledge, the preceding section emphasizes that cardiomyocytes of the healthy heart have distinctive patterns of gap junction organization and connexin expression. These gap junction features are known to form the cell-to-cell pathways for the precisely orchestrated patterns of current flow that govern the normal heart rhythm. From this perspective, a key question that follows is whether alterations of gap junction organization and connexin expression play a role in abnormal impulse propagation and arrhythmia in heart disease (Smith et al. 1991; Green and Severs 1993). Arrhythmias are multifactorial in origin, involving an interplay between gap-junctional coupling, membrane excitability and cell and tissue architecture (Rohr et al. 1997, Shaw and Rudy 1997, Spach et al. 2000); moreover, gap-junctional coupling is itself determined by a combination of factors including channel gating, the assembly/disassembly of functional gap junction plaques and the pattern, amount and types of connexin expressed. Alterations to gap junction organization and connexin expression thus represent one potential facet of a constellation of factors that could contribute to the life-threatening pro-arrhythmogenic substrates that beset human heart disease.

Dissecting out any role that gap junctions might play in human heart disease in vivo is thus a complex issue. With all this in mind, a fundamental starting point is to determine what changes in gap junctions and connexins, if any, are we actually able to see in human heart disease. Having identified any such changes, the next step involves developing and applying approaches to analyze their functional significance. The following sections explore progress in these areas both with respect to ventricular myocardium and atrial myocardium. Conceptually, it is helpful to use two headings in summarizing disease-related gap junction changes: (1) structural remodeling, which defines any alteration in the arrangement, distribution or organization of gap junctions; and (2) remodeling of connexin expression, involving alteration in the amount and/or types of connexin expressed. As we shall see, these remodeling patterns are, in practice, commonly closely linked.

4.4 Ventricular Myocardium in Disease

Two principal gap junction-related alterations are consistently seen in the diseased ventricle; disturbances in the distribution of gap junctions, and reduced levels of their major component, connexin43. However, the precise nature of this gap junction and connexin remodeling varies according to the stage and type of myocardial disease.

4.4.1 Structural Remodeling

Disease-related alteration in the distribution of connexin43 gap junctions was first reported in the myocardial zone bordering infarct scar tissue in the ventricles of explanted hearts from patients undergoing cardiac transplantation because of end-stage ischemic heart disease (Smith et al. 1991). Connexin43 immunolabeling in the border zone myocytes is typically scattered in disordered fashion over the lateral surfaces of the cells rather than in the polar, intercalated disk arrays characteristic of normal myocardium (Fig. 4.3). Electron microscopy reveals that both true laterally disposed gap junctions that connect adjacent cells, and annular profiles of apparently internalized and hence nonfunctional gap-junctional membrane, contribute

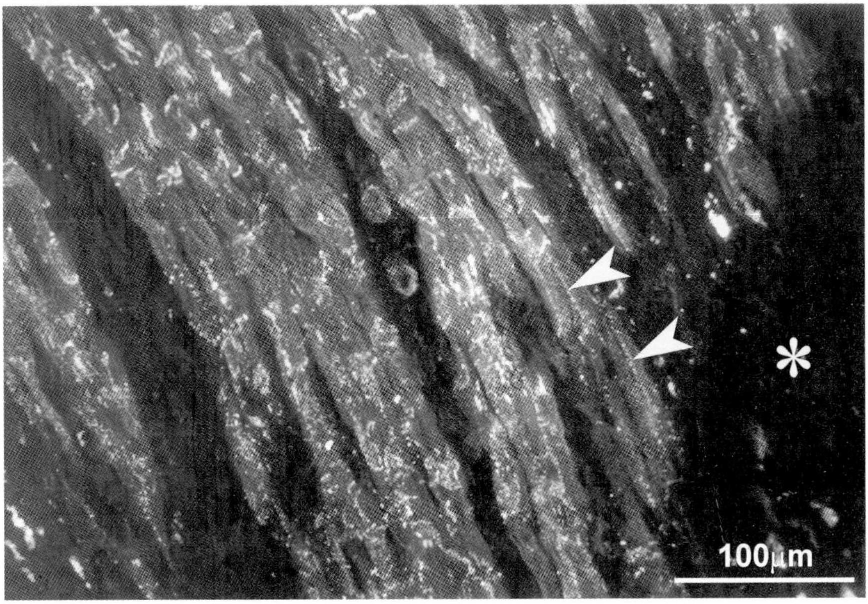

Fig. 4.3. Disordered distribution of connexin43 gap junctions in border zone ventricular myocardium facing infarcted tissue from a patient with ischemic heart disease (Green and Severs 1993). Positively labeled myocardial tissue is seen approaching the connexin43-immunonegative infarct (*asterix*). The normal ordered arrangement of connexin43 gap junctions (illustrated in Fig. 4.1) is severely disrupted in cardiac muscle cells abutting the infarct scar where many of the immunolabeled gap-junctional spots appear distributed in disordered fashion along the lateral borders of the myocytes (*arrowheads*)

to the nonordered connexin43 immunolabeling patterns in these human infarct border-zone myocytes (Smith et al. 1991). Gap junction disarray is not only associated with established infarct scar tissue in the human ventricle, but is initiated rapidly in response to ventricular ischemia and infarction in experimental animals (Matsushita et al. 1999; Daleau et al. 2001). In a canine model at 4-days post-infarction, zones of gap junction lateralization in border zone myocytes extending across the full thickness of the epicardial layer have been shown to correlate positionally with electrophysiologically identified figure-of-eight reentrant circuits (Peters et al. 1997). Longer term features of remodeling in myocardium distant from the infarct in the canine ventricle include reduction in the size and the number of gap junctions per unit length of intercalated disk, and fewer side-to-side, but relative preservation of end-to-end connections between myocytes (Luke and Saffitz 1991).

Lateralization of connexin43 gap junctions in a pattern resembling that in human infarct border zone myocytes has more recently been reported in the ventricles of patients with compensated hypertrophy due to valvular aortic stenosis, but is not apparent in decompensated hypertrophy from the same cause; instead, a heterogeneous distribution is seen in which the gap junctions are organized largely in normal intercalated disk arrays, but with patches of tissue in which the junctions are not detectable or markedly fewer (Kostin et al. 2004). Lateralization is found in some (though not necessarily all) models of ventricular hypertrophy in the rat (Uzzaman et al. 2000; Emdad et al. 2001), and has been shown to be associated, in one model, with reduction in longitudinal conduction velocity (Uzzaman et al. 2000). Spectacularly disordered arrangements of ventricular connexin43 gap junctions are apparent in human hypertrophic cardiomyopathy; this may largely reflect the haphazard myocyte organization characteristic of this condition, the most common cause of sudden cardiac death due to arrhythmia in young adults (Sepp et al. 1996). Some focal disordering of connexin43 gap junctions is also found in small areas of the explanted ventricle in transplant patients with heart failure due to idiopathic dilated cardiomyopathy and myocarditis (Kostin et al. 2003).

A rather different form of structural remodeling is associated with human hibernating myocardium (Kaprielian et al. 1998). The term "hibernating myocardium" is used in patients with ischemic heart disease to refer to a region of ventricular myocardium which shows impaired contraction, but which has the capacity to recover contractile function after a coronary artery by-pass operation has been carried out (Camici et al. 1997; Heusch 1998). In human hibernating myocardium from patients with ischemic heart disease, the large connexin43 gap junctions typically found at the periphery of the intercalated disk are smaller in size, and the overall amount of connexin43 immunolabeling per intercalated disk is reduced, compared with normally perfused (and reversibly ischemic) segments of the same heart (Kaprielian et al. 1998). From these observations, a possible link between connexin43 gap junction remodeling and impaired ventricular contraction in human heart disease was first proposed (Kaprielian et al. 1998).

4.4.2 Remodeling of Connexin43 Expression

As emphasized by the above observations on hibernating myocardium, disease-related structural remodeling is often accompanied by alterations in the amount of connexin expressed. The most thoroughly and widely documented disease-related alteration in ventricular connexin expression involves down-regulation of connexin43. A significant reduction in connexin43 transcript and protein levels is apparent in the left ventricles of transplant patients with end-stage congestive heart failure (Dupont et al. 2001b). Several studies have confirmed that such a reduction of ventricular connexin43 expression is consistently found irrespective of whether heart failure is due to idiopathic dilated cardiomyopathy, ischemic heart disease, valvular aortic stenosis or other aetiologies (Kitamura et al. 2002; Kostin et al. 2003, 2004, Yamada et al. 2003). The reduction in ventricular connexin43 appears to develop progressively during the course of disease, as suggested by the pattern of change in connexin amounts in pressure-overloaded hearts with valvular aortic stenosis classified according to ejection fraction (Kostin et al. 2004) and the finding of reduced connexin43 in the nonfailing ventricles of patients with ischemic heart disease undergoing coronary artery by-pass surgery (Peters et al. 1993). Intriguingly, the study of Kostin et al. (2004) raises the possibility that the long-term disease-related decline in connexin43 may be preceded by an adaptive increase in the connexin during the early stage of compensated hypertrophy.

The possible functional significance of reduced connexin43 levels in the diseased human ventricle continues to be actively debated. In any such discussion, it is important to bear in mind that while it may, with some justification, be argued that total connexin levels reflect the *potential* capacity for cell-to-cell communication or coupling, they give no indication of the quantity of functional (open) channels; this, combined with predictions from computer modeling that reductions of up to 40% in gap junction content (without change in junction size) would have no significant effect on conduction velocity (Jongsma and Wilders 2000), suggests that reduction of connexin43 alone would in theory hold no adverse functional consequences for the diseased ventricle. However, reduction of connexin levels is neither a sole nor uniform change, even within the context of gap junction remodeling. In view of the complex relationship between passive and active membrane properties (Rudy and Shaw 1997; Shaw and Rudy 1997; Viswanathan et al. 1999), the multiplicity of structural and functional alterations in the diseased heart and the assumptions inherent in computer modeling, in vivo-extrapolation of the effects of a single change (i.e. reduced connexin43 levels), in isolation from other factors, may not give the full picture. Studies on experimental animals and the intact heart are therefore also instructive in gaining further insight.

In intact isolated hearts of transgenic mice expressing half the normal level of connexin43, experimental ischemia reportedly leads to a marked increase in incidence, frequency and duration of ventricular tachycardias (Lerner et al. 2000) even though there may only be a modest reduction in conduction velocities (Guerrero et al. 1997; Morley et al. 1999). Moreover, in a transgenic mouse model of juvenile dilated cardiomyopathy, reduced connexin43 and conduction defects become apparent at 4 weeks after birth, with contractile dysfunction and heart failure following at 12 weeks (Hall et al. 2000). On the other hand, in transgenic mice generated to give cardiac specific loss of connexin43, the magnitude of connexin43 reduction associated with sudden death due to spontaneous ventricular arrhythmia is in the order of 86–95% (Gutstein et al. 2001a), much lower than the average reduction found in the diseased human ventricle (~50%). In the cardiac restricted connexin43 knock-out model, reduction in connexin43 is accompanied by reduction in gap-junctional conductance during the period prior to ventricular fibrillation when heart function appears normal (Yao et al. 2002). Outbred cardiac-restricted knock-outs show a more gradual decline in connexin43 levels with a progressive increase in inducible ventricular arrhythmias (Gutstein et al. 2002).

In relating these findings to the failing human ventricle, an important feature to emphasize is that connexin43 reduction is not uniform (Fig. 4.4); a considerable variation in the extent of connexin43 reduction is found between and, in particular, within hearts, some regions of some diseased hearts reaching a reduction of > 90% of control values (Dupont et al. 2001b; i.e. similar to the levels at which fatal cardiac arrhythmia occurs in the cardiac restricted connexin43 mouse knock-out). Thus, average values for the overall reduction in ventricular connexin43 in the diseased human heart disguise considerable spatial heterogeneity in the extent of the reduction. It is the existence of this heterogeneity, arising from a combination of altered gap junction arrangement and reduction of connexin43 that is now considered critical. Heterogeneous reduction of connexin43 could quite plausibly lead to exaggeration of inhomogeneities in resting potential and action potential upstroke velocity and duration, affecting individual cell excitability and refractory period, dispersion of which is a key pro-arrhythmic factor. Inhomogeneous wave front propagation could in turn, it is thought, lead to asynchronous myocyte contraction and poor ventricular force development.

Evidence that heterogeneity of cardiac connexin43 expression may indeed be causally linked to disturbances in electromechanical function came from work by Gutstein et al. (2001b) using chimeric mice created from connexin43-deficient stem cells and blastocysts. These experimental mice have a patchy expression of ventricular connexin43, mimicking some of the remodeling patterns found in the diseased human ventricle, and were demonstrated to have abnormal conduction and marked contractile dysfunction, just as originally hypothesized in the human studies (Smith et al. 1991; Kaprielian et al. 1998; Dupont et al. 2001b). Thus, the possibility that spatially heterogeneous reduction of connexin43 of the magnitude and nature observed in the diseased human ventricle could similarly predispose to arrhythmia and contractile dysfunction remains very much on the research agenda.

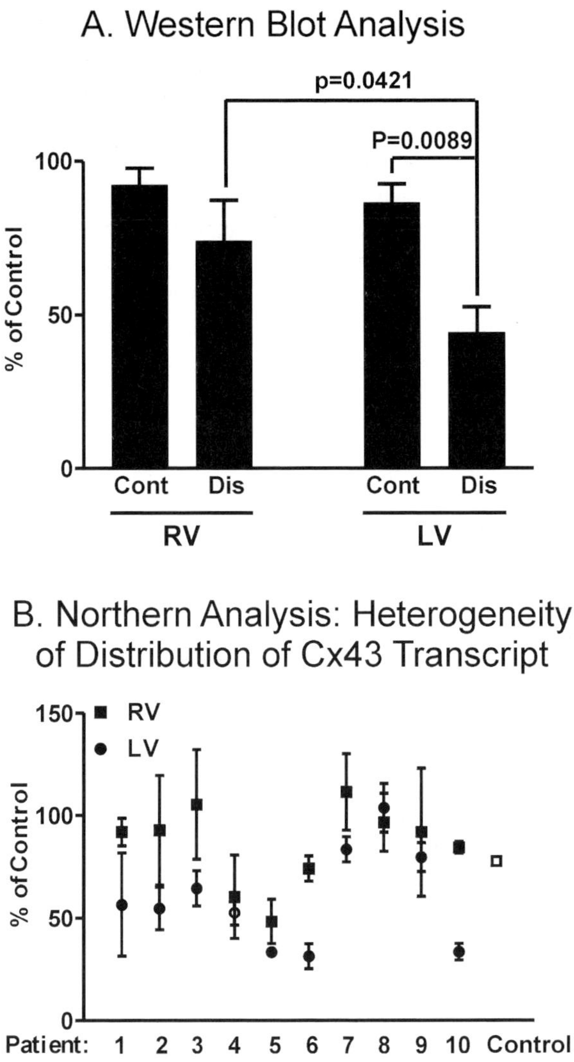

Fig. 4.4A, B. Data illustrating heterogeneous reduction of ventricular connexin43 levels in human congestive heart failure. The Western blot analysis (**A**) shows significant reduction in connexin43 in samples of failing ventricle from explanted hearts of patients undergoing cardiac transplantation. Note significant reduction of connexin43 levels in failing left ventricle compared with control samples (*cont*) from normal hearts (intended for transplantation, but not used for technical reasons). That the reduction in connexin43 does not uniformly affect all regions of the heart is shown in the northern blot data in **B**. Multiple samples taken from the same diseased heart in many (though not all) instances reveal a considerable spread of values, shown by the large standard deviations. In control hearts, however, the standard deviation is very small (*open square*). (Dupont et al. 2001b)

Precisely how reduction in connexin43 expression and gap junction size and number are brought about in the setting of human cardiac disease remains unclear; among candidate mechanisms are association of the PDZ-MAGUK protein ZO-1 with connexin43 (Barker et al. 2002), growth factors such as VEGF (Pimentel et al. 2002), and activation of the stress-activated protein kinase, c-Jun N-terminal kinase (JNK; Petrich et al. 2002, 2004). JNK activation in a transgenic mouse model leads to an impressive combination of loss of connexin43, slowing of ventricular conduction, contractile dysfunction and congestive heart failure (Petrich et al. 2004). For further information on possible regulatory mechanisms of connexin43 down-regulation, the reader is referred to a recent review by Saffitz and Kléber (2004).

4.4.3 Remodeling of Expression of Connexin45 and Connexin40

Fewer studies have documented alterations in expression of connexins other than connexin43 in the diseased human ventricle, though the available evidence suggests that elevated levels of connexin45, and in some instances also connexin40, may feature. Up-regulation of connexin45 in parallel with the reduction in connexin43 in the failing human ventricle was recently reported by Yamada et al. (2003). As in the human ventricle, connexin43 and connexin45 commonly appear co-localized in the same gap junctional plaque (Vozzi et al. 1999, Yamada et al. 2003), the possibility exists that these two connexins may be assembled into heteromeric connexons and channels. Connexin43/connexin45 heteromeric channels are reported to have reduced unitary conductance compared with that of homomeric connexin43 channels (Martinez et al. 2002). Thus, even though the overall levels of connexin45 are low, elevated connexin45 in the failing ventricle might, by altering the stoichiometry of connexins 43 and 45 at the level of the individual channel, exert a significant effect on channel properties, perhaps reducing and/or creating microheterogeneities in coupling (Yamada et al. 2003).

Increased levels of connexin40 transcript have been detected in the ventricles of patients with congestive heart failure due to ischemic heart disease, but not that due to idiopathic dilated cardiomyopathy (Dupont et al. 2001b). This elevated connexin40 expression correlates with an increased depth of connexin40 expressing myocytes from the endocardial surface, adjacent to the Purkinje myocytes. Whether this change represents an expansion of the Purkinje system or a transformation of working myocytes to become more Purkinje-like is unclear. The significance of this expanded zone of connexin40 expression, which shows some similarity to that reported in ventricular hypertrophy in the rat (Bastide et al. 1993), is unknown. One speculation is that it might represent some form of adaptive response in the face of declining connexin43 levels. Such a spatially localized effect could augment local heterogeneities in coupling brought about by heterogeneous reduction of connexin43, increasing yet further the susceptibility to arrhythmia.

4.5 Short-Term Effects of Ischemia

Apart from gap junction and connexin remodeling during the long-term development of heart disease, a number of studies have examined short-term changes in connexin expression in response to brief ischemic episodes, especially in relation to pre-conditioning (i.e. brief, repetitive episodes of ischemia). In experimental animals, reduced levels of connexin43 have been reported in response to such episodes of ischemia and reperfusion though this appears to be a temporary effect (Daleau et al. 2001). In addition, rapid dephosphorylation of connexin43 and translocation of connexin43 from surface gap junctions into the cytosol has been reported when electrical uncoupling is induced by acute ischemia in the Langendorff-perfused rat heart (Beardslee et al. 2000). These processes are also reversible upon reperfusion (Beardslee et al. 2000) and substantially reduced with ischemic preconditioning (Jain et al. 2003; Schulz et al. 2003). In transgenic mice expressing half the normal level of connexin43, preconditioning apparently does not afford protection from prolonged ischemia as it does in mice with the normal level of connexin43 (Schwanke et al. 2002). However, while this last study found that infarct size was not reduced by ischemic preconditioning in the transgenic animals, another report has concluded that these animals do develop smaller infarcts after coronary ligation than do their wild-type counterparts (Kanno et al. 2003). For further discussion of gap junctions and preconditioning, the reader is referred to a recent review by Schulz and Heusch (2004).

Ethical considerations preclude corresponding studies of the short term effects of ischemia in the human heart. However, during cardiopulmonary by-pass, the human heart may be subject to stress resembling the challenge of ischemia. The accessibility of right atrial appendage samples during cardiac surgery has enabled investigation of temporal changes in connexins and gap junctions during cardiopulmonary by-pass which may reflect changes in the heart as a whole, i.e. including the ventricles which, for ethical reasons, cannot be sampled (Yeh et al. 2002). Connexin43 expression and gap junctions appear reduced during cardiopulmonary by-pass, with coronary artery disease patients showing a greater reduction than other patients. This suggests that despite the application of hypothermia and cardioplegic solution, protection of the heart may in some instances be inadequate during the operation, especially in patients with coronary artery disease. Whether these changes are sufficient to contribute to the common occurrence of post-operative ventricular dysfunction is yet to be determined (Yeh et al. 2002)

4.6 Remodeling of Gap Junctions and Connexin Expression in Diseased Atrial Myocardium

Our discussion so far has examined altered gap junction and connexin expression in relation to electromechanical dysfunction of the ventricles. Arrhythmia also commonly afflicts the atria; indeed, atrial fibrillation, a condition in which wavelets of electrical activity propagate in multiple directions leading to disorganized depolarization and ineffective atrial contraction, is the most common sustained arrhythmia in humans (Zipes 1997). The incidence of atrial fibrillation increases with age and has now reached epidemic proportions. The condition is associated with progressive electrical, contractile and structural remodeling (Allessie et al. 2002), including altered cell size and shape, interstitial fibrosis, and altered expression of ion channels, connexins and gap junction organization. Remodeling, resulting from atrial fibrillation itself, exacerbates the condition, so that once established it tends to persist (Wijffels et al. 1995). There is now considerable literature on gap junction remodeling in atrial fibrillation, but in contrast to the data on ventricular dysfunction, the picture remains somewhat confusing. A variety of factors may underlie the apparent disparities in the findings, including species and age differences, manner of atrial fibrillation induction, different clinical subsets of patients and associated pathological factors, and flawed techniques or interpretation.

A marked increase in connexin43 expression was reported in atrial fibrillation induced in dogs after 10–14 weeks of right atrial pacing (Elvan et al. 1997), a change abated by pre(pacing)-treatment with the angiotensin-converting enzyme-inhibitor, enalapril (Sakabe et al. 2004). No corresponding alteration was apparent in pacing-induced atrial fibrillation in goats, however; instead, a heterogeneous distribution of connexin40 label was found with an apparent reduction in connexin40/connexin43 ratio (van der Velden et al. 2000; Ausma et al. 2003). Studies on human atrial samples have variously reported a net increase (Polontchouk et al. 2001) or decrease (Kostin et al. 2002, Nao et al. 2003) in connexin40 expression in patients with chronic atrial fibrillation, with redistribution of connexin40 labeling to predominate at the lateral borders of the myocytes (Polontchouk et al.

2001; Kostin et al. 2002). Other junctional proteins such as N-cadherin and desmoplakin show similar changes in distribution (Kostin et al. 2002), suggesting a spatial association between gap junctions and adhesive junctions during their re-organization at the cell surface, as occurs in the maturing heart (Peters et al. 1994; Angst et al. 1997). The lateralization reported in these studies is quite distinct from the heterogeneity of connexin40 seen in the goat model (van der Velden et al. 1998). The available literature suggests fundamental differences in the normal distribution of connexin40 gap junctions in the goat and human atrial samples studied, goats in sinus rhythm apparently having a homogeneous distribution (van der Velden et al. 1998, 2000) while humans show a heterogeneous distribution (Vozzi et al. 1999; Dupont et al. 2001a). Indeed, the heterogeneity of connexin40 distribution reported as an atrial fibrillation-induced change in the goat model resembles that naturally seen in the human atrium (Dupont et al. 2001a, Kostin et al. 2002), though this feature may become more marked in patients with chronic atrial fibrillation (Kostin et al. 2002).

Emerging evidence emphasizes the importance of differing patient subgroups and related pathology as further confounding factors in attempting to draw generalizations on gap junction and connexin changes in human chronic atrial fibrillation. One line of work suggests that a reduced ratio of connexin40 to total connexin may be specifically associated with complex activation patterns rather than more simple activation patterns (Kanagaratnam et al. 2004). Another emphasizes the potentially critical importance of the degree of atrial dilatation (Kaba et al. 2003). The effects of atrial fibrillation are difficult to isolate from those of atrial dilatation, as the two conditions are so closely associated, particularly in patients undergoing cardiac surgery. When the contribution from atrial dilatation is minimal, chronic atrial fibrillation appears to be associated with an increase in connexin43, rather than a reduction in connexin40, compared to patients in sinus rhythm (Kaba et al. 2003). However, with severe atrial dilatation connexin43 appears to diminish in patients with atrial fibrillation compared to well-matched sinus rhythm counterparts (Kostin et al. 2002), raising the possibility of temporal changes during the course of disease.

Apart from gap junction changes identified in samples with established atrial fibrillation, it is also of interest to consider whether any features of gap junction organization or connexin expression may predispose the initiation of atrial fibrillation. Samples of right atrial appendage from patients in sinus rhythm undergoing coronary artery bypass show a range of connexin40 levels. Of the patients who subsequently develop post-operative atrial fibrillation, the majority have higher levels of connexin40 than those who do not develop the condition (Dupont et al. 2001a). In a substantial number of patients, the initiating foci of atrial fibrillation are not located in the atrium itself, but in proximal portions of the thoracic veins which have a sleeve of myocardium continuous with that of the atria (Jalife 2003). Gap junctions in the myocardial sleeve of the canine superior vena cava show areas of atypical expression, in which regions of diffusely distributed connexin43 gap junctions are surrounded by areas in which tiny connexin40 gap junctions predominate (Yeh et al. 2001). This, together with the spatial orientation of the myocytes, endows the myocardial sleeve with a heterogeneous structure that may predispose to ectopic activation.

4.7 Significance of Connexin Co-Expression: New Tools

As the work discussed in this chapter has shown, specific patterns of connexin co-expression characterize functionally specialized subsets of myocytes in the healthy heart, and alterations in connexin expression levels are a key feature in at least some categories of arrhythmic heart disease. While electrophysiological studies on transfected cells have emphasized the distinctiveness of the channel properties conferred by different connexin types, work on genetically engineered mice indicates that at the level of the whole organ a considerable capacity apparently exists for one connexin type to substitute for the function of another (Plum et al. 2000). To advance further our understanding of the functional correlates of connexin co-expression at the level of groups of cells, we are currently developing an in vitro model, designed to express the same combinations of connexins in the same relative quantities as those found in vivo. This involves use of an inducible plasmid system that allows the co-expression of two different connexins under the control of Ecdysone and Tetracycline inducible promoters (Halliday et al. 2003). By using an IRES sequence to link the connexin to an antibiotic resistance gene (i.e. producing a bi-cistronic mRNA transcript), preliminary problems of heterogeneity of expression in this model were overcome (Halliday et al. 2003).

RLE (rat liver epithelial) cells are particularly suitable for use in our model as they give a uniform monolayer, endogenously express high levels of connexin43 and have a high level of intercellular communication. We have transfected the cells with one connexin under the control of the Ecdysone promoter to generate fully inducible connexin40 and connexin45 clones. The expression levels of connexin40 can be accurately controlled and all cells express similar levels of the endogenous and transfected connexin (Fig. 4.5). Initial double fluorescence-labeling experiments show that although expression is homogeneous, some cell interfaces may have more of one or the other connexin, raising the possibility that different ratios of connexins within a connexon may impose specificity of compatibility (Fig. 4.6). Transfection with the Tet-off system gives similar results. At present, double

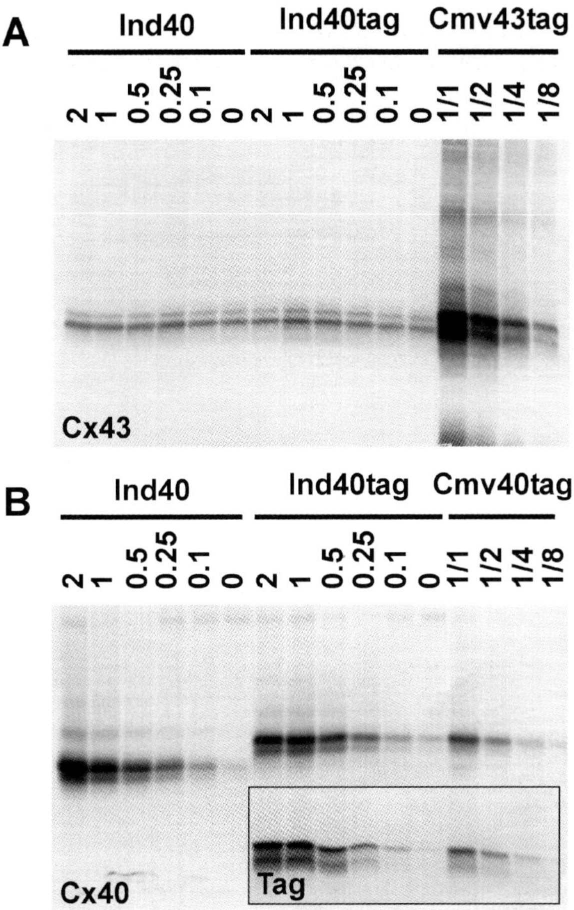

Fig. 4.5A, B. Western blot analysis of connexin expression response under the control of the ecdysone system. **A** Analysis of endogenous connexin43 in the transfected connexin40 (Ind40) and connexin40tag (Ind40tag) cell lines cultured at different levels of induction (indicated by the concentration of ponasterone A above given in μM). **B** Western blot for connexin40 in the same samples. Note progressive reduction in the amount of connexin40 expressed as the ponasterone A concentration is decreased. The set of blots under *Cmvtag* show data from which a standard curve is produced (the Cmv promoter gives a constant expression level of connexin; the fractions above the blots give the protein concentrations of the cells used) The introduction of the V5 tag to the connexin40 (inset denoted by *Tag*) gives a common point throughout, allowing the relative abundance of different connexins to be quantified in relation to each other

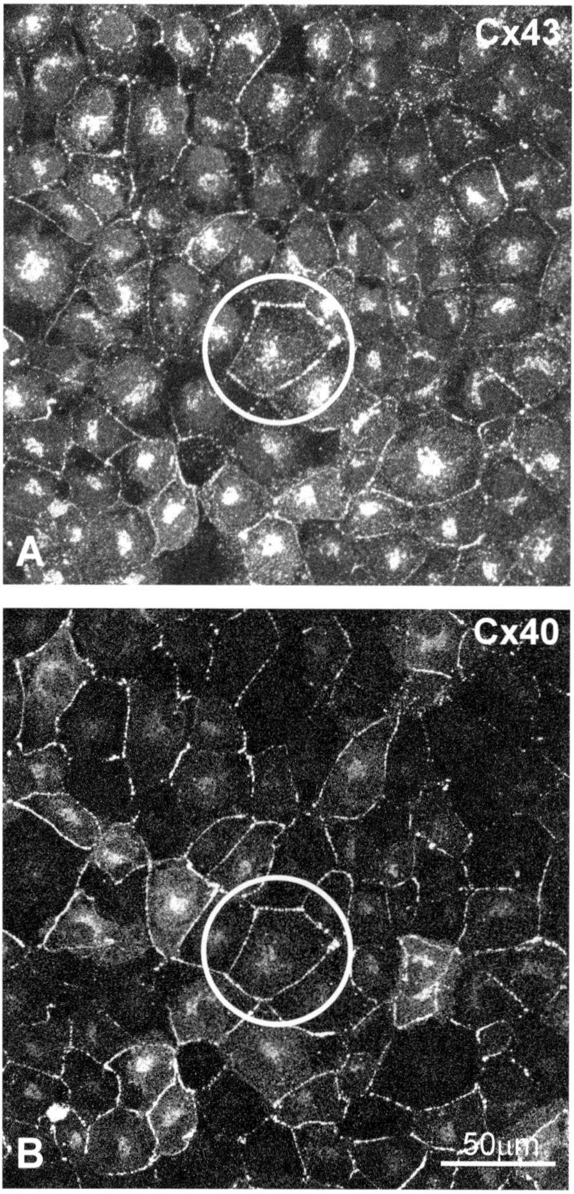

Fig. 4.6A, B. Homogeneous connexin expression using the Ecdysone system. **A, B** Immuno-confocal imaging of transfected connexin40 and endogenous connexin43 in the *same* area of connexin40 transfectants. Expression is high and homogeneous throughout the monolayer, but distribution is less uniform at individual cell interfaces. In the example encircled, connexin40 expression is similar on each cell interface though distribution of connexin43 can be seen to be abundant on one side of the cell, but sparse on the opposite side. This micro-heterogeneity at the level of the individual interacting cell raises the possibility of differences in the connexin make-up of channels between a given cell and its various neighbors

transfection with both systems is underway to give a cell line with three connexins, two of which will be under the control of inducible promoters.

In order to obtain relative quantification and therefore an accurate ratio of connexin expression in the co-expressing cell lines or in the cardiac tissue that the cell lines are designed to mimic, we introduced a V5 6xhistidine tag at the C-terminal of the connexins. The V5 immunoreactivity can be used to calibrate the specific reaction with anti-connexin antibodies in Western blots (Fig. 4.5). Using this technique we have accurately measured levels of connexin40 within the cell line (Fig. 4.7). At the higher levels of induction it should be noted that expression of the transfected connexin can be manipulated to levels both above and below that of endogenous connexin43. This flexibility allows us to imitate not only those patterns observed in the healthy heart, but also those of pathological conditions.

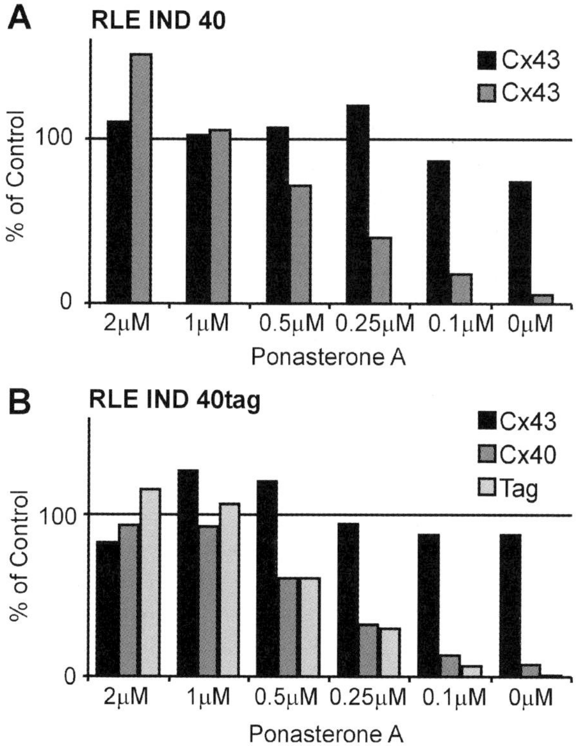

Fig. 4.7. Quantification of connexin expression from Western blot analysis. *Above* Analysis of connexin40 transfectants. *Below* Analysis of cells transfected with connexin40 including the 6x histidine (V5) tag. Levels of induction are indicated on the horizontal axis as concentration of ponasterone A used in culture media. The immunoreaction was normalized to the mean endogenous expression of connexin43 of the samples as shown on the vertical axis

With this model, it will be possible to establish the influence of varying ratios of co-expressed connexins on (1) intracellular resistance, a major determinant of conduction velocity in myocardium, and (2) intercellular transfer of small molecules (using direct microinjection of fluorescent dyes such as Lucifer yellow and ethidium bromide). The 6xhistidine stretch can be used to carry out affinity purification, determine the stoichiometry of the heteromeric connexons and relate it to specific electrical and diffusional properties. Finally, applying a similar approach to HL1 cells (Claycomb et al. 1998), a cardiac muscle cell line that continuously undergoes contraction and division, will permit the direct measurement of conduction velocities in relation to various levels of connexin co-expression.

4.8 Concluding Comment

To conclude, remodeling of myocyte gap junctions and connexins – notably disordering in the pattern of junctional distribution and reduced levels of ventricular connexin43 – feature prominently in defined categories of human heart disease, and in at least some instances similar alterations correlate with electrophysiologically identified pro-arrhythmic changes and contractile dysfunction in animal models. Our insights into how and why arrhythmia may occur have thus been significantly advanced, but whether therapeutic approaches based on this knowledge will ever become feasible presents a formidable challenge for the future (Spach and Starmer 1995; Severs 1999; van der Velden and Jongsma 2002b; Kanno et al. 2003).

Acknowledgements. This work was supported by the British Heart Foundation (PG/02/083). We thank Stephen Rothery, Deborah Halliday and Steven Coppen for their help and contributions.

References

Alcolea S, Theveniau-Ruissy M, Jarry-Guichard T, Marics I, Tzouanacou E, Chauvin JP, Briand JP, Moorman AF, Lamers WH, Gros DB (1999) Downregulation of connexin 45 gene products during mouse heart development. Circ Res 84:1365–1379

Allessie M, Ausma J, Schotten U (2002) Electrical, contractile and structural remodeling during atrial fibrillation. Cardiovasc Res 54:230–246

Angst BD, Khan LUR, Severs NJ, Whitely K, Rothery S, Thompson RP, Magee AI, Gourdie RG (1997) Dissociated spatial patterning of gap junctions and cell adhesion junctions during postnatal differentiation of ventricular myocardium. Circ Res 80:88–94

Ausma J, van der Velden HM, Lenders MH, van Ankeren EP, Jongsma HJ, Ramaekers FC, Borgers M, Allessie MA (2003) Reverse structural and gap-junctional remodeling after prolonged atrial fibrillation in the goat. Circulation 107:2051–2058

Barker RJ, Price RL, Gourdie RG (2002) Increased association of ZO-1 with connexin43 during remodeling of cardiac gap junctions. Circ Res 90:317–324

Bastide B, Neyses L, Ganten D, Paul M, Willecke K, Traub O (1993) Gap junction protein connexin40 is preferentially expressed in vascular endothelium and conductive bundles of rat myocardium and is increased under hypertensive conditions. Circ Res 73:1138–1149

Beardslee MA, Lerner DL, Tadros PN, Laing JG, Beyer EC, Yamada KA, Kléber AG, Schuessler RB, Saffitz JE (2000) Dephosphorylation and intracellular redistribution of ventricular connexin43 during electrical uncoupling induced by ischemia. Circ Res 87:656–662

Beyer E, Seul KH, Larson DM (1997) Cardiovascular gap junction proteins: molecular characterization and biochemical regulation. In: De Mello WC, Janse MJ, Norwell MA (eds) Heart Cell Communication in Health and Disease. Kluwer, New York, pp 45–51

Bukauskas FF, Elfgang C, Willecke K, Weingart R (1995) Biophysical properties of gap junction channels formed by mouse connexin40 in induced pairs of transfected human HeLa cells. Biophys J 68:2289–2298

Camici PG, Wijns W, Borgers M, DeSilva R, Ferrari R, Knuuti J, Lammertsma AA, Liedtke AJ, Paternostro G, Vatner SF (1997) Pathophysiological mechanisms of chronic reversible left ventricular dysfunction due to coronary artery disease (hibernating myocardium). Circulation 96:3205–3214

Claycomb WC, Lanson NA, Stallworth BS, Egeland DB, Delcarpio JB, Bahinski A, Izzo NJ (1998) HL-1 cells: A cardiac muscle cell line that contracts and retains phenotypic characteristics of the adult cardiomyocyte. Proc Natl Acad Sci USA 95:2979–2984

Coppen SR, Dupont E, Rothery S, Severs NJ (1998) Connexin45 expression is preferentially associated with the ventricular conduction system in mouse and rat heart. Circ Res 82:232–243

Coppen SR, Gourdie RG, Severs NJ (2001) Connexin45 is the first connexin to be expressed in the central conduction system of the mouse heart. Exp Clin Cardiol 6:17–23

Coppen SR, Kaba RA, Halliday D, Dupont E, Skepper JN, Elneil S, Severs NJ (2003) Comparison of connexin expression patterns in the developing mouse heart and human foetal heart. Mol Cell Biochem 242:121–127

Coppen SR, Kodama I, Boyett MR, Dobrzynski H, Takagishi Y, Honjo H, Yeh H-I, Severs NJ (1999a) Connexin45, a major connexin of the rabbit sinoatrial node, is co-expressed with connexin43 in a restricted zone at the nodal-crista terminalis border. J Histochem Cytochem 47:907–918

Coppen SR, Severs NJ (2002) Diversity of connexin expression patterns in the atrioventricular node: vestigial consequence or functional specialization? J Cardiovasc Electrophysiol 13:625–626

Coppen SR, Severs NJ, Gourdie RG (1999b) Connexin45 (a6) expression delineates an extended conduction system in the embryonic and mature rodent heart. Dev Genet 24:82–90

Daleau P, Boudriau S, Michaud M, Jolicoeur C, Kingma JG, Jr. (2001) Preconditioning in the absence or presence of sustained ischemia modulates myocardial Cx43 protein levels and gap junction distribution. Can J Physiol Pharmacol 79:371–378

Davis LM, Rodefeld ME, Green K, Beyer EC, Saffitz JE (1995) Gap junction protein phenotypes of the human heart and conduction system. J Cardiovasc Electrophysiol 6:813–822

Dupont E, Ko YS, Rothery S, Coppen SR, Baghai M, Haw M, Severs NJ (2001a) The gap-junctional protein, connexin40, is elevated in patients susceptible to post-operative atrial fibrillation. Circulation 103:842–849

Dupont E, Matsushita T, Kaba R, Vozzi C, Coppen SR, Khan N, Kaprielian R, Yacoub MH, Severs NJ (2001b) Altered connexin expression in human congestive heart failure. J Mol Cell Cardiol 33:359–371

Elvan A, Huang X, Pressler ML, Zipes DP (1997) Radiofrequency catheter ablation of the atria eliminates pacing-induced sustained atrial fibrillation and reduces connexin43 in dogs. Circulation 96:1675–1685

Emdad L, Uzzaman M, Takagishi Y, Honjo H, Uchida T, Severs NJ, Kodama I, Murata Y (2001) Gap junction remodelling in hypertrophied left ventricles of aortic-banded rats: prevention by angiotensin II type 1 receptor blockade. J Mol Cell Cardiol 33:219–231

Gourdie RG, Green CR, Severs NJ (1991) Gap junction distribution in adult mammalian myocardium revealed by an antipeptide antibody and laser scanning confocal microscopy. J Cell Sci 99:41–55

Gourdie RG, Severs NJ, Green CR, Rothery S, Germroth P, Thompson RP (1993) The spatial distribution and relative abundance of gap-junctional connexin40 and connexin43 correlate to functional properties of the cardiac atrioventricular conduction system. J Cell Sci 105:985–991

Green CR, Severs NJ (1993) Distribution and role of gap junctions in normal myocardium and human ischemic heart disease. Histochemistry 99:105–120

Gros D, Jarry-Guichard T, ten Velde I, De Mazière AMGL, Van Kempen MJA, Davoust J, Briand JP, Moorman AFM, Jongsma HJ (1994) Restricted distribution of connexin40, a gap junctional protein, in mammalian heart. Circ Res 74:839–851

Guerrero PA, Schuessler RB, Davis LM, Beyer EC, Johnson CM, Yamada KA, Saffitz JE (1997) Slow ventricular conduction in mice heterozygous for connexin43 null mutation. J Clin Invest 99:1991–1998

Gutstein DE, Liu FY, Zhang J, Chen FL, Ghodsi N, Kini AS, Fishman GI (2002) Progressive loss of connexin43 in the heart yields a phenotypic shift from ventricular dysfunction to fatal arrhythmias. Circulation 106:767

Gutstein DE, Morley GE, Tamaddon H, Vaidya D, Schneider MD, Chen J, Chien KR, Stuhlmann H, Fishman GI (2001a) Conduction slowing and sudden arrhythmic death in mice with cardiac-restricted inactivation of connexin43. Circ Res 88:333–339

Gutstein DE, Morley GE, Vaidya D, Liu F, Chen FL, Stuhlmann H, Fishman GI (2001b) Heterogeneous expression of gap junction channels in the heart leads to conduction defects and ventricular dysfunction. Circulation 104:1194–1199

Hall DG, Morley GE, Vaidya D, Ard M, Kimball TR, Witt SA, Colbert MC (2000) Early onset heart failure in transgenic mice with dilated cardiomyopathy. Pediatr Res 48:36–42

Heusch G (1998) Hibernating myocardium. Physiol Rev 78:1055–1085

Honjo H, Boyett MR, Coppen SR, Takagishi Y, Severs NJ, Kodama I (2002) Heterogeneous expression of connexins in rabbit sinoatrial node cells: correlation between connexin isoform and cell size. Cardiovasc Res 50:89–96

Hoyt RH, Cohen ML, Saffitz JE (1989) Distribution and three-dimensional structure of intercellular junctions in canine myocardium. Circ Res 64:563–574

Jain SK, Schuessler RB, Saffitz JE (2003) Mechanisms of delayed electrical uncoupling induced by ischemic preconditioning. Circ Res 92:1138–1144

Jalife J (2003) Rotors and spiral waves in atrial fibrillation. J Cardiovasc Electrophysiol 14:776–780

Jongsma HJ, Wilders R (2000) Gap junctions in cardiovascular disease. Circ Res 86:1193–1197

Kaba RA, Coppen SR, Dupont E, Skepper JN, Elneil S, Haw MP, Pepper JR, Yacoub MH, Rothery S, Severs NJ (2001) Comparison of connexin 43, 40 and 45 expression patterns in the developing human and mouse hearts. Cell Communication and Adhesion 8:339–343

Kaba RA, Coppen SR, Rothery S, Dupont E, Yacoub MH, De Souza AC, Pepper JR, Severs N (2003) Relevance of cardiac gap junctions to the clinical course of patients with chronic atrial fibrillation. Eur Heart J 24:508

Kanagaratnam P, Cherian A, Stanbridge RD, Glenville B, Severs NJ, Peters NS (2004) Relationship between connexins and atrial activation during human atrial fibrillation. J Cardiovasc Electrophysiol 15:206–216

Kanno S, Kovacs A, Yamada KA, Saffitz JE (2003) Connexin43 as a determinant of myocardial infarct size following coronary occlusion in mice. J Am Coll Cardiol 41:681–686

Kanno S, Saffitz JE (2001) The role of myocardial gap junctions in electrical conduction and arrhythmogenesis. Cardiovasc Pathol 10:169–177

Kaprielian RR, Gunning M, Dupont E, Sheppard MN, Rothery SM, Underwood R, Pennell DJ, Fox K, Pepper J, Poole-Wilson PA, Severs NJ (1998) Down-regulation of immunodetectable connexin43 and decreased gap junction size in the pathogenesis of chronic hibernation in the human left ventricle. Circulation 97:651–660

Kitamura H, Ohnishi Y, Yoshida A, Okajima K, Azumi H, Ishida A, Galeano EJ, Kubo S, Hayashi Y, Itoh H, Yokoyama M (2002) Heterogeneous loss of connexin43 protein in nonischemic dilated cardiomyopathy with ventricular tachycardia. J Cardiovasc Electrophysiol 13:865–870

Ko YS, Yeh HI, Ko YL, Hsu YC, Chen CF, Wu S, Lee YS, Severs NJ (2004) Three-dimensional reconstruction of the rabbit atrioventricular conduction axis by combining histological, desmin, and connexin mapping data. Circulation 109:1172–1179

Kostin S, Dammer S, Hein S, Klovekorn WP, Bauer EP, Schaper J (2004) Connexin 43 expression and distribution in compensated and decompensated cardiac hypertrophy in patients with aortic stenosis. Cardiovasc Res 62:426–436

Kostin S, Klein G, Szalay Z, Hein S, Bauer EP, Schaper J (2002) Structural correlate of atrial fibrillation in human patients. Cardiovasc Res 54:361–379

Kostin S, Rieger M, Dammer S, Hein S, Richter M, Klovekorn WP, Bauer EP, Schaper J (2003) Gap junction remodeling and altered connexin43 expression in the failing human heart. Mol Cell Biochem 242:135–144

Lerner DL, Beardslee MA, Saffitz JE (2001) The role of altered intercellular coupling in arrhythmias induced by acute myocardial ischemia. Cardiovasc Res 50:263–269

Lerner DL, Yamada KA, Schuessler RB, Saffitz JE (2000) Accelerated onset and increased incidence of ventricular arrhythmias induced by ischemia in Cx43-deficient mice. Circulation 101:547–552

Luke RA, Saffitz JE (1991) Remodeling of ventricular conduction pathways in healed canine infarct border zones. J Clin Invest 87:1594–1602

Martinez AD, Hayrapetyan V, Moreno AP, Beyer EC (2002) Connexin43 and connexin45 form heteromeric gap junction channels in which individual components determine permeability and regulation. Circ Res 90:1100–1107

Matsushita T, Oyamada M, Fujimoto K, Yasuda Y, Masuda S, Wada Y, Oka T, Takamatsu T (1999) Remodeling of cell-cell and cell-extracellular matrix interactions at the border zone of rat myocardial infarcts. Circ Res 85:1046–1055

Moreno AP, Laing JG, Beyer EC, Spray DC (1995) Properties of gap junction channels formed of connexin 45 endogenously expressed in human hepatoma (SKHep1) cells. Am J Physiol 268:C356–C365

Morley GE, Vaidya D, Samie FH, Lo C, Delmar M, Jalife J (1999) Characterization of conduction in the ventricles of normal and heterozygous Cx43 knockout mice using optical mapping. J Cardiovasc Electrophysiol 10:1361–1375

Nao T, Ohkusa T, Hisamatsu Y, Inoue N, Matsumoto T, Yamada J, Shimizu A, Yoshiga Y, Yamagata T, Kobayashi S, Yano M, Hamano K, Matsuzaki M (2003) Comparison of expression of connexin in right atrial myocardium in patients with chronic atrial fibrillation versus those in sinus rhythm. Am J Cardiol 91:678–683

Peters NS, Green CR, Poole-Wilson PA, Severs NJ (1993) Reduced content of connexin43 gap junctions in ventricular myocardium from hypertrophied and ischaemic human hearts. Circulation 88:864–875

Peters NS, Severs NJ, Coromilas J, Wit AL (1997) Disturbed connexin43 gap junction distribution correlates with the location of reentrant circuits in the epicardial border zone of healing canine infarcts that cause ventricular tachycardia. Circulation 95:988–996

Peters NS, Severs NJ, Rothery SM, Lincoln C, Yacoub MH, Green CR (1994) Spatiotemporal relation between gap junctions and fascia adherens junctions during postnatal development of human ventricular myocardium. Circulation 90:713–725

Petrich BG, Eloff BC, Lerner DL, Kovacs A, Saffitz JE, Rosenbaum DS, Wang Y (2004) Targeted activation of c-Jun N-terminal kinase in vivo induces restrictive cardiomyopathy and conduction defects. J Biol Chem 279:15330–15338

Petrich BG, Gong X, Lerner DL, Wang X, Brown JH, Saffitz JE, Wang Y (2002) c-Jun N-terminal kinase activation mediates downregulation of connexin43 in cardiomyocytes. Circ Res 91:640–647

Pimentel RC, Yamada KA, Kléber AG, Saffitz JE (2002) Autocrine regulation of myocyte Cx43 expression by VEGF. Circ Res 90:671–677

Plum A, Hallas G, Magin T, Dombrowski F, Hagendorff A, Schumacher B, Wolpert C, Kim J-S, Lamers WH, Evert M, Meda P, Traub O, Willecke K (2000) Unique and shared functions of different connexins in mice. Curr Biol 10:1083–1091

Polontchouk L, Haefliger J-A, Ebelt B, Schafer T, Stuhlmann D, Mehlhorn U, Kuhn-Regnier F, Rainer De Vivie E, Dhein S (2001) Effects of chronic atrial fibrillation on gap junction distribution in human and rat atria. J Am Coll Cardiol 38:883–891

Rohr S, Kucera JP, Fast VG, Kléber AG (1997) Paradoxical improvement of impulse conduction in cardiac tissue by partial cellular uncoupling. Science 275:841–844

Rudy Y, Shaw RM (1997) Cardiac excitation: an interactive process of ion channels and gap junctions. Adv Exp Med Biol 430:269–279

Saffitz JE, Kléber AG (2004) Effects of mechanical forces and mediators of hypertrophy on remodeling of gap junctions in the heart. Circ Res 94:585–591

Sakabe M, Fujiki A, Nishida K, Sugao M, Nagasawa H, Tsuneda T, Mizumaki K, Inoue H (2004) Enalapril prevents perpetuation of atrial fibrillation by suppressing atrial fibrosis and over-expression of connexin43 in a canine model of atrial pacing-induced left ventricular dysfunction. J Cardiovasc Pharmacol 43:851–859

Schulz R, Gres P, Skyschally A, Duschin A, Belosjorow S, Konietzka I, Heusch G (2003) Ischemic preconditioning preserves connexin 43 phosphorylation during sustained ischemia in pig hearts in vivo. FASEB J 17:1355–1357

Schulz R, Heusch G (2004) Connexin 43 and ischemic preconditioning. Cardiovasc Res 62:335–344

Schwanke U, Konietzka I, Duschin A, Li X, Schulz R, Heusch G (2002) No ischemic preconditioning in heterozygous connexin43-deficient mice. Am J Physiol (Heart Circ Physiol) 283:H1740–H1742

Sepp R, Severs NJ, Gourdie RG (1996) Altered patterns of cardiac intercellular junction distribution in hypertrophic cardiomyopathy. Heart 76:412–417

Severs NJ (1985) Intercellular junctions and the cardiac intercalated disk. In: Harris P, Poole-Wilson PA (eds) Advances in myocardiology. Plenum, New York, pp 223–242

Severs NJ (1989) Constituent cells of the heart and isolated cell models in cardiovascular research. In: Piper HM, Isenberg G (Eds) Isolated adult cardiomyocytes. volume 1. CRC, Boca Raton, pp 3–41

Severs NJ (1999) Cardiovascular disease. In: Cardew G (Ed.) Gap junction-mediated intercellular signalling in health and disease. Wiley, New York, pp 188–206

Severs NJ (2001) Gap junction remodeling and cardiac arrhythmogenesis: cause or coincidence? J Cell Mol Med 5:355–366

Severs NJ (2002) Gap junctions and connexin expression in human heart disease. In: De Mello WC, Janse MJ (eds) Heart cell coupling and impulse propagation in health and disease. Kluwer, Boston, pp 321–334

Severs NJ (2003) Communicating junctions, connexins and the cardiomyocyte: from cell biology to cardiology. In: Singal PK, Dixon IMC, Kirshenbaum LA, Dhalla NS (Eds) Cardiac remodeling and failure. Kluwer, Boston, pp 417–434

Severs NJ, Coppen SR, Dupont E, Yeh HI, Ko YS, Matsushita T (2004a) Gap junction alterations in human cardiac disease. Cardiovasc Res 62:368–377

Severs NJ, Dupont E, Coppen SR, Halliday D, Inett E, Baylis D, Rothery S (2004b) Remodelling of gap junctions and connexin expression in heart disease. Biochim Biophys Acta 1662:138–148

Severs NJ, Rothery S, Dupont E, Coppen SR, Yeh H-I, Ko Y-S, Matsushita T, Kaba R, Halliday D (2001) Immunocytochemical analysis of connexin expression in the healthy and diseased cardiovascular system. Microsc Res Tech 52:301–322

Shaw RM, Rudy Y (1997) Ionic mechanisms of propagation in cardiac tissue – roles of the sodium and L-type calcium currents during reduced excitability and decreased gap junction coupling. Circ Res 81:727–741

Smith JH, Green CR, Peters NS, Rothery S, Severs NJ (1991) Altered patterns of gap junction distribution in ischemic heart disease. An immunohistochemical study of human myocardium using laser scanning confocal microscopy. Am J Pathol 139:801–821

Spach MS, Heidlage JF, Dolber PC, Barr RC (2000) Electrophysiological effects of remodeling cardiac gap junctions and cell size. Circ Res 86:302–311

Spach MS, Starmer CF (1995) Altering the topology of gap junctions, a major therapeutic target for atrial fibrillation. Cardiovasc Res 30:337–344

Uzzaman M, Honjo H, Takagishi Y, Emdad L, Magee AI, Severs NJ, Kodama I (2000) Remodeling of gap-junctional coupling in hypertrophied right ventricles of rats with monocrotaline-induced pulmonary hypertension. Circ Res 86:871–878

van der Velden HM, Jongsma HJ (2002a) Cardiac gap junctions and connexins: their role in atrial fibrillation and potential as therapeutic targets. Cardiovasc Res 54:270–279

van der Velden HM, van Kempen MJ, Wijffels MC, van Zijverden M, Groenewegen WA, Allessie MA, Jongsma HJ (1998) Altered pattern of connexin40 distribution in persistent atrial fibrillation in the goat. J Cardiovasc Electrophysiol 9:596–607

van der Velden HMW, Ausma J, Rook MB, Hellemons AJCGM, Van Veen TAAB, Allessie MA, Jongsma HJ (2000) Gap junctional remodeling in relation to stabilization of atrial fibrillation in the goat. Cardiovasc Res 46:476–486

van der Velden HMW, Jongsma HJ (2002b) Cardiac gap junctions and connexins: their role in atrial fibrillation and potential as therapeutic targets. Cardiovasc Res 54:270–279

Van Kempen MJA, ten Velde I, Wessels A, Oosthoek PW, Gros D, Jongsma HJ, Moorman AFM, Lamers WH (1995) Differential connexin distribution accommodates cardiac function in different species. Microsc Res Tech 31:420–436

Van Kempen MJA, Vermeulen JLM, Moorman AFM, Gros D, Paul DL, Lamers WH (1996) Developmental changes of connexin40 and connexin43 messenger RNA. Cardiovasc Res 32:886–900

van Veen TA, van Rijen HV, Jongsma HJ (2000) Electrical conductance of mouse connexin45 gap junction channels is modulated by phosphorylation. Cardiovasc Res 46:496–510

Veenstra RD, Wang HZ, Beyer EC, Brink PR (1994) Selective dye and ionic permeability of gap junction channels formed by connexin45. Circ Res 75:483–490

Viswanathan PC, Shaw RM, Rudy Y (1999) Effects of IKr and IKs heterogeneity on action potential duration and its rate dependence: a simulation study. Circulation 99:2466–2474

Vozzi C, Dupont E, Coppen SR, Yeh H-I, Severs NJ (1999) Chamber-related differences in connexin expression in the human heart. J Mol Cell Cardiol 31:991–1003

Wijffels MCEF, Kirchhof CJHJ, Dorland R, Allessie MA (1995) Atrial fibrillation begets atrial fibrillation: a study in awake chronically instrumented goats. Circulation 92:1954–1968

Yamada KA, Rogers JG, Sundset R, Steinberg TH, Saffitz JE (2003) Up-regulation of connexin45 in heart failure. J Cardiovasc Electrophysiol 14:1205–1212

Yao JA, Gutstein DE, Liu FY, Wit AL, Fishman GI (2002) Dissociation of heart function from electrophysiological properties of gap junctions in cardiac-restricted connexin43 knockout mice. Circulation 106:638

Yeh H-I, Lai Y-J, Lee S-H, Lee Y-N, Ko Y-S, Chen S-A, Severs NJ, Tsai C-H (2001) Heterogeneity of myocardial sleeve morphology and gap junctions in canine superior vena cava. Circulation 104:3152–3157

Yeh HI, Hou SH, Hu HR, Lee YN, Li JY, Dupont E, Coppen SR, Ko YS, Severs NJ, Tsai CH (2002) Alteration of gap junctions and connexins in the right atrial appendage during cardiopulmonary bypass. J Thorac Cardiovasc Surg 124:1106–1112

Zipes DP (1997) Specific arrhythmias: diagnosis and treatment. In: Braunwald E (ed) Heart disease. WB Saunders, Philadelphia, pp 640–704

5 Gap Junction Expression in Brain Tissues with Focus on Development

Rolf Dermietzel[1], Carola Meier[1]

5.1 Gap Junctions and Neurogenesis: Some Basic Aspects

The transfer of information between cells appears to be an essential step in the evolution of multicellular organisms. Prior to the evolution of communicative structures such as chemical and electrical synapses (gap junctions), primitive eukaryotes secreted and recognized signal molecules and possessed the potential to associate via cell adhesion, thereby assembling into social complexes of higher order. The onset of differentiation necessitated direct communication between homotypic members and the creation of compartmental boundaries. The invention of direct pathways in the form of communicative structures, which allowed the establishment of compartmental boundaries, e.g., the creation of diffusion gradients and directed signal exchange, was undoubtedly a major step in the evolution of complex multicellular organisms. The early invention of gap junctions during evolutionary processes is therefore not surprising (Revel 1988).

During embryogenesis this phylogenetic trait is recapitulated. Direct transfer of molecules can be detected from the eight-cell stage embryo onwards (McLachlin and Kidder 1986), and gap junction forming proteins (i.e., connexins, Cx) are already present at the zygote stage (Barron et al. 1989). Gap junctions are equipped with all the molecular features necessary for a direct exchange of signaling molecules. They are constituted of transmembrane channels that allow the intercellular exchange of ions (ionic coupling) and transmission of biologically active molecules (metabolic coupling) with a molecular mass below 1200 Da between adjacent cells in multicellular organisms. In the central nervous system, gap junctions are prevalent during the early phase of neurogenesis (prenatal phase), while chemical synapses emerge in the postnatal phase when constitution of appropriate circuits is required in order to process information from the outside world to the brain or within the brain. Apparently, both modes of communication serve different functions, and it is no surprise to find

[1] Institut für Anatomie, Abteilung Neuroanatomie/Molekulare Hirnforschung, Medizinische Fakultät der Ruhr-Universität, 44780 Bochum, Germany, Tel.: +49-234-3225002, Fax: +49-234-3214655, E-mail: rolf.dermietzel@ruhr-uni-bochum.de

Gap Junctions in Development and Disease
(ed. by E. Winterhager)

them in different places or at different times. During early neurogenesis (embryonic day (E) 12–16), the problem that neuroblasts have to solve is primarily regulation of proliferation. Thus, gap junctions at the early stages of neurogenesis may enable neuroblasts to communicate with each other to fulfill these particular tasks, probably in a more metabolic than electrical way. A long-standing concept of gap junction developmental biology is that these junctions are prone to convey intercellular signals through second messengers and low molecular weight morphogenes. They have been suggested to comprise a suitable means to create local gradients of morphogenes, which could provide cells with positional information to control proliferation and differentiation according to their localization within the embryo (Wolpert 1969, 1994). In the developing amphibian neural plate, for instance, in which cells have been proven to be electrically coupled, the cells at different positions along the neural plate have been found at different membrane potentials (Warner 1985). This indicates an electrical gradient between regions of different membrane potentials, creating gradients of ion concentrations along the plate. Although essentially theoretical, the idea of gap junctions being involved in cueing positional information has gained considerable interest over recent years. A major caveat in obtaining experimental clues for this model is that definite data on the interplay between intercellular propagation of signals and external inputs via receptor-mediated signal transmission have rarely been obtained. Such data are of utmost importance since the dissection of external and internal cues and/or the unraveling of their interplay is necessary to define the part that gap junctions contribute to intercellular signaling during morphogenesis. A new twist regarding functional implications of gap junctions during brain development has recently been added to this issue by the discovery that gap junction hemichannels are required for the initiation of calcium waves which apparently play a crucial role in regulating the proliferation of neuronal progenitor cells in the ventricular zone (VZ; Weissman et al. 2004). Such a mechanism in which components of gap junctions in the form of hemichannels may be involved is essentially new and should be discussed in more detail.

The issue of physiological functions of gap junction hemichannels in brain tissues has gained considerable interest over the last few years (for review, see Goodenough and Paul 2003). Hemichannels consist of connexons (hexameres of connexins) which assemble in the trans-Golgi complex from where they are delivered to the plasma membrane (Musil and Goodenough 1993). A fraction of hemichannels seems to be constitutively present in the plasma membrane (Beyer and Steinberg 1991; Li et al. 1996; Hofer and Dermietzel 1998), waiting to be paired with their complementary partners in the adjacent cell. It was long thought that hemichannels are silent and lack any function per se. However, in vitro data exploiting the *Xenopus* oocyte expression system indicated that hemichannels of certain connexin isoforms can occur in open conformation, i.e., Cx46 (Paul et al. 1991), or can be forced into an open state by incubation in low calcium media (Bruzzone et al. 1996; Li et al. 1996; Pfahnl and Dahl 1999). Further detailed studies of transfected cell lines and primary cells including astrocytes indicated that hemichannels are present in mammalian cells as well, and a detailed description of the physiology and pharmacology of hemichannels has been made (Goodenough and Paul 2003).

In brain tissues, hemichannels have been characterized preferentially in astrocytes, and multiple functions have been attributed to this gap junction constituent. They have been suggested to, firstly, account for ATP release (Cotrina et al. 1998), and secondly, to be responsible for the nonexocytotic, Ca^{2+}-independent release of glutamate from astrocytes (Ye et al. 2003). The connexin isoform responsible for hemichannel formation in astrocytes is most likely Cx43 (Dermietzel et al. 1991; Giaume et al. 1991), which has been described as the major gap junction protein in this neural cell type. Interestingly, Cx43 has been found to lack open hemichannel conformation after functional expression in *Xenopus* oocytes under physiological calcium conditions (Bennett et al. 1991), suggesting that the channel when exposed to its natural environment occurs in a closed state, and requires regulative mechanisms to fulfill proper physiological functioning. Since the current experimental paradigm for hemichannel opening relies on low or zero Ca^{2+} conditions in the external milieu (which are hardly achieved under physiological conditions), the functional cues for hemichannel opening in vivo are still enigmatic (but see Bennett et al. 2003, for recent discussion). In the context of developmental processes, this account on hemichannel physiology seems to be a necessary prerequisite for a better understanding of the functional implications in which this gap junction component might be involved.

In terms of the positional information model, which implies a more or less linear diffusion gradient of a morphogenetic substance in a tissue, proper transient modulation in the form of oscillations and/or regenerative waves has become a more favorable paradigmatic factor. Waves can be regarded as transient gradients in time and space. In this way waves carry, besides local differences in concentration (amplitude), time-related information in the form of fluctuations (frequency) of a given substrate. Frequency coding of cytosolic Ca^{2+} has attracted considerable attention (Berridge and Galione 1988; Tsien and Tsien 1990) and the demonstration that gene expression of Ca^{2+}-dependent transcription factors is related to frequency modulation of intracellular Ca^{2+} fluctuations (Dolmetsch et al. 1998; Li et al. 1998) and can discriminate among different transcriptional pathways (Dolmetsch et al. 1998) has considerably improved our understanding of signal transmission encoded by oscillatory transients of this important second messenger.

Synchronous Ca^{2+} oscillations in the developing brain were first described by LoTurco and Kriegstein (1991) in clusters of coupled precursor cells of the ventricular zone in embryonic neocortex, and the contribution that gap junctions may make to the coordination of these events has been discussed (Owens and Kriegstein 1998). The early expression of Cx43 in periventricular precursor cells has been documented, suggesting that this connexin might be responsible for intercellular wave propagation at this early stage of neurogenesis, although Cx26 has also been found in the ventricular zone (Dermietzel et al. 1989; Bittman and LoTurco 1999) indicating a co-expression of both proteins in precursor cells. Interestingly, the expression of both connexins in undifferentiated neuroepithelial cells was found to be closely related to the cell cycle and a dynamic process of coupling and uncoupling as a result of regulated connexin expression during the cell cycle of proliferating neocortical cells was suggested (Bittman and LoTurco 1999). This idea is very much in context with earlier observations of a cell cycle-dependent expression of connexins in nonneural (Dermietzel et al. 1987) and neural tissues (Miragall et al. 1997). With a set of elegant studies the Kriegstein group (Weissman et al. 2004) was able to correlate Ca^{2+} wave initiation and propagation with cortical neuron proliferation. While the contribution of gap junctions to Ca^{2+} wave propagation seems to be of a subsidiary order prevailed on by external transmission through $P2Y_1$ receptor-mediated ATP signaling, hemichannels are rendered essential for the initiation of waves, since the onset of spontaneous waves could be blocked by the common gap junction blocker carbenoxolone leaving wave propagation only slightly affected. The function of hemichannels during the initial phase of wave generation is suggested to be involved in ATP release which generates subsequent waves. Since the spontaneous opening of hemichannels appears to occur in a cell-cycle-specific manner, and the disruption of the Ca^{2+} wave signaling pathway reduces proliferation of progenitor cells in the VZ this pathway is suggested to contribute to regulation of cortical neuron proliferation. The minor effect of gap junction inhibition on Ca^{2+} wave propagation does not necessarily mean that gap junction coupling is not an essential means in coordinating developmental activities during early neurogenesis, especially since dye coupling between cohorts of VZ cells have frequently been encountered (LoTurco and Kriegstein 1991), rather it is an indication that they contribute to additional mechanisms that need to be coordinated among the proliferating cell clusters. In this context, it is of interest to discuss nonjunctional effects of connexins, besides initiation and coordination of calcium waves that may play a role in regulating proliferation. Several lines of evidence suggest that gap junctions convey functions not directly associated with their channel-forming capabilities. Since several tumor cell lines express connexins without establishing communication-competent junctions (Musil et al. 1990; Hofer et al. 1996), it was suggested that junctional communication might influence proliferation, and many studies have correlated the suppression of growth in transformed cells with restoration of communication, typically by connexin transfection (see Mesnil 2002, for recent review). Paradoxically, it appears that in some cases connexin expression alone, without establishment of intercellular channels, might be sufficient to achieve this goal (Zhang et al. 2003). For instance, retroviral delivery of Cx43 or Cx26 to MDA-MB-231 cells does not restore intercellular communication or even cause the assembly of gap junction plaques, but does dramatically suppress tumor growth when these cells are implanted in nude mice (Qin et al. 2002). The idea of a regulative effect of connexins on proliferation apart from their gap junction forming capabilities has gained even more support by the demonstration that the transcription factor NOV associates with the carboxy-terminus of Cx43 thereby imposing a negative growth control on cell proliferation (Fu et al. 2004; Gellhaus et al. 2004). How this effect is achieved is a matter of extensive investigation.

Interestingly, in Cx43 knock-out animals subtle defects in the lamination of the cortical layers have been described (Fushiki et al. 2003) and were attributed to deficits in migration. In the context described above, the lack of Cx43 and consequently the lack of hemichannels could also account for impairment of Ca^{2+} wave initiation and thus lead to defects in progenitor cell proliferation which ultimately would generate perturbations of cortical pattern formation. In addition to this indirect effect of hemichannels on pattern formation, a direct role on migratory activity of gap junction and/or their hemichannel components cannot be excluded. Data from transgenic mice overexpressing Cx43, Cx43 knock-out mice (KO), and transgenic mice expressing a dominant negative Cx43 gene support this idea (Ewart et al. 1997; Huang et al. 1998). According to these data, overexpression of Cx43 results in an increase in migratory activity of cardiac neural crest cells, whereas the absence of Cx43 or its functional inhibition by the dominant negative fusion protein leads to a decrease in migration. It is noteworthy that the perturbation of Cx43 expression exerts also an influence on the proliferation of myocardial cells, with an increase in the overexpressing animals and a decrease in the KO mice. Apparently, the timely presence of cardiac neural crest cells in certain regions of the developing heart seems necessary for correct developmental modeling of myocardial cell derivatives; in heart organogenesis, the modeling of the conotruncal cardiac outflow (Reaume et al. 1995). It is tempting to extrapolate the essence of these findings to neocorticogenesis since both processes depend on similar basic mechanisms: (1) local symmetric proliferation of neural precursor cells, (2) guided migration, and (3) differentiation distant from the source of origin.

5.2 Segregation of Connexin Expression During Brain Development

5.2.1 Prenatal Expression

Biophysical, electrophysiological and tracer studies have shown that the different connexin isoforms confer different functional properties to gap junction coupling including unitary conductance, pH dependence and ion selectivity (Veenstra 2001). In this context it is of interest to elaborate a little more on the expression profiles of connexins during brain development. Unfortunately, this undertaking is hampered by the fact that divergent data exist with respect to cell-specific expression. We will thus draw a preliminary outline of the expression profiles which have found general acceptance in the literature before discussing the various findings below.

During the early phase of development (E12), premature neuroblasts co-express Cx43 and Cx26 (Fig. 5.1). Immunocytochemical studies show that the highest level of expression occurs in the ventricular zone (Dermietzel

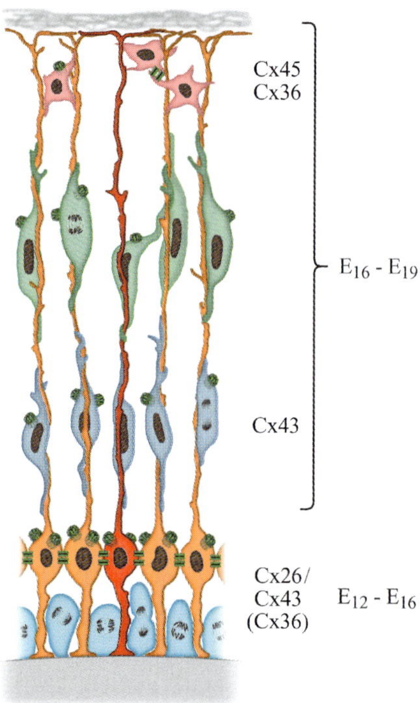

Fig. 5.1. Schematic representation of connexin expression pattern during neocorticogenesis

et al. 1989; Bittman et al. 1997; Bittman and LoTurco 1999). This finding is consistent with the high degree of dye coupling between precursor cells in the periventricular germinal layer (LoTurco and Kriegstein 1991). However, RT-PCR analysis on an immortalized mouse embryonic hippocampal progenitor cell line (MK31) in an uncommitted state indicated a more complex expression which included Cx26, Cx33, Cx36, Cx43 and Cx45 (Rozental et al. 2000). The expression of messenger RNA alone is, of course, not indicative for the presence of the translational product, especially since a variety of connexin mRNAs was described without adequate translation in several tissues (Traub et al. 1998; Condorelli et al. 2003). In addition, the RT-PCR method is sensitive enough to detect traces of mRNA which could account for signals from "leaky promotor" activity having no functional meaning. In the case of Cx45 Western blot analyses confirmed that this protein is present at prenatal stage E18.5 revealing this gap junction protein as a further embryonic connexin (Sohl et al. 2004). This stage, however, is too late to discern expression in uncommitted precursor cells. Connexin31 has also been described in early mouse embryos, thereby revealing a transient locally restricted pattern between E8 and E11 to rhombomeres r3 and r5 (Jungbluth et al. 2002) of the hindbrain. At later stages (from E10.5–E11.5), Cx31 is expressed in the ventral rhombomere r4. A further site of expression occurs at E11.5 in the boundary cap cells in the entry/exit points of all mixed sensory/motor and entry points of pure sensory nerves. This particular pattern shows that some connexins are extremely restricted in terms of their spatial and temporal expression during embryogenesis, a fact that should be taken into account when developmental profiles of connexin genes are considered.

Segregation of connexin expression becomes evident when lineage restriction of the undifferentiated neuroblasts commences. Although a definite schedule of the expression profiles of connexins along the specific differentiation pathways in situ have not been deciphered yet, studies on neural precursor cells lines indicate that commitment of the cells to a neuronal pathway coincides with a shift in expression from Cx43 to Cx36 (Fig. 5.1), while cells following the glial lineage sustain Cx43 expression (for details see below). Corroborative data were reported by Belliveau et al. (1997), who demonstrated an expression of Cx43 in P19 cells, a pluripotent embryonal carcinoma cell line. Upon exposure to retinoic acid and subsequent lineaging along the glial pathway Cx26 was found within 3 days following differentiation into immature neurons while Cx36 was not investigated in that study because of a lack of information on this connexin at that time. Interestingly, Cx32 was not detected in cells of glial restriction at this time point which confirmed that Cx32 is a late protein in brain development (Dermietzel et al. 1989). With respect to early embryonic expression of Cx36 it is remarkable that this connexin, which in adult stages is exclusively expressed in neuronal tissues including their lineage derivatives, i.e., beta-cells of the pancreas, adrenal chromaffine cells (Serre-Beinier et al. 2000; Degen et al. 2004; Li et al. 2004), reveals highly restricted expression in chick somites around E2, which express MYOD and FGF8, indicative for myotome fate. RT-PCR studies on Cx36 expression in mouse embryos indicate that traces of Cx36 mRNA can be detected around E7.5 and in situ hybridization together with Northern blot analysis provide evidence of restricted neural expression at E9.5 (Gulisano et al. 2000). Again, the restriction of Cx36 expression in the early embryo, similar to the expression of Cx31, is remarkable, exhibiting a confinement to boundaries which play a specific function during brain morphogenesis. A feasible explanation for the differential expression of connexins during early neurogenesis is that individual connexins fulfill specific functions which fit into the developmental scheme of brain morphogenesis.

5.2.2 Postnatal Expression

Most studies related to developmental expression of gap junction proteins focus on postnatal stages. An obvious reason for this preference is that functional data on slice preparations have indicated that neuronal coupling within the first two postnatal weeks in rodents is exceptionally high and declines thereafter. Single-cell injection of gap junction permeable dyes like Lucifer Yellow, biocytin, neurobiotin as well as Ca^{2+} imaging studies revealed extensive coupling between principal neurons as well as between interneurons of the neocortex (Connors et al. 1983; Yuste et al. 1992; Peinado et al. 1993a; Rorig et al. 1995; Yuste et al. 1995; Kandler and Katz 1998) and in other brain areas (Goodman and Spitzer 1979; Spitzer 1991; Christie and Jelinek 1993). Interneuronal coupling peaks before and during the major period of synapse formation at a time when the principal synaptic circuits are established (Miller 1988; Rorig and Sutor 1996a; Kandler and Katz 1998). This coincidence was suggestive for a functional link of gap junction coupling and the emergence of chemical synapses and thus the primary findings were soon encapsulated in a theoretical frame (Kandler and Katz 1995). In slices of neocortex, Yuste et al. (1992, 1995) described groups of neurons that spontaneously increased their Ca^{2+} in synchrony, defining locally correlated areas which they termed "neuronal domains" (Yuste et al. 1992). The size, shape, and organization of these events mirror the structure of vertical horizontal domains of the mature cortex. It was therefore suggested that they might represent an early manifestation of the cortical functional architecture of the mature neocortex. Since blocking of gap junctions with halothane and octanol abolished the domain-oriented Ca^{2+} waves, it was concluded that gap junction coupling is responsible for the co-activation in the cortical domains. Whether the co-activation in the postnatal neuronal domains is responsible for the functional wiring and stabilization of the modular architecture of the mature cortex, and how this may happen is still a matter of debate. Since electrical coupling between neurons in the neocortex seems to be too weak to be responsible for the synchronization of neuronal activities during the occurrence of neuronal domains (Connors et al. 1983; Peinado et al. 1993b), metabolic coupling was suggested to play a role in this event (Kandler and Katz 1998). Based on their observation that Ca^{2+} spread in neuronal postnatal domains involves the intercellular diffusion and probably the partial regeneration of IP_3 and that the increase of IP_3 is related to glutamate release, Kandler and Katz (1998) suggested a model for initiation and propagation of calcium waves that includes neurotransmitter interaction. In this model glutamate, probably deriving from developing synapses, is assumed to work through metabotropic glutamate receptors (mGluR) and G-protein coupled phospholipase C, resulting in the production of IP_3 that can be transmitted via gap junctions throughout the neuronal domains. This concept is of particular interest since the maximum amount of gap junction coupling and its decline in the postnatal developing brain is inversely correlated to the structural and functional evolution of chemical synapses (see Kandler and Katz 1998, for review). Whether an external signaling via ATP transmission is involved in the postnatal propagation of calcium waves has not been considered yet and remains a subject for future studies. Nevertheless, the concept of a close correlation of the postnatal evolution of the metabotropic glutamate system and coupling through gap junctions has gained additional support by the finding that mice lacking functional NMDA receptors show a remarkable decrease of spontaneous calcium domains by 90% and a decrease of about 20% of neuronal domains after IP_3 injection (Okada et al. 2003). Further support in favor of this concept comes from studies on hypothalamic neurons with chronically blocked glutamate receptors and knock-out mice with a deletion of the NMDA NR1 receptor subunit (Belousov et al. 2004). Chronic NMDA receptor blockade in vivo prevented developmental gap junction uncoupling and reduced the developmental decrease in Cx36 expression similar to the NMDA NR1 knock-out, thereby favoring the idea that a functional relationship of both systems during postnatal development exists. In addition, the paper provided evidence that this regulation depends on signal transmission pathways in which protein-kinase C (PKC), CamK II/IV and CREB are involved.

Other neurotransmitters have also been described to exhibit a modulatory effect on gap junction communication in the developing brain.

Serotoninergic innervation is present in the sensory cortex when there is widespread junctional coupling, and incubation of postnatal brain slices with serotonin results in a reduction of tracer-coupled cells (Rorig and Sutor 1996b). The effect appears to be mediated through 5-HT_2 receptors and includes further signal transmission pathways among which IP_3-dependent calcium release and activation of PKC seem to be major players. Similar effects were reported for dopamine which induces a reduction in tracer coupling of about 50% when applied to prefrontal and frontal cortices during the second postnatal week. Both D_1 and D_2 receptors seem to participate in this effect (Rorig et al. 1995), which is at least partially mediated by protein-kinase A. Interestingly, functional interaction of neurotransmitters like dopamine with gap junction coupling is also a significant feature in the mature brain where it has been extensively studied in horizontal cells of the retina. Here, dopamine regulates the size of the receptive fields in response to light exposure (Piccolino et al. 1984; Dong and McReynolds 1991). Upon illumination the number of coupled cells decreases dramatically which leads to a sharpening of the spatial resolution by focusing signal propagation from the receptor cells to the bipolar cells. PKA has been described to be essential for this interaction and it seems that mechanisms similar to those described for neurons of mature brains also account for the effects in the developing brain.

5.3 Drawing a General Scheme of Gap Junction Function in the Developing Brain

In order to compose a scheme of gap junction function during brain development, one can follow the general ideas posted by Montoro and Yuste 2004. A basic feature seems to be that gap junctions play differential roles during neurogenesis. It appears that in the early prenatal stages of neurogenesis there is a necessity for strong coordination in groups of cells. Apparently, Ca^{2+} waves may confer signals from cell to cell, which could influence the coordination of cell proliferation as well as early differentiation mechanisms. The pathway of Ca^{2+} transmission entails an initial trigger event that seems to be dependent on gap junction hemichannels followed by a subsequent cell-to-cell propagation. It seems likely that nonjunctional effects of connexins, besides initiation and coordination of calcium waves, play a role in regulating proliferation events. Transmission of Ca^{2+} waves is composed of a major extracellular ATP-dependent component and a direct intercellular component via gap junctions. The main effect of Ca^{2+} is thought to be synchronization of S-phase entry generating cohorts of coordinately dividing cells which then exit from the VZ into the subventricular

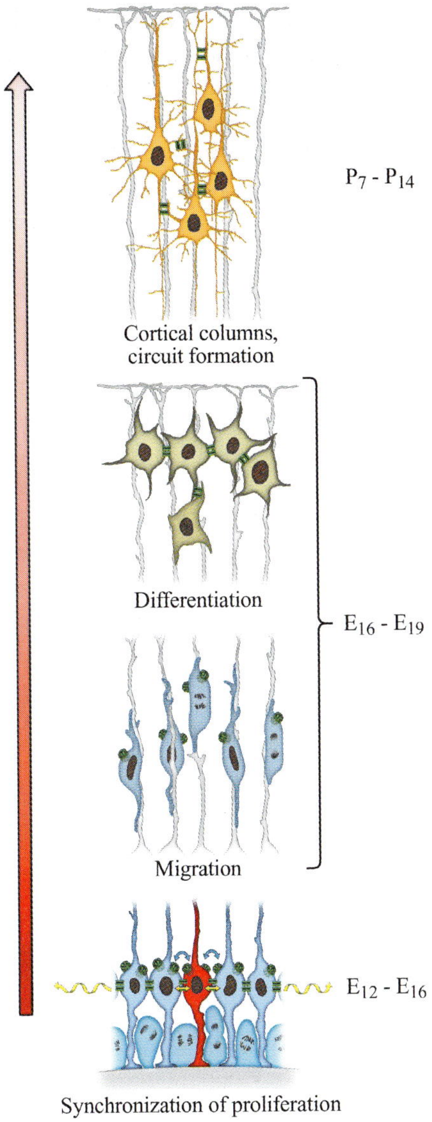

Fig. 5.2. Schematic representation of putative functions of gap junctions and gap junction hemichannels during neocorticogenesis. During early neurogenesis (E12–E16), hemichannels trigger the initiation of calcium waves which are suggested to coordinate proliferation of precursor cells through S-phase synchronization. Cohorts of synchronized cells migrate from the ventricular zone to the developing cortical plate where asymmetric division and differentiation commences (E16–E19). Gap junctions and/or hemichannels may be involved in the process of migration and coordination of differentiation. During postnatal maturation of the neocortex (P7–P14), postmitotic neurons assemble in cortical domains which reveal coactivation in the form of calcium transients. Interneuronal gap junction coupling is suggested to play a role in stabilizing and defining the mature neuronal circuits at this stage

zone from which isochronic division and migration of cohorts of precursor cells from the VZ to the cortical plate happen. In addition, migration may be coordinated by hemichannels and/or gap junction interaction.

In late neurogenesis asymmetric division and differentiation lead to the pattern formation of the cortical plate. At this stage, gap junctions in concert with extracellular cues may transmit morphogenetic signals for local differentiation of postmitotic immature neurons.

In the postnatal brain interneuronal gap junctional coupling increases between the first and second postnatal week. This peak of interneuronal coupling correlates with a maximum of Cx36 and Cx45 expression, the latter being a putative second neuronal connexin (see below). Since coupling in the postnatal phase occurs in vertical domains which penetrate the cortical layers, it is thought that the domains may represent an early manifestation of the cortical functional architecture of the mature neocortex, and that junctional coupling plays a role in stabilizing and defining the mature neuronal circuits (see Figs. 5.1, 5.2).

It is worthwhile mentioning that in the maturing cortex the expression of Cx36 segregates preferentially to GABAergic interneurons with low or even absence of expression in principal cells and coupling seems to be involved in the generation of gamma frequency network oscillations (Deans et al. 2001; Hormuzdi et al. 2001, 2004).

One has to consider that several aspects of this principal scheme are not so well supported and require further experimental work, especially since targeted deletion of connexin genes expressed early in neurogenesis reveals only minor defects of corticogenesis (Fushiki et al., 2003). Here, compensatory mechanisms may account for the lack of function of a particular gene. Further genetic approaches utilizing double knock-out animals or transgenic strains expressing proteins with dominant negative function may offer better insights into the developmental capabilities of gap junctions and/or their basic constituents, the connexins.

5.4
Pitfalls in Defining Cell-Specific Expression of Connexins in Brain Tissues

A definite attribution of some connexins to specific cell types is obscured by methodological problems. The discovery that about 20 different connexin isoforms exist (20 in mouse and 21 in humans; Willecke et al. 2002; Sohl et al. 2004) has made the detection by common substrate-based techniques, i.e., immunocytochemistry or in situ hybridization, a major handicap because of cross-reactivity between the different isoforms. This problem becomes even more pronounced when developmental studies are considered, since the segregation of connexins during cell fate decision and the unequivocal determination of cell types by specific marker proteins can rarely be achieved in tissue samples. The generation of knock-out animals carrying reporter genes like lacZ driven by connexin-specific promoters has been an advantage in defining the expression profiles of connexins better, and a reassignment has been achieved in some cases in adult tissues (Sohl et al. 2004). This is the case for Cx45 and Cx47. The former is now attributed to neurons (Maxeiner et al. 2003), while it was first described as an oligodendrocytic connexin (Dermietzel et al. 1997; Kunzelmann et al. 1997). The misleading attribution of Cx45 to oligodendrocytes has in part been due to cross-reactivity of the applied antibody with the homologous Cx47, which is now found to be expressed in oligodendrocytes (Odermatt et al., 2003) instead of neurons where it was first described (Teubner et al. 2001). Despite this reassignment of Cx45 expression in adult tissue, its developmental expression is not completely clarified. Apparently, Cx45, besides segregating to the neuronal lineage, remains expressed in cells which lack neuronal markers (Condorelli et al. 2003), indicating a nonneuronal expression site as well. A similar situation can be found in the case of Cx26 expression. A general consensus has been reached that Cx26 is expressed in early neurogenesis in the progenitor cells of the ventricular zone. It was further described to become restricted to leptomeningeal and ependymal cells in the postnatal brain (Dermietzel et al. 1989; Bittman and LoTurco 1999). However, expression in astrocytes has also frequently been reported (Nagy et al. 2001). Recent studies exploiting transgenic techniques with connexin-specific promoter and reporter genes indicate that postnatal expression of Cx26 is confined to nonneural cells, i.e., leptomeningeal cells and ependymal cells (Filippov et al. 2003). One major argument against the transgenic technique is that in weak promoter activity or in heterozygous animals, cells with a low abundance of expression can not be detected. The only way to overcome the described difficulties is to apply a broad spectrum of techniques, including mutant animals (Meier et al. 2002), and approaches that combine micro-sampling of tissue with molecular biological methods (Weickert et al. 2005).

5.5 Segregation of Connexins During Glial Lineaging

5.5.1 The Oligodendrocytic Lineage

Although neurons and glia fulfill very different functions in the mature brain, their common developmental origin, i.e., the same population of multipotent stem cells (Rao 1999), implies a connection between their gap junction make-up. Like neurons, astrocytes and oligodendrocytes are both of neurectodermal origin and develop from the ventricular zone. Prior to maturation, oligodendrocytes and type II astrocytes have a common progenitor, the O2A cell, which is characterized by bipolar morphology and expression of the NG2 marker protein. These progenitor cells were shown to express Cx45, both in cortical as well as cerebellar areas of the brain, as demonstrated by expression of a Cx45-LacZ reporter gene (Maxeiner et al. 2003). However, Cx45 expression is only transient as with maturation, the progeny of O2A cells loses Cx45 expression (Maxeiner et al., 2003; Kleopa et al., 2004), except for a small population of O2A spindle-shaped NG2-positive cells (Maxeiner et al. 2003).

In addition to lineage progression towards mature oligodendrocytes and type II astrocytes, O2A cells also give rise to oligodendrocyte precursor cells (Noble and Wolswijk 1992). These cells types are still present in the adult rodent brain, and display characteristic patterns of connexins expression; oligodendrocyte precursor cells, characterized by small cell bodies with stellate processes and NG2 protein expression, were demonstrated to express Cx32 (Melanson-Drapeau et al. 2003). In general, expression of Cx32 – both spatial and temporal – was shown to coincide with maturation of oligodendrocytes (Belliveau et al. 1991; Belliveau and Naus 1995). However, Cx32 expression in precursor cells pointed to a distinctly different functional set-up. When investigating the implications of Cx32 deletion on oligodendrocyte development, the observed effects were manifold; in Cx32-deficient mice (Nelles et al. 1996), proliferation of nestin- and NG2-positive oligodendrocyte progenitor cells in the gyrus dentatus were enhanced, however, there was also an increase in apoptosis of this cell population, resulting in an increase in progenitor cell turnover (Melanson-Drapeau et al. 2003). Expression of Cx32 therefore influences the number of adult oligodendrocyte progenitor cells by affecting proliferation, survival and differentiation of a subset of oligodendrocyte precursor cells (Melanson-Drapeau et al. 2003). As no significant levels of Cx32 have been detected in embryonic brain (Parnavelas et al. 1983; Dermietzel et al. 1989; Belliveau et al. 1991; Nadarajah et al. 1997), an impact of Cx32-deficiency on the retention of embryonic progenitor populations was not expected. Accordingly, lineage progression of the main proportion of cells developing from O2A progenitor cells (i.e., oligodendrocytes and type II astrocytes) was not affected following Cx32-deficiency (Melanson-Drapeau et al. 2003).

Connexin32 expression, as described for oligodendrocyte progenitor cells, is maintained throughout oligodendrocyte maturation (Dermietzel et al. 1989; Micevych and Abelson 1991; Belliveau and Naus 1995; Giaume and Venance 1995; Scherer et al. 1995; Spray and Dermietzel 1995). In mature oligodendrocytes, two additional connexin proteins are present: Cx29 and Cx47 (Altevogt et al. 2002; Li et al. 2002; Menichella et al. 2003; Nagy et al. 2003; Odermatt et al. 2003; Kleopa et al. 2004; Nagy et al. 2004). Data on the expression of Cx45 (see above), are more controversial as studies on Cx45-transgenic animals seem to exclude its expression in mature oligodendrocytes (Maxeiner et al. 2003). Cx36 expression has also been described in cultured oligodendrocytes (Parenti et al. 2002), however, culture conditions might have changed the molecular make-up of cells.

Gap junctions between oligodendrocytes themselves are rarely if ever observed (Massa and Mugnaini 1982; Waxman and Black 1984; Rash et al. 2001a). However, oligodendrocytes are integrated into the astrocytic network by gap junctional communication, creating a "panglial syncytium" (Massa and Mugnaini 1982) consistent with heterotypic channels. It has been suggested that oligodendrocyte Cx32 associates mainly with astrocytic Cx26, whereas oligodendrocyte Cx47 associates with astrocytic Cx43 and Cx30 (Rash et al. 2001b; Rash et al. 2001a; Altevogt and Paul 2004).

The lack of Cx32, causing symptoms of Charcot-Marie-Tooth X-linked neuropathy in the peripheral nervous system, affects the central nervous system only very mildly, leading to reduced myelin volume and increased excitability (Sutor et al. 2000). Moreover, deletion of Cx47 resulted in minor impairment of central myelin, i.e., slight degeneration and vacuolation of myelin, particularly in the optic tract (Odermatt et al. 2003). However, combined deletions of Cx32 *and* Cx47 exhibited severe defects in CNS myelin and led to the premature death of animals (Menichella et al. 2003; Odermatt et al. 2003), indicating that (1) either connexin might possibly compensate for the other, leading to a comparably mild phenotype in each of the individual knock-outs and (2) both connexins play a role in myelination. This is also supported by the expression patterns of Cx32 and Cx47, which closely parallel that of myelin proteins, with a peak of mRNA expression between P10 and P30 (Menichella et al. 2003).

In comparison, the role of Cx29 is less identified: Immunohistochemically, Cx29 staining does not show any colocalization with any known glial connexin in the central nervous system (Altevogt and Paul 2004) – pointing to a role independent of that in the establishment of the pan–glial network. In addition, immunolabeling for Cx29 appears more diffuse and irregular in shape than for the other gap junction proteins described, leading to the speculation that Cx29 might not be generally incorporated into gap junctional plaques (Altevogt and Paul 2004). This notion was supported by freeze-fracture studies on the peripheral nervous system which suggest that Cx29 connexons might not even be assembled in common gap junction plaques (Li et al. 2002). However, the developmental expression profile of Cx29 parallels that of proteolipid protein (PLP; Altevogt et al. 2002), showing a temporal regulation similar to other myelin proteins and, therefore, also coinciding with Cx32 and Cx47 expression (Sohl et al. 2001).

In addition to intercellular gap junctions, oligodendrocytic gap junctions also exist between different parts of one cell, named 'reflexive' gap junctions. Immunohistochemical and freeze-fracture replica immunolabeling (FRIL) data suggest that both Cx29 and Cx32 are also localized at intracellular membranes, as has also been demonstrated for Schwann cells of the peripheral nervous system (Li et al. 2002; Meier et al. 2004). The function of inner cellular connexins in oligodendrocytes was considered to be primarily metabolic, allowing the passage of nutrients and ions from the soma to all layers of myelin (Paul 1995).

5.5.2 The Astrocytic Lineage

As demonstrated in vitro, type II astrocytes are, in addition to oligodendrocytes, the second mature cell maturing from O2A progenitors (Noble and Wolswijk 1992). The characteristics of astrocytic subpopulations (type I and type II astrocytes) have been investigated in culture (Raff et al. 1983). Strikingly, there is no evidence for type II astrocytes to be coupled by gap junctions: type II astrocytes – at least in the optic nerve – do not express Cx26, Cx32, or Cx43 (Sontheimer et al. 1990), nor do they establish functional channels (Belliveau and Naus 1994).

Most recent data support the notion that the astrocytic cell population in vivo (at least in the hippocampus) might indeed be diverse, and comprise coupled as well as noncoupled astrocytes. The most prominent differences are the morphology of cells as well as the expression of the glutamate transporter GluR (Matthias et al. 2003). Wallraff et al. (2004) were able to demonstrate that GluR-negative astrocyte populations lack gap junctional tracer coupling, whereas GluR expression is associated with gap junctional coupling. Therefore, the GluR-negative population is not participating in the astrocytic and pan-glial networks of the brain, pointing to a different role of these cells (Wallraff et al., 2004). Morphologically different astrocytes have also been identified in brain and spinal cord, i.e., fibrous and protoplastic astrocytes. In view of this data, subpopulations of astrocytes may account for regional differences observed for the distribution of astrocytic connexins described below.

However, many astrocytes form extensive gap junction contacts, and mature cells were shown to express a number of connexin proteins, including Cx43, Cx30 and presumably Cx26 (reviewed by Nagy and Rash 2000). These three connexins have been identified in the adult rat brain and spinal cord and were ultrastructurally identified as astrocytic gap junctions by immunolabeling on freeze-fracture replicas (reviewed by Nagy and Rash 2000). Studies on Cx43-deficient astrocytes also imply that the pattern of connexin expression can be modulated: In Cx43-deficient cells, the additional expression of Cx40, Cx45 and Cx46 was detected (Dermietzel et al. 2000) by RT-PCR. Although there are still some controversies on the occurrence of some of these connexins in astrocytes, those issues have been reviewed in detail by Nagy et al. (2004). We will therefore focus on their expression during brain development.

In general, the expression of astrocytic gap junction proteins is developmentally regulated and displays regional distribution (Nagy and Rash 2000; Rouach et al. 2000). In developing astrocytes, Cx43 is expressed during prenatal development and in abundance postnatally (Dermietzel et al. 1989). Interestingly, the implication of Cx43 on astrocyte maturation seem to be negligible as ablation of Cx43 did not influence astrocytic development (Naus et al. 1997). However, Cx43 expression was shown to commence prior to astrocyte generation; in rats, Cx43 protein appears already at embryonic day (E) 12 (Dermietzel et al. 1989; Nadarajah et al. 1997) as the neural tube is closing. At this timepoint, the telencephalic vesicles first appear and the subventricular epithelium is tightly packed with dividing stem cells. With the onset of cortical neurogenesis (E14), Cx43 protein was distributed throughout the ventricular zone and at E19, its association with radial glia cells was demonstrated (Nadarajah et al. 1997; Bittman and LoTurco 1999). As radial glia cells are involved in neuronal cortical migration (Rakic 2002), these observations indicate that connexins might also contribute to that process.

Connexin43 protein levels increase during the prenatal period and expression peaks around birth (Dermietzel et al. 1989; Nadarajah et al. 1997). At this time, Cx43 immunoreactivity was detected throughout the cortical thickness, however, it was concentrated in the lower cortical layers (Nadarajah et al. 1997). The characteristic pattern of punctate immunostaining was most often reached during the second postnatal week, however, the timepoint can vary depending on the functional maturation of brain areas. Once the mature pattern of Cx43, i.e., predominant location in the infragranular layers at P28 (Nadarajah et al. 1997) was established, there were no changes in the amount of protein observed over time (Dermietzel et al. 1989).

For Cx26, another gap junction protein thought to be present in mature astrocytes, the spatial and temporal expression was distinctly different. In general, the labeling intensity of Cx26 was weaker than that of Cx30 (see below) and Cx43, both during development and in adult brain (Dermietzel et al. 1989; Nadarajah et al. 1997; Altevogt and Paul 2004). Connexin26 expression also commenced at E12 (Dermietzel et al. 1989), the protein being located throughout the neuroepithelium (Nadarajah et al. 1997). Between E14 and E16, Cx26 showed a peak of expression being distributed throughout the teleencephalic wall, whereas from E19 onwards, Cx26 was concentrated in the proliferative (ventricular and subventricular) zones (Nadarajah et al. 1997). Intriguingly, the temporal expression of Cx26 protein closely matches the period of neurogenesis (E14–E16) as well as its decline (>E19). However, data obtained with mice that express the β-galactosidase gene under the Cx26 promoter are quite controversial, indicating that cells of the ventricular zone as well as astrocytes are devoid of Cx26 (Filippov et al. 2003).

As demonstrated for Cx43 (see above), Cx26 protein was also documented in radial glial cells of the ventricular zone (Bittman and LoTurco 1999). Postnatally, the overall amount of Cx26 gradually decreased (Dermietzel et al. 1989; Nadarajah et al. 1997) and thereby paralleled the expression of Cx45 (Maxeiner et al. 2003). However, Cx26 protein was clearly maintained in pinealocytes, cells of the leptomeninges and the ependyma (Dermietzel et al. 1989). Despite initial conflicting observations concerning Cx26 expression in the mature brain, the view has cristallized out that Cx26 is the third astrocyte gap junction protein (reviewed by Nagy et al. 2004). Nevertheless, Cx26 appears least represented and seems to be restricted to subpopulations (Mercier and Hatton 2001; Nagy et al. 2001; Nagy et al. 2004), a fact that might explain the negative results in the transgenic animals (see above).

Contrasting the expression patterns of Cx43 and Cx26 in astrocytes, Cx30 displays a delayed onset in appearance and has been described in mature astrocytes only (Dahl et al. 1996; Nakase and Naus 2004). In the rat brain, Cx30 mRNA becomes detectable during the second postnatal week (Dahl et al. 1996; Condorelli et al. 2002). Connexin30 protein is first detected at P16, and after a peak of expression during the fourth postnatal week, remains at high levels throughout adulthood (Kunzelmann et al. 1999).

Comparing all three astrocytic gap junction proteins, their distribution as well as expression levels display striking regional differences in adult animals (Altevogt and Paul 2004; Nagy et al. 2004). In general, immunostaining for these three connexins was far more prominent in gray matter than in white (Altevogt and Paul 2004), and this distribution was particularly evident for Cx30 which was virtually absent in white matter astrocytes (Nagy et al. 1999). Regional differences of astrocytic connexin expression, which have also been reported by other authors (Lynn et al., 2001; Condorelli et al., 2002), may therefore reflect the morphological and functional differences of astrocytes described above.

5.6 Upstream Events Regulating Connexin Expression

In addition to the description of temporal and spatial expression of connexins as well as their action, another important topic is beginning to evolve; the upstream events, regulating connexin expression and function itself, are currently being elucidated. Briefly, in the nervous system these involve transcriptional regulators like Sox10 (Bondurand et al. 2001) and p38/SAPK2 kinase (Zvalova et al. 2004), protein phosphatases (Herve and Sarrouilhe 2002) and protein kinases (Cruciani and Mikalsen 2002; Warn-Cramer and Lau 2004), neuronal apoptotic death following kainate-induced seizures (Condorelli et al. 2003), signaling molecules like "bone morphogenetic proteins" (Bani-Yaghoub et al. 2000) and Wnt (van der Heyden et al. 1998), as well as neuroinflammation (reviewed by Kielian and Esen 2004). In neuroinflammation, brain macrophages/microglial cells and astrocytes are cooperatively generating the local inflammatory response of the brain (Benveniste 1992; McGeer and McGeer 1995; Benveniste 1998). Contact between these two cell populations was shown to reduce connexin expression and gap junctional coupling in astrocytes (Rouach et al. 2002; Faustmann et al. 2003), an effect which was mimicked by astrocytic treatment with inflammatory cytokines (John et al. 1999; Martinez and Saez 2000). In addition to transcriptional and junctional regulation of gap junction proteins, the issue of hemichannel involvement has been discussed recently (reviewed by Contreras et al. 2004).

In contrast to neurectodermal astrocytes and oligodendrocytes, microglial cells are mesodermal derivatives. Microglia are ubiquitous in brain and are the main immune effector in the central nervous system (Streit et al. 1988). In response to "injury", a term comprising any injury that can trigger inflammatory response in the brain, microglia cells are recruited to the site of injury and commence proliferation (Gehrmann et al. 1995). Recently, microglia was also shown to undergo an increase in connexin expression and dye coupling upon activation, an effect which was also obtained by treatment with inflammatory cytokines (Eugenin et al. 2001) or Ca^{2+} ionophores (Martinez et al. 2002) in vitro.

Since connexins were identified as functional targets for signaling molecules like Wnt and bone morphogenetic proteins (BMPs), developmental expression patterns have to be viewed under functional aspects. Studies on the embryonal mouse P19 cell line (Bani-Yaghoub et al. 2000) reveal an increase of Cx43 mRNA and protein expression upon BMP treatment. In parallel, neuronal differentiation was inhibited, whereas formation of astrocytes was enhanced. In the P19 cell line, but also in a PC12 cell system, Wnt also caused an increase in gap junctional communication, implying that Cx43 is a functional target of Wnt (van der Heyden et al. 1998). Intriguingly, the expression pattern of Wnt in E10.5 mouse embryos closely resembles that of Cx43, both being distributed as a gradient in regions spanning the mid-/hindbrain border (Wilkinson et al. 1987). Although the spatial and temporal regulation of connexin expression has been implicated in the control of differentiation and growth for decades (Caveney 1985), the underlying cellular mechanisms are only currently being unraveled.

Whereas Cx43 expression seems to be increased in cells becoming astrocytes, commitment to a neuronal fate seems to have an adverse effect, i.e., the downregulation of Cx43 with differentiation in vivo (Leung et al. 2002) and in vitro (Rozental et al. 1998).

These observations correspond to those of neuroblastic development: Neuroblasts of the ventricular zone are coupled via Cx43 gap junctions (Lo Turco and Kriegstein 1991). In later neuroblastic development, Cx43 expression is discontinued. In contrast, expression of the neuronal Cx36 commences at E16 and is increased with a peak in expression between E18 and postnatal day (P0; Rozental et al. 2000).

The data compiled in this review indicate that gap junctions represent an essential tool involved in morphogenetic mechanisms. With respect to neurogenesis, it became clear that gap junctions and their molecular constituents, the connexins, do not serve a monofunctional role. They rather seem to convey different stage-dependent functions that fit into the respective phase of the developmental program. An important issue that has emerged during recent years is whether gap junctions, beyond their classical role as mediators for cell-to-cell communication, account for additional functions which do not involve direct ionic or metabolic coupling between cells, but exploit truncated parts of the junctional complex in the form of hemichannels and/or portions of the entire molecular complex – a concept that provides a scientific challenge for the years to come.

References

Altevogt BM, Paul DL (2004) Four classes of intercellular channels between glial cells in the CNS. J Neurosci 24:4313–4323

Altevogt BM, Kleopa KA, Postma FR, Scherer SS, Paul DL (2002) Connexin29 is uniquely distributed within myelinating glial cells of the central and peripheral nervous systems. J Neurosci 22:6458–6470

Bani-Yaghoub M, Felker JM, Sans C, Naus CC (2000) The effects of bone morphogenetic protein 2 and 4 (BMP2 and BMP4) on gap junctions during neurodevelopment. Exp Neurol 162:13–26

Barron DJ, Valdimarsson G, Paul DL, Kidder GM (1989) Connexin32, a gap junction protein, is a persistent oogenetic product through preimplantation development of the mouse. Dev Genet 10:318–323

Belliveau DJ, Naus CC (1994) Cortical type 2 astrocytes are not dye coupled nor do they express the major gap junction genes found in the central nervous system. Glia 12:24–34

Belliveau DJ, Naus CC (1995) Cellular localization of gap junction mRNAs in developing rat brain. Dev Neurosci 17:81–96

Belliveau DJ, Kidder GM, Naus CC (1991) Expression of gap junction genes during postnatal neural development. Dev Genet 12:308–317

Belliveau DJ, Bechberger JF, Rogers KA, Naus CC (1997) Differential expression of gap junctions in neurons and astrocytes derived from P19 embryonal carcinoma cells. Dev Genet 21:187–200

Belousov A, Arumugam H, Denisova J (2004) Signal transduction in the developmental regulation of gap junctions. FENS Forum Abstracts 2:A006.002

Bennett MV, Barrio LC, Bargiello TA, Spray DC, Hertzberg E, Saez JC (1991) Gap junctions: new tools, new answers, new questions. Neuron 6:305–320

Bennett MV, Contreras JE, Bukauskas FF, Saez JC (2003) New roles for astrocytes: Gap junction hemichannels have something to communicate. Trends Neurosci 26:610–617
Benveniste EN (1992) Inflammatory cytokines within the central nervous system: sources, function, and mechanism of action. Am J Physiol 263:C1–16
Benveniste EN (1998) Cytokine actions in the central nervous system. Cytokine Growth Factor Rev 9:259–275
Berridge MJ, Galione A (1988) Cytosolic calcium oscillators. Faseb J 2:3074–3082
Beyer EC, Steinberg TH (1991) Evidence that the gap junction protein connexin-43 is the ATP-induced pore of mouse macrophages. J Biol Chem 266:7971–7974
Bittman K, Owens DF, Kriegstein AR, LoTurco JJ (1997) Cell coupling and uncoupling in the ventricular zone of developing neocortex. J Neurosci 17:7037–7044
Bittman KS, LoTurco JJ (1999) Differential regulation of connexin 26 and 43 in murine neocortical precursors. Cereb Cortex 9:188–195
Bondurand N, Girard M, Pingault V, Lemort N, Dubourg O, Goossens M (2001) Human Connexin 32, a gap junction protein altered in the X-linked form of Charcot-Marie-Tooth disease, is directly regulated by the transcription factor SOX10. Hum Mol Genet 10:2783–2795
Bruzzone R, White TW, Goodenough DA (1996) The cellular Internet: on-line with connexins. Bioessays 18:709–718
Caveney S (1985) The role of gap junctions in development. Annu Rev Physiol 47:319–335
Christie MJ, Jelinek HF (1993) Dye-coupling among neurons of the rat locus coeruleus during postnatal development. Neuroscience 56:129–137
Condorelli DF, Mudo G, Trovato-Salinaro A, Mirone MB, Amato G, Belluardo N (2002) Connexin-30 mRNA is up-regulated in astrocytes and expressed in apoptotic neuronal cells of rat brain following kainate-induced seizures. Mol Cell Neurosci 21:94–113
Condorelli DF, Trovato-Salinaro A, Mudo G, Mirone MB, Belluardo N (2003) Cellular expression of connexins in the rat brain: neuronal localization, effects of kainate-induced seizures and expression in apoptotic neuronal cells. Eur J Neurosci 18:1807–1827
Connors BW, Benardo LS, Prince DA (1983) Coupling between neurons of the developing rat neocortex. J Neurosci 3:773–782
Contreras JE, Sanchez HA, Veliz LP, Bukauskas FF, Bennett MV, Saez JC (2004) Role of connexin-based gap junction channels and hemichannels in ischemia-induced cell death in nervous tissue. Brain Res Brain Res Rev 47:290–303
Cotrina ML, Lin JH, Alves-Rodrigues A, Liu S, Li J, Azmi-Ghadimi H, Kang J, Naus CC, Nedergaard M (1998) Connexins regulate calcium signaling by controlling ATP release. Proc Natl Acad Sci USA 95:15735–15740
Cruciani V, Mikalsen SO (2002) Connexins, gap junctional intercellular communication and kinases. Biol Cell 94:433–443
Dahl E, Manthey D, Chen Y, Schwarz HJ, Chang YS, Lalley PA, Nicholson BJ, Willecke K (1996) Molecular cloning and functional expression of mouse connexin-30, a gap junction gene highly expressed in adult brain and skin. J Biol Chem 271:17903–17910
Deans MR, Gibson JR, Sellitto C, Connors BW, Paul DL (2001) Synchronous activity of inhibitory networks in neocortex requires electrical synapses containing connexin36. Neuron 31:477–485
Degen J, Meier C, Van Der Giessen RS, Sohl G, Petrasch-Parwez E, Urschel S, Dermietzel R, Schilling K, De Zeeuw CI, Willecke K (2004) Expression pattern of lacZ reporter gene representing connexin36 in transgenic mice. J Comp Neurol 473:511–525
Dermietzel R, Yancey SB, Traub O, Willecke K, Revel JP (1987) Major loss of the 28-kD protein of gap junction in proliferating hepatocytes. J Cell Biol 105:1925–1934
Dermietzel R, Traub O, Hwang TK, Beyer E, Bennett MV, Spray DC, Willecke K (1989) Differential expression of three gap junction proteins in developing and mature brain tissues. Proc Natl Acad Sci USA 86:10148–10152

Dermietzel R, Hertzberg EL, Kessler JA, Spray DC (1991) Gap junctions between cultured astrocytes: immunocytochemical, molecular, and electrophysiological analysis. J Neurosci 11:1421–1432

Dermietzel R, Farooq M, Kessler JA, Althaus H, Hertzberg EL, Spray DC (1997) Oligodendrocytes express gap junction proteins connexin32 and connexin45. Glia 20:101–114

Dermietzel R, Gao Y, Scemes E, Vieira D, Urban M, Kremer M, Bennett MV, Spray DC (2000) Connexin43 null mice reveal that astrocytes express multiple connexins. Brain Res Brain Res Rev 32:45–56

Dolmetsch RE, Xu K, Lewis RS (1998) Calcium oscillations increase the efficiency and specificity of gene expression. Nature 392:933–936

Dong CJ, McReynolds JS (1991) The relationship between light, dopamine release and horizontal cell coupling in the mudpuppy retina. J Physiol 440:291–309

Eugenin EA, Eckardt D, Theis M, Willecke K, Bennett MV, Saez JC (2001) Microglia at brain stab wounds express connexin 43 and in vitro form functional gap junctions after treatment with interferon-gamma and tumor necrosis factor-alpha. Proc Natl Acad Sci USA 98:4190–4195

Ewart JL, Cohen MF, Meyer RA, Huang GY, Wessels A, Gourdie RG, Chin AJ, Park SM, Lazatin BO, Villabon S, Lo CW (1997) Heart and neural tube defects in transgenic mice overexpressing the Cx43 gap junction gene. Development 124:1281–1292

Faustmann PM, Haase CG, Romberg S, Hinkerohe D, Szlachta D, Smikalla D, Krause D, Dermietzel R (2003) Microglia activation influences dye coupling and Cx43 expression of the astrocytic network. Glia 42:101–108

Filippov MA, Hormuzdi SG, Fuchs EC, Monyer H (2003) A reporter allele for investigating connexin 26 gene expression in the mouse brain. Eur J Neurosci 18:3183–3192

Fu CT, Bechberger JF, Ozog MA, Perbal B, Naus CC (2004) CCN3 (NOV) interacts with connexin43 in C6 glioma cells: possible mechanism of connexin-mediated growth suppression. J Biol Chem 279:36943–36950

Fushiki S, Perez Velazquez JL, Zhang L, Bechberger JF, Carlen PL, Naus CC (2003) Changes in neuronal migration in neocortex of connexin43 null mutant mice. J Neuropathol Exp Neurol 62:304–314

Gehrmann J, Matsumoto Y, Kreutzberg GW (1995) Microglia: intrinsic immuneffector cell of the brain. Brain Res Brain Res Rev 20:269–287

Gellhaus A, Dong X, Propson S, Maass K, Klein-Hitpass L, Kibschull M, Traub O, Willecke K, Perbal B, Lye SJ, Winterhager E (2004) Connexin43 interacts with NOV: a possible mechanism for negative regulation of cell growth in choriocarcinoma cells. J Biol Chem 279:36931–36942

Giaume C, Venance L (1995) Gap junctions in brain glial cells and development. Perspect Dev Neurobiol 2:335–345

Giaume C, Fromaget C, el Aoumari A, Cordier J, Glowinski J, Gros D (1991) Gap junctions in cultured astrocytes: single-channel currents and characterization of channel-forming protein. Neuron 6:133–143

Goodenough DA, Paul DL (2003) Beyond the gap: functions of unpaired connexon channels. Nat Rev Mol Cell Biol 4:285–294

Goodman CS, Spitzer NC (1979) Embryonic development of identified neurones: differentiation from neuroblast to neurone. Nature 280:208–214

Gulisano M, Parenti R, Spinella F, Cicirata F (2000) Cx36 is dynamically expressed during early development of mouse brain and nervous system. Neuroreport 11:3823–3828

Hofer A, Dermietzel R (1998) Visualization and functional blocking of gap junction hemichannels (connexons) with antibodies against external loop domains in astrocytes. Glia 24:141–154

Hofer A, Saez JC, Chang CC, Trosko JE, Spray DC, Dermietzel R (1996) C-erbB2/neu transfection induces gap junctional communication incompetence in glial cells. J Neurosci 16:4311–4321
Hormuzdi SG, Pais I, LeBeau FE, Towers SK, Rozov A, Buhl EH, Whittington MA, Monyer H (2001) Impaired electrical signaling disrupts gamma frequency oscillations in connexin 36-deficient mice. Neuron 31:487–495
Hormuzdi SG, Filippov MA, Mitropoulou G, Monyer H, Bruzzone R (2004) Electrical synapses: a dynamic signaling system that shapes the activity of neuronal networks. Biochim Biophys Acta 1662:113–137
Huang GY, Wessels A, Smith BR, Linask KK, Ewart JL, Lo CW (1998) Alteration in connexin 43 gap junction gene dosage impairs conotruncal heart development. Dev Biol 198:32–44
John GR, Scemes E, Suadicani SO, Liu JS, Charles PC, Lee SC, Spray DC, Brosnan CF (1999) IL-1beta differentially regulates calcium wave propagation between primary human fetal astrocytes via pathways involving P2 receptors and gap junction channels. Proc Natl Acad Sci USA 96:11613–11618
Jungbluth S, Willecke K, Champagnat J (2002) Segment-specific expression of connexin31 in the embryonic hindbrain is regulated by Krox20. Dev Dyn 223:544–551
Kandler K, Katz LC (1995) Neuronal coupling and uncoupling in the developing nervous system. Curr Opin Neurobiol 5:98–105
Kandler K, Katz LC (1998) Coordination of neuronal activity in developing visual cortex by gap junction-mediated biochemical communication. J Neurosci 18:1419–1427
Kielian T, Esen N (2004) Effects of neuroinflammation on glia-glia gap junctional intercellular communication: a perspective. Neurochem Int 45:429–436
Kleopa KA, Orthmann JL, Enriquez A, Paul DL, Scherer SS (2004) Unique distributions of the gap junction proteins connexin29, connexin32, and connexin47 in oligodendrocytes. Glia 47:346–357
Kunzelmann P, Blumcke I, Traub O, Dermietzel R, Willecke K (1997) Coexpression of connexin45 and -32 in oligodendrocytes of rat brain. J Neurocytol 26:17–22
Kunzelmann P, Schroder W, Traub O, Steinhauser C, Dermietzel R, Willecke K (1999) Late onset and increasing expression of the gap junction protein connexin30 in adult murine brain and long-term cultured astrocytes. Glia 25:111–119
Leung DS, Unsicker K, Reuss B (2002) Expression and developmental regulation of gap junction connexins cx26, cx32, cx43 and cx45 in the rat midbrain-floor. Int J Dev Neurosci 20:63–75
Li H, Liu TF, Lazrak A, Peracchia C, Goldberg GS, Lampe PD, Johnson RG (1996) Properties and regulation of gap junctional hemichannels in the plasma membranes of cultured cells. J Cell Biol 134:1019–1030
Li W, Llopis J, Whitney M, Zlokarnik G, Tsien RY (1998) Cell-permeant caged InsP3 ester shows that Ca^{2+} spike frequency can optimize gene expression. Nature 392:936–941
Li X, Olson C, Lu S, Nagy JI (2004) Association of connexin36 with zonula occludens-1 in HeLa cells, betaTC-3 cells, pancreas, and adrenal gland. Histochem Cell Biol 122:485–498
Li X, Lynn BD, Olson C, Meier C, Davidson KG, Yasumura T, Rash JE, Nagy JI (2002) Connexin29 expression, immunocytochemistry and freeze-fracture replica immunogold labelling (FRIL) in sciatic nerve. Eur J Neurosci 16:795–806
Lo Turco JJ, Kriegstein AR (1991) Clusters of coupled neuroblasts in embryonic neocortex. Science 252:563–566
Lynn BD, Rempel JL, Nagy JI (2001) Enrichment of neuronal and glial connexins in the postsynaptic density subcellular fraction of rat brain. Brain Res 898:1–8
Martinez AD, Saez JC (2000) Regulation of astrocyte gap junctions by hypoxia-reoxygenation. Brain Res Brain Res Rev 32:250–258

Martinez AD, Eugenin EA, Branes MC, Bennett MV, Saez JC (2002) Identification of second messengers that induce expression of functional gap junctions in microglia cultured from newborn rats. Brain Res 943:191–201
Massa PT, Mugnaini E (1982) Cell junctions and intramembrane particles of astrocytes and oligodendrocytes: a freeze-fracture study. Neuroscience 7:523–538
Matthias K, Kirchhoff F, Seifert G, Huttmann K, Matyash M, Kettenmann H, Steinhauser C (2003) Segregated expression of AMPA-type glutamate receptors and glutamate transporters defines distinct astrocyte populations in the mouse hippocampus. J Neurosci 23:1750–1758
Maxeiner S, Kruger O, Schilling K, Traub O, Urschel S, Willecke K (2003) Spatiotemporal transcription of connexin45 during brain development results in neuronal expression in adult mice. Neuroscience 119:689–700
McGeer PL, McGeer EG (1995) The inflammatory response system of brain: implications for therapy of Alzheimer and other neurodegenerative diseases. Brain Res Brain Res Rev 21:195–218
McLachlin JR, Kidder GM (1986) Intercellular junctional coupling in preimplantation mouse embryos: effect of blocking transcription or translation. Dev Biol 117:146–155
Meier C, Dermietzel R, Davidson KG, Yasumura T, Rash JE (2004) Connexin32-containing gap junctions in Schwann cells at the internodal zone of partial myelin compaction and in Schmidt-Lanterman incisures. J Neurosci 24:3186–3198
Meier C, Petrasch-Parwez E, Habbes HW, Teubner B, Guldenagel M, Degen J, Sohl G, Willecke K, Dermietzel R (2002) Immunohistochemical detection of the neuronal connexin36 in the mouse central nervous system in comparison to connexin36-deficient tissues. Histochem Cell Biol 117:461–471
Melanson-Drapeau L, Beyko S, Dave S, Hebb AL, Franks DJ, Sellitto C, Paul DL, Bennett SA (2003) Oligodendrocyte progenitor enrichment in the connexin32 null-mutant mouse. J Neurosci 23:1759–1768
Menichella DM, Goodenough DA, Sirkowski E, Scherer SS, Paul DL (2003) Connexins are critical for normal myelination in the CNS. J Neurosci 23:5963–5973
Mercier F, Hatton GI (2001) Connexin 26 and basic fibroblast growth factor are expressed primarily in the subpial and subependymal layers in adult brain parenchyma: roles in stem cell proliferation and morphological plasticity? J Comp Neurol 431:88–104
Mesnil M (2002) Connexins and cancer. Biol Cell 94:493–500
Micevych PE, Abelson L (1991) Distribution of mRNAs coding for liver and heart gap junction proteins in the rat central nervous system. J Comp Neurol 305:96–118
Miller RF (1988) Are single retinal neurons both excitatory and inhibitory? Nature 336:517–518
Miragall F, Albiez P, Bartels H, de Vries U, Dermietzel R (1997) Expression of the gap junction protein connexin43 in the subependymal layer and the rostral migratory stream of the mouse: evidence for an inverse correlation between intensity of connexin43 expression and cell proliferation activity. Cell Tissue Res 287:243–253
Montoro RJ, Yuste R (2004) Gap junctions in developing neocortex: a review. Brain Res Brain Res Rev 47:216–226
Musil LS, Goodenough DA (1993) Multisubunit assembly of an integral plasma membrane channel protein, gap junction connexin43, occurs after exit from the ER. Cell 74:1065–1077
Musil LS, Cunningham BA, Edelman GM, Goodenough DA (1990) Differential phosphorylation of the gap junction protein connexin43 in junctional communication-competent and -deficient cell lines. J Cell Biol 111:2077–2088
Nadarajah B, Jones AM, Evans WH, Parnavelas JG (1997) Differential expression of connexins during neocortical development and neuronal circuit formation. J Neurosci 17:3096–3111

Nagy JI, Rash JE (2000) Connexins and gap junctions of astrocytes and oligodendrocytes in the CNS. Brain Res Brain Res Rev 32:29–44

Nagy JI, Dudek FE, Rash JE (2004) Update on connexins and gap junctions in neurons and glia in the mammalian nervous system. Brain Res Brain Res Rev 47:191–215

Nagy JI, Patel D, Ochalski PA, Stelmack GL (1999) Connexin30 in rodent, cat and human brain: selective expression in gray matter astrocytes, co-localization with connexin43 at gap junctions and late developmental appearance. Neuroscience 88:447–468

Nagy JI, Ionescu AV, Lynn BD, Rash JE (2003) Connexin29 and connexin32 at oligodendrocyte and astrocyte gap junctions and in myelin of the mouse central nervous system. J Comp Neurol 464:356–370

Nagy JI, Li X, Rempel J, Stelmack G, Patel D, Staines WA, Yasumura T, Rash JE (2001) Connexin26 in adult rodent central nervous system: demonstration at astrocytic gap junctions and colocalization with connexin30 and connexin43. J Comp Neurol 441:302–323

Nakase T, Naus CC (2004) Gap junctions and neurological disorders of the central nervous system. Biochim Biophys Acta 1662:149–158

Naus CC, Bechberger JF, Zhang Y, Venance L, Yamasaki H, Juneja SC, Kidder GM, Giaume C (1997) Altered gap junctional communication, intercellular signaling, and growth in cultured astrocytes deficient in connexin43. J Neurosci Res 49:528–540

Nelles E, Butzler C, Jung D, Temme A, Gabriel HD, Dahl U, Traub O, Stumpel F, Jungermann K, Zielasek J, Toyka KV, Dermietzel R, Willecke K (1996) Defective propagation of signals generated by sympathetic nerve stimulation in the liver of connexin32-deficient mice. Proc Natl Acad Sci USA 93:9565–9570

Noble M, Wolswijk G (1992) Development and regeneration in the O-2A lineage: studies in vitro and in vivo. J Neuroimmunol 40:287–293

Odermatt B, Wellershaus K, Wallraff A, Seifert G, Degen J, Euwens C, Fuss B, Bussow H, Schilling K, Steinhauser C, Willecke K (2003) Connexin 47 (Cx47)-deficient mice with enhanced green fluorescent protein reporter gene reveal predominant oligodendrocytic expression of Cx47 and display vacuolized myelin in the CNS. J Neurosci 23:4549–4559

Okada H, Miyakawa N, Mori H, Mishina M, Miyamoto Y, Hisatsune T (2003) NMDA receptors in cortical development are essential for the generation of coordinated increases in $[Ca^{2+}](i)$ in "neuronal domains". Cereb Cortex 13:749–757

Owens DF, Kriegstein AR (1998) Patterns of intracellular calcium fluctuation in precursor cells of the neocortical ventricular zone. J Neurosci 18:5374–5388

Parenti R, Campisi A, Vanella A, Cicirata F (2002) Immunocytochemical and RT-PCR analysis of connexin36 in cultures of mammalian glial cells. Arch Ital Biol 140:101–108

Parnavelas JG, Luder R, Pollard SG, Sullivan K, Lieberman AR (1983) A qualitative and quantitative ultrastructural study of glial cells in the developing visual cortex of the rat. Philos Trans R Soc Lond B Biol Sci 301:55–84

Paul DL (1995) New functions for gap junctions. Curr Opin Cell Biol 7:665–672

Paul DL, Ebihara L, Takemoto LJ, Swenson KI, Goodenough DA (1991) Connexin46, a novel lens gap junction protein, induces voltage-gated currents in nonjunctional plasma membrane of *Xenopus* oocytes. J Cell Biol 115:1077–1089

Peinado A, Yuste R, Katz LC (1993a) Gap junctional communication and the development of local circuits in neocortex. Cereb Cortex 3:488–498

Peinado A, Yuste R, Katz LC (1993b) Extensive dye coupling between rat neocortical neurons during the period of circuit formation. Neuron 10:103–114

Pfahnl A, Dahl G (1999) Gating of cx46 gap junction hemichannels by calcium and voltage. Pflugers Arch 437:345–353

Piccolino M, Neyton J, Gerschenfeld HM (1984) Decrease of gap junction permeability induced by dopamine and cyclic adenosine 3':5'-monophosphate in horizontal cells of turtle retina. J Neurosci 4:2477–2488

Qin H, Shao Q, Curtis H, Galipeau J, Belliveau DJ, Wang T, Alaoui-Jamali MA, Laird DW (2002) Retroviral delivery of connexin genes to human breast tumor cells inhibits in vivo tumor growth by a mechanism that is independent of significant gap junctional intercellular communication. J Biol Chem 277:29132–29138

Raff MC, Abney ER, Cohen J, Lindsay R, Noble M (1983) Two types of astrocytes in cultures of developing rat white matter: differences in morphology, surface gangliosides, and growth characteristics. J Neurosci 3:1289–1300

Rakic P (2002) Evolving concepts of cortical radial and areal specification. Prog Brain Res 136:265–280

Rao MS (1999) Multipotent and restricted precursors in the central nervous system. Anat Rec 257:137–148

Rash JE, Yasumura T, Dudek FE, Nagy JI (2001a) Cell-specific expression of connexins and evidence of restricted gap junctional coupling between glial cells and between neurons. J Neurosci 21:1983–2000

Rash JE, Dudek FE, Yasumura T, Davidson K, Furman CS, Nagy JI (2001b) Identification of cells expressing Cx43, Cx30, Cx26, Cx32 and Cx36 in gap junctions of rat brain and spinal cord. Cell Commun Adhes 8:315–320.

Reaume AG, de Sousa PA, Kulkarni S, Langille BL, Zhu D, Davies TC, Juneja SC, Kidder GM, Rossant J (1995) Cardiac malformation in neonatal mice lacking connexin43. Science 267:1831–1834

Revel JP (1988) The oldest multicellular animal and its junctions. In: Hertzberg EL, Johnson RG (eds) Gap Junctions Allan R Liss, New York, pp 135–149

Rorig B, Sutor B (1996a) Regulation of gap junction coupling in the developing neocortex. Mol Neurobiol 12:225–249

Rorig B, Sutor B (1996b) Serotonin regulates gap junction coupling in the developing rat somatosensory cortex. Eur J Neurosci 8:1685–1695

Rorig B, Klausa G, Sutor B (1995) Dye coupling between pyramidal neurons in developing rat prefrontal and frontal cortex is reduced by protein kinase A activation and dopamine. J Neurosci 15:7386–7400

Rouach N, Glowinski J, Giaume C (2000) Activity-dependent neuronal control of gap-junctional communication in astrocytes. J Cell Biol 149:1513–1526

Rouach N, Calvo CF, Glowinski J, Giaume C (2002) Brain macrophages inhibit gap junctional communication and downregulate connexin 43 expression in cultured astrocytes. Eur J Neurosci 15:403–407

Rozental R, Morales M, Mehler MF, Urban M, Kremer M, Dermietzel R, Kessler JA, Spray DC (1998) Changes in the properties of gap junctions during neuronal differentiation of hippocampal progenitor cells. J Neurosci 18:1753–1762

Rozental R, Srinivas M, Gokhan S, Urban M, Dermietzel R, Kessler JA, Spray DC, Mehler MF (2000) Temporal expression of neuronal connexins during hippocampal ontogeny. Brain Res Brain Res Rev 32:57–71

Scherer SS, Deschenes SM, Xu YT, Grinspan JB, Fischbeck KH, Paul DL (1995) Connexin32 is a myelin-related protein in the PNS and CNS. J Neurosci 15:8281–8294

Serre-Beinier V, Le Gurun S, Belluardo N, Trovato-Salinaro A, Charollais A, Haefliger JA, Condorelli DF, Meda P (2000) Cx36 preferentially connects beta-cells within pancreatic islets. Diabetes 49:727–734

Sohl G, Eiberger J, Jung YT, Kozak CA, Willecke K (2001) The mouse gap junction gene connexin29 is highly expressed in sciatic nerve and regulated during brain development. Biol Chem 382:973–978

Sohl G, Odermatt B, Maxeiner S, Degen J, Willecke K (2004) New insights into the expression and function of neural connexins with transgenic mouse mutants. Brain Res Brain Res Rev 47:245–259

Sontheimer H, Minturn JE, Black JA, Waxman SG, Ransom BR (1990) Specificity of cell–cell coupling in rat optic nerve astrocytes in vitro. Proc Natl Acad Sci USA 87:9833–9837
Spitzer NC (1991) A developmental handshake: neuronal control of ionic currents and their control of neuronal differentiation. J Neurobiol 22:659–673
Spray DC, Dermietzel R (1995) X-linked dominant Charcot-Marie-Tooth disease and other potential gap-junction diseases of the nervous system. Trends Neurosci 18:256–262
Streit WJ, Graeber MB, Kreutzberg GW (1988) Functional plasticity of microglia: a review. Glia 1:301–307
Sutor B, Schmolke C, Teubner B, Schirmer C, Willecke K (2000) Myelination defects and neuronal hyperexcitability in the neocortex of connexin 32-deficient mice. Cereb Cortex 10:684–697
Teubner B, Odermatt B, Guldenagel M, Sohl G, Degen J, Bukauskas F, Kronengold J, Verselis VK, Jung YT, Kozak CA, Schilling K, Willecke K (2001) Functional expression of the new gap junction gene connexin47 transcribed in mouse brain and spinal cord neurons. J Neurosci 21:1117–1126
Traub O, Hertlein B, Kasper M, Eckert R, Krisciukaitis A, Hulser D, Willecke K (1998) Characterization of the gap junction protein connexin37 in murine endothelium, respiratory epithelium, and after transfection in human HeLa cells. Eur J Cell Biol 77:313–322
Tsien RW, Tsien RY (1990) Calcium channels, stores, and oscillations. Annu Rev Cell Biol 6:715–760
van der Heyden MA, Rook MB, Hermans MM, Rijksen G, Boonstra J, Defize LH, Destree OH (1998) Identification of connexin43 as a functional target for Wnt signalling. J Cell Sci 111 (Pt 12):1741–1749
Veenstra RD (2001) Determining ionic permeabilities of gap junction channels. Methods Mol Biol 154:293–311
Wallraff A, Odermatt B, Willecke K, Steinhauser C (2004) Distinct types of astroglial cells in the hippocampus differ in gap junction coupling. Glia 48:36–43
Warn-Cramer BJ, Lau AF (2004) Regulation of gap junctions by tyrosine protein kinases. Biochim Biophys Acta 1662:81–95
Warner AE (1985) The role of gap junctions in amphibian development. J Embryol Exp Morphol 89 Suppl:365–380
Waxman SG, Black JA (1984) Freeze-fracture ultrastructure of the perinodal astrocyte and associated glial junctions. Brain Res 308:77–87
Weickert S, Ray A, Zoidl G, Dermietzel R (2005) Expression of neural connexins and pannexin1 in the hippocampus and inferior olive: a quantitative approach. Brain Res Mol Brain Res 133:102–109
Weissman TA, Riquelme PA, Ivic L, Flint AC, Kriegstein AR (2004) Calcium waves propagate through radial glial cells and modulate proliferation in the developing neocortex. Neuron 43:647–661
Wilkinson DG, Bailes JA, McMahon AP (1987) Expression of the proto-oncogene int-1 is restricted to specific neural cells in the developing mouse embryo. Cell 50:79–88
Willecke K, Eiberger J, Degen J, Eckardt D, Romualdi A, Guldenagel M, Deutsch U, Sohl G (2002) Structural and functional diversity of connexin genes in the mouse and human genome. Biol Chem 383:725–737
Wolpert L (1969) Positional information and the spatial pattern of cellular differentiation. J Theor Biol 25:1–47
Wolpert L (1994) Positional information and pattern formation in development. Dev Genet 15:485–490
Ye ZC, Wyeth MS, Baltan-Tekkok S, Ransom BR (2003) Functional hemichannels in astrocytes: a novel mechanism of glutamate release. J Neurosci 23:3588–3596
Yuste R, Peinado A, Katz LC (1992) Neuronal domains in developing neocortex. Science 257:665–669

Yuste R, Nelson DA, Rubin WW, Katz LC (1995) Neuronal domains in developing neocortex: mechanisms of coactivation. Neuron 14:7–17

Zhang YW, Kaneda M, Morita I (2003) The gap junction-independent tumor-suppressing effect of connexin 43. J Biol Chem 278:44852–44856

Zvalova D, Cordier J, Mesnil M, Junier MP, Chneiweiss H (2004) p38/SAPK2 controls gap junction closure in astrocytes. Glia 46:323–333

6 Connexins Responsible for Hereditary Deafness – The Tale Unfolds

Martine Cohen-Salmon, Francisco J. del Castillo, Christine Petit[1]

6.1 Introduction

In recent years, connexins, the major proteins of gap junctions, have been demonstrated to be crucial for audition. Genetic studies have shown the involvement of several connexin-encoding genes in deafness. However, the physiopathological mechanisms leading to deafness when connexins are absent or mutated are far from being understood. In this review, we have tried to integrate the actual genetic, molecular and functional data reported so far (Sect. 6.2), and provide an update on the hypotheses proposed for the role of gap junctions in the inner ear (Sect. 6.3).

Hearing impairment (deafness) is the most common sensory disorder, affecting approximately 1 in 1000 individuals at birth or during early childhood. In developed countries, about 60% of the cases are of genetic origin (Petit et al. 2001). This proportion might even be underestimated. Inherited hearing impairment is either syndromic, i.e., associated with other clinical features, in 30% of the cases, or nonsyndromic, in 70%. Conductive deafness is due to defects in the outer or middle ear, whereas sensorineural deafness is caused by anomalies located either in the inner ear or, more rarely, in the neural pathways, transmitting sound signals to the cortical auditory centers of the brain. Four modes of inheritance have been described: maternal (mitochondrial), X-linked (DFN forms), autosomal dominant (DFNA forms) and autosomal recessive (DFNB forms). DFNB forms are generally congenital and severe, thereby impeding speech acquisition (prelingual deafness).

Inherited deafness is genetically heterogeneous, with about 40 currently identified genes involved, and about other 40 loci reported for which the involved genes are still unknown (hereditary hearing loss homepage: http://www.uia.ac.be/dnalab/hhh). Over 50 causative mutations in *CX26* (*GJB2*), encoding the gap junction protein connexin 26, have been shown to account for a large proportion of prelingual hearing impairment in most populations (the connexin-deafness homepage: http://www.crg.es/deafness;

[1] Unité de Génétique des Déficits Sensoriels, INSERM U587, Institut Pasteur, 25 rue du Dr Roux, 75724 Paris, cedex 15, France, E-mail: cpetit@pasteur.fr

Gap Junctions in Development and Disease
(ed. by E. Winterhager)

Kenneson et al. 2002; see Sect. 6.2.1). The implication of gap junctions in audition has been further highlighted by the discovery of deafness-causing mutations in three other connexin genes, *CX30* (*GJB6*), *CX31* (*GJB3*), and *CX43* (*GJA1*).

6.2 Connexin Genes and Hearing Impairment

Table 6.1 summarizes the implication of connexin genes in various forms of deafness.

6.2.1 *CX26* (*GJB2*) and Autosomal Recessive Deafness DFNB1

Mutations in the *CX26* gene at the DFNB1 locus on chromosome 13q12 are the major cause of inherited hearing impairment; they are responsible for up to 50% of prelingual, recessive, non-syndromic deafness cases in the Mediterranean basin, 30% in northern Europe, 40% in the USA, 22% in Brazil, 25% in the Far East, 17% in Ghana, and 15% in Australia (for a detailed listing, see review by Kenneson et al. 2002; Oliveira et al. 2002). In these countries, the molecular analysis of *CX26* has therefore become a standard of care for diagnosis and counseling of patients. DFNB1 deafness is prelingual and shows a variable degree of severity (from mild to profound), even between siblings harboring the same mutations, which indicates the possible modulation of the phenotype by environmental factors or modifying genes acting in concert with the *CX26* mutations (Cryns et al. 2004).

A great variety of mutations have been found to be responsible for DFNB1 (the Cx26-deafness homepage at http://www.crg.es/deafness/cx26mut.php contains a full listing), including missense and nonsense mutations, as well as deletions or insertions. Among these mutations, five have been found at a very high frequency in different ethnic groups: **35delG** (originally reported as 30delG) in European-Mediterranean populations (Estivill et al. 1998; Denoyelle et al. 1999); **167delT** in the Ashkenazi Jewish population (Morell et al. 1998; Sobe et al. 1999); **235delC** (originally reported as 233delC) in eastern Asian (Japanese, Chinese, and Korean) populations (Fuse et al. 1999; Abe et al. 2000; Kudo et al. 2000; Park et al. 2000; Liu et al. 2002; Wang et al. 2002); **R143 W** in Ghana (Brobby et al. 1998; Hamelmann et al. 2001); and **W24X** in Indian and Romany (gypsy) populations (Maheshwari et al. 2003; Minarik et al. 2003; RamShankar et al. 2003). Because of their restricted geographical distribution, these mutations most likely

Table 6.1. Connexin genes involved in various forms of deafness

Gene	Recessive isolated deafness	Dominant isolated deafness	Dominant syndromic deafness
CX26 GJB2	**DFNB1**: Prelingual, mild to profound	**DFNA3**: Prelingual, severe to profound	Profound with diffuse keratoderma Moderate with mutilating keratoderma (Vohwinkel's syndrome) Profound with palmoplantar keratoderma Severe with keratitis-ichthyosis-deafness syndrome (KID) Profound with palmoplantar keratoderma, knuckle pads, leukonychia (Bart-Pumphrey syndrome)
CX26/CX30 GJB2/GJB6	**DFNB1**: Prelingual, severe to profound		
CX30 GJB6		**DFNA3'**: Profound, adult onset	Prelingual, mild to moderate with keratitis-ichthyosis-deafness syndrome (KID)
CX31 GJB3	Late onset[a]	Late onset[a]	Mild, associated with neuropathy
CX43 GJA1			Conductive, associated with dominant oculo-dento-digital dysplasia (ODDD)

[a] Mutations showing partial penetrance

arose by independent founder effects. Analysis of the association of each mutation with a specific haplotype (in genetic markers closely linked to *CX26* on chromosome 13q12) has indeed provided convincing evidence for founder effects in the cases of 35delG (Van Laer et al. 2001; Rothrock et al. 2003), 167delT (Sobe et al. 1999, 2000), and 235delC (Ohtsuka et al. 2003; Yan et al. 2003).

The frequency of these mutations in the general population is exceptionally high (e.g., for 35delG, carrier frequencies are 1/35 and 1/79 in southern and northern Europe, respectively; Gasparini et al. 2000), and as a direct consequence, there is a high proportion of sporadic DFNB1 cases. An explanation would be that these mutations may provide carriers with phenotypic advantages, resulting in positive selection. Recently, it was shown that R143 W/R143 W individuals have higher sodium and chloride concentrations in sweat, which could protect the skin from microbial colonization (Meyer et al. 2002). In addition, wt/R143 W and R143 W/R143 W individuals have a significantly thicker epidermis than their wt/wt relatives (Meyer et al. 2002). Moreover, cultured keratinocytes expressing deafness-associated mutant Cx26 alleles (including R143 W) display reduced cell death compared to cells expressing either wild-type Cx26 or skin disease-associated Cx26 alleles (Common et al. 2004). Thus, the extended cell lifespan could result in the observed thickening of the epidermis, which may enhance protection against pathogen invasion or trauma, and therefore confer a selective advantage on carriers of the mutation.

Several studies have been carried out to understand the molecular mechanisms underlying DFNB1 pathogenesis, either by resorting to creation of murine models or by analyzing mutant connexin 26 variants in heterologous expression systems. Because full ablation of *CX26* in mouse results in embryonic lethality due to a placental defect (Gabriel et al. 1998), the absence of Cx26 in the inner ear, caused by null mutations such as 35delG, can only be reproduced by conditional knockout tools or by trying to rescue the placental defect. We generated a murine cochlear-specific knockout of Cx26 by using the Cre-*loxP* system, where Cx26 is only deleted in one of the cellular networks formed by gap junctions in the inner ear (see Sect. 6.3.3), the *epithelial* gap junction network in the inner ear, leaving the *connective-tissue* gap junction network intact (Cohen-Salmon et al. 2002). This experiment demonstrated that Cx26 is not required for inner ear maturation, but crucial for the survival of the organ of Corti. In mutants, the organ of Corti degenerated upon sound stimulation, leading to deafness. Hence, it has been hypothesized that Cx26 epithelial gap junctions play a role in the K^+ homeostasis in the organ of Corti. A complete model for DFNB1, where Cx26 is deleted in both inner ear gap junction networks, is now necessary to fully elucidate the pathogenesis of DFNB1.

Concerning missense mutations causing DFNB1 deafness, their recessive mode of inheritance implies that the mutated protein does not critically interfere with the channel activity of the wild-type protein in heterozygous subjects, and that half the normal amount of the wild-type protein is sufficient for the inner ear to function. The majority of these mutations have been tested for membrane targeting, electrical coupling, and dye transfer to adjacent cells in different heterologous expression systems, such as *Xenopus* oocytes or HeLa cells, which are naturally devoid of gap junctions (Martin et al. 1999; D'Andrea et al. 2000; Thonnissen et al. 2002; Bruzzone et al. 2003; Wang et al. 2003). These experiments demonstrated a loss of function due to: (1) impaired translation of the mRNAs, or enhanced protein degradation (e.g., M1 V, F161S, P173R); (2) altered sorting of the proteins to the plasma membrane (e.g., G12 V, S19T, W77R, L90P, F161S); (3) deficient oligomerization into hemichannels (e.g., M34T, L90P, R184P); or (4) inability to form functional homotypic gap-junction channels (e.g., V37I, S113R, ΔE120, R127H, M163 V).

In some cases, there are also difficulties in reconciling in vitro results with observations made in patients, or even between data collected in different heterologous expression systems. For instance, the mutation R127H has been reported to be either a polymorphism or a recessive mutation in patients, and described as being harmless for the channel function in HeLa cells, while it abolished ionic conductance and neurobiotin transfer in N2A cells (Thonnissen et al. 2002; Wang et al. 2003). The M34T Cx26 mutated form has been successively reported to be dominant, recessive, or non-pathogenic (Kelsell et al. 1997; White et al. 1998; Martin et al. 1999; Thonnissen et al. 2002; Feldmann et al. 2004). Such discrepancies would be due to the fact that the effect of M34T on channel function depends on its expression level. Indeed, a dominant negative effect on wt Cx26 is only observed when the M34T variant is overexpressed (Skerrett et al. 2004). Therefore, it is now widely accepted that M34T is not a dominant mutation, as firstly reported. Finally, deafness-causing mutant Cx26 forms V84L and R143W show functional properties very similar to wild-type Cx26 (Bruzzone et al. 2003; Wang et al. 2003), including normal ionic conductance and the ability to transfer dye to neighboring cells, which indicate that such mutations must have very subtle effects on intercellular communication, which may not be readily spotted in heterologous systems. Beltramello et al. (2005) have indeed shown this to be the case for V84L by analyzing the effect of the mutation in transfected HeLa cells (Beltramello et al. 2005). They found a direct effect of the mutation on the transfer of inositol 1,4,5-trisphosphate (IP_3) between cells. The consequent inhibition of Ca^{2+} waves that would occur in supporting cells of the organ of Corti may prevent their coordinated activity, leading to hearing impairment (see Sect. 6.3.3).

Another striking particularity for DFNB1 is the high proportion of patients (10–42% in different populations) carrying only one mutant *CX26* allele, as well as the existence of some familial cases with evidence of linkage to DFNB1, but harboring no *CX26* mutations (del Castillo et al. 2002). It was thus postulated that another gene close to *CX26* might be responsible for these cases. The most likely candidate was *CX30*, located just 35 kb telomeric to *CX26* (Kelley et al. 1999), and known to be involved in a dominant form of hearing loss (see Sect. 6.2.3). Although no recessive, deafness-causing *CX30* point mutations have been found so far, this hypothesis finally received experimental support when several laboratories (Lerer et al. 2001; del Castillo et al. 2002; Pallares-Ruiz et al. 2002) identified a deletion of about 309 kb, named *Δ*(*GJB6*-D13S1830), not affecting the *CX26* coding region, but truncating *CX30* (Fig. 6.1). This deletion ac-

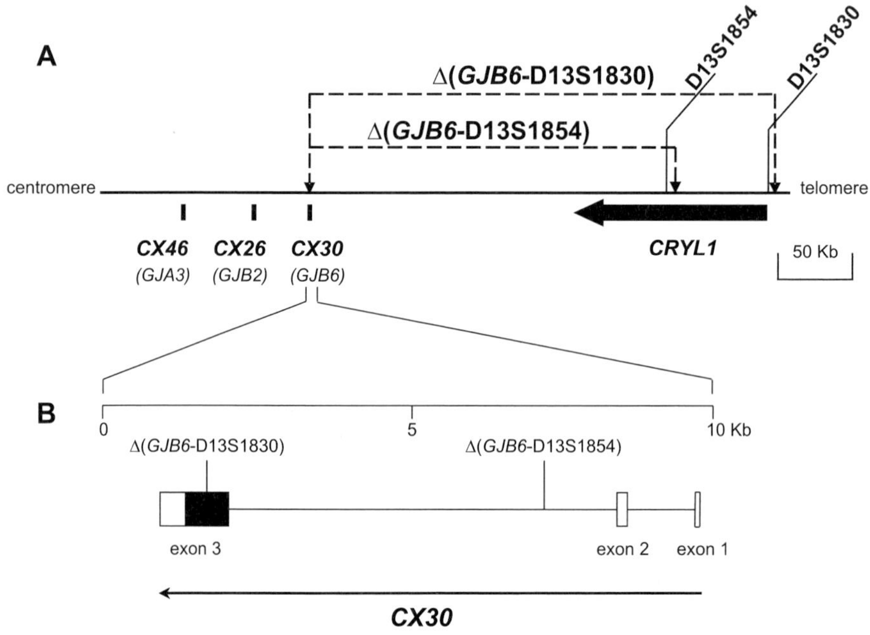

Fig. 6.1 A, B. Deletions of the *CX30* gene in DFNB1 hearing-impaired patients. **A** Map of the region of human chromosome 13q12 that harbors the DFNB1 locus. Genes are indicated by *thick bars*. Genes *CX26* and *CX30* are transcribed from telomere to centromere (*right* to *left*). The other known genes located in this region code for Cx46 (*GJA3*) and *λ*-crystallin (*CRYL1*). *Vertical arrows* and *dashed lines* indicate the extension of *Δ*(*GJB6*-D13S1830) and *Δ*(*GJB6*-D13S1854) deletions. The precise location of both deletion breakpoints within the *CX30* gene is shown in **B**. *Empty boxes* denote 5'- and 3'-untranslated regions; the *filled box* indicates the Cx30 coding sequence. The DNA segment lost in deletion *Δ*(*GJB6*-D13S1854) is completely contained within the segment lost in deletion *Δ*(*GJB6*-D13S1830)

companied a *CX26* mutation in *trans* (i.e., double heterozygosity) in about 50% of the cases of unexplained *CX26* deafness (del Castillo et al. 2002). The Δ(*GJB6*-D13S1830) deletion has been found so far in populations of European descent, being most frequent in Spain, France, the United Kingdom, Israel and Brazil, where it accounts for 5.9–9.7% of all DFNB1 alleles (del Castillo et al. 2003). Subsequently, a novel 232-kb deletion truncating *CX30*, termed Δ(*GJB6*-D13S1854), was identified, accounting for 1.9–25.5% of European *CX26* heterozygous patients that remained uncharacterized after screening for Δ(*GJB6*-D13S1830; del Castillo et al. 2005; Fig. 6.1). Though deletions affecting *CX30* are clearly involved in DFNB1, it is still unknown whether the *CX30* gene itself, in concert with *CX26*, is implicated in this form of deafness (digenic inheritance). This hypothesis is supported by several facts. Firstly, Cx26 and Cx30 colocalize and form heteromeric gap junctions in the inner ear (see Sect. 6.3.2). In addition, ablation of *Cx30* in mice results in severe constitutive hearing impairment (Teubner et al. 2003; see Sect. 6.3.3). Finally, whereas simple *Cx26*$^{+/-}$ or *Cx30*$^{+/-}$ heterozygous mice are asymptomatic, *Cx26*$^{+/-}$/*Cx30*$^{+/-}$ double heterozygous mice display a moderate hearing loss, showing that **in mice** the functional loss of one allele for each gene results in hearing impairment (Michel et al. 2003). Even so, the contribution of *CX30* to DFNB1 in humans remains to be determined, since no recessive deafness-causing point mutation in *CX30* has been found so far. An alternative hypothesis to explain the phenotype of double heterozygous patients is that deletions Δ(*GJB6*-D13S1830) and Δ(*GJB6*-D13S1854) eliminate a putative, as-yet-unknown *cis*-regulatory element crucial for the expression of *CX26* in the inner ear. These hypotheses are not necessarily mutually exclusive, however. Indeed, it can be argued that the difference in the severity of the phenotype (moderate vs. profound deafness) between *Cx26*$^{+/-}$/*Cx30*$^{+/-}$ mice and double heterozygous human patients could be due to the removal of such a *cis*-regulatory element.

Since individuals harboring the same mutations sometimes display different severity of symptoms, it has not been possible to draw a clear correlation between the various *CX26* mutations and the clinical characteristics of the associated hearing loss. However, after systematic analyses of the audiometric data of a large cohort of DFNB1 patients, a consistent pattern emerges (Azaiez et al. 2004; Cryns et al. 2004). Individuals harboring two null mutations (stop mutations or frameshifts such as 35delG) are much more likely to have severe-to-profound deafness than individuals with a null/missense or missense/missense *CX26* genotype. Similarly, individuals with a null/missense genotype are more likely to have severe-to-profound deafness than individuals harboring two missense mutations. Interestingly, individuals carrying Δ(*GJB6*-D13S1830) with any null mutation in *CX26* are consistently the most affected of all patients studied (Azaiez et al. 2004). Expansion of these studies to include larger numbers of patients will probably make possible accurate predictions of the severity of hearing impairment for some of the most common DFNB1 genotypes. On the other hand, identifying those families that represent exceptions to the indicated rules may help in discovering the putative modifier genes thought to be responsible for DFNB1 phenotypic variations.

6.2.2 *CX26* (*GJB2*) and Autosomal Dominant Deafness DFNA3

Some rare mutations in *CX26* have been shown to cause autosomal dominant deafness DFNA3, as well as syndromic forms of deafness associated with skin symptoms. Mutation W44C results in prelingual, high-frequency, severe to profound sensorineural hearing impairment (Denoyelle et al. 1998). Functional analysis of this mutation revealed the inability of mutated channels to transfer Lucifer yellow when expressed in HeLa cells (Martin et al. 1999). When co-expressed with wild-type Cx26, W44C mutants altered channel activity by reducing both the magnitude of the trans-junctional current and the voltage dependence of wild-type Cx26 channels. A dominant negative effect is also hypothesized for the C202F mutation which results in postlingual, predominantly high-frequency, progressive hearing impairment with a variable degree of severity (Morle et al. 2000). Several missense mutations and a short deletion have been reported to cause dominant forms of deafness associated with skin anomalies: (1) R75W is responsible for prelingual, profound deafness sometimes associated with diffuse keratoderma. Partial penetrance could be invoked since R75W has also been found in an unaffected member of the same family (Richard et al. 1998). Kudo et al. (2003) generated a transgenic insertion mouse model expressing mutant R75W Cx26 ubiquitously. The transgenic mice displayed profound hearing loss and degeneration of the organ of Corti shortly after birth, demonstrating that the R75W mutation is indeed deleterious for inner ear function (see also Sect. 6.3.3). (2) D66H is responsible for moderate sensorineural deafness with mutilating keratoderma (Vohwinkel's syndrome; Maestrini et al. 1999). (3) G59A is responsible for high frequency, moderate to severe progressive deafness with palmoplantar keratoderma (Heathcote et al. 2000). (4) DE42 causes profound hearing loss and palmoplantar keratoderma (Rouan et al. 2001). (5) N54K is responsible for profound congenital deafness with palmoplantar keratoderma, knuckle pads, and leukonychia (Bart-Pumphrey syndrome; Richard et al. 2004). (6) Finally, three missense mutations, G12R, S17F, and D50N have been found to cause keratitis–ichthyosis–deafness syndrome (KID), associating severe prelingual deafness with a broad spectrum of cutaneous features, as well as corneal symptoms (Richard et al. 2002).

Although Cx26 is probably implicated in both skin differentiation and keratinocyte proliferation, for instance during wound healing (Brandner et al. 2004), the majority of *CX26* mutations do not cause any epidermal phenotype, suggesting that Cx26 alone is not crucial for normal epidermal differentiation. A possible explanation for the effect of the mutant forms of Cx26 causing skin symptoms is that they display a dominant negative effect on other connexins expressed in the skin. It has been shown that mutated Cx26 forms DE42, D66H and R75W indeed exert a dominant negative effect, not only on wild-type Cx26, as demonstrated for other mutations causing dominant deafness, but also on wild-type Cx30 (Marziano et al. 2003) and Cx43 (Rouan et al. 2001). This *trans*-dominant effect was also demonstrated in transfected HeLa cells with G59A mutant Cx26 on wild-type Cx30 (Marziano et al. 2003).

6.2.3
CX30 (*GJB6*) and Autosomal Dominant Deafness DFNA3′

No recessively inherited, disease-causing point mutations in *CX30* have been reported so far, although, as previously discussed, Cx30 might be involved in DFNB1. Some rare cases of patients homozygotes for Δ(GJB6-D13S1830) as well as one Δ(GJB6-D13S1830)/Δ(GJB6-D13S1854) compound heterozygote have been described (del Castillo et al. 2002, 2003, 2005). However, as reported in Section 6.2.1, the effect of the *CX30* truncation, which is mostly found in CX26 heterozygous patients, has not yet been established. Nevertheless, ablation of *Cx30* in mice results in severe hearing loss (Teubner et al. 2003).

On the other hand, missense mutations in *CX30* with autosomal dominant inheritance have been shown to cause either nonsyndromic hearing loss DFNA3' (a single mutation, T5M) or hidrotic ectodermal dysplasia (Clouston syndrome), a skin disorder with no auditory symptoms (Lamartine et al. 2000). T5M heterozygotes display profound hearing loss in middle to high frequencies, with adult onset (Grifa et al. 1999). Functional studies on *Xenopus* paired oocytes demonstrated the dominant negative effect of this mutation on Cx30 channel activity (Grifa et al. 1999). More recently, a different missense mutation, V37E, was found in a patient with keratitis–ichthyosis–deafness (KID) syndrome (Jan et al. 2004), a disease usually caused by mutations in *CX26*. Surprisingly, this mutation had been previously found in a Clouston syndrome patient without any reported hearing defect (Smith et al. 2002). This observed variability illustrates again the complexities of connexin interactions, and underscores the possible influence of genetic and epigenetic factors which modify the clinical phenotype.

6.2.4 Other Connexin Genes and Deafness

CX31 (GJB3)

Although the implication of *CX31* in skin pathology is well established (see Chap. 7 by G. Richard), its involvement in hearing loss remains uncertain. Three different studies have reported mutations in *CX31* causing either late-onset dominant (Xia et al. 1998), or recessive (Liu et al. 2000), nonsyndromic deafness, as well as a dominant, syndromic form of deafness that presented with peripheral neuropathy (Lopez-Bigas et al. 2001). However, these claims are problematic for several reasons: (1) the small size of the families and paucity of affected individuals precluded the use of linkage analysis in all three studies to confirm the association between *CX31* and the disease; (2) some of the mutations showed partial penetrance, and in one case, an affected member of the family harbored no *CX31* mutations; (3) no functional analyses supported the pathogenic role of the mutations, with the exception of deafness- and neuropathy-associated *Δ*D66 mutation (Lopez-Bigas et al. 2001) that resulted in impaired intracellular trafficking of the mutant connexin (Di et al. 2002). Furthermore, *Cx31* knockout mice have normal hearing, without any inner ear structural defects (Plum et al. 2001). Thus, the role of *CX31* in hearing impairment is still not convincingly established.

CX43 (GJA1)

Liu et al. (2001) implicated *CX43* in nonsyndromic sensorineural deafness among African Americans. They reported four hearing-impaired subjects that were homozygous for either L11F or V24A substitutions. However, segregation of the mutations within the families was not checked, either because some patients were sporadic cases or because other deaf relatives were not available for testing. Subsequently, these mutations have been shown in fact to involve the *CX43* pseudogene located on chromosome 5 (Paznekas et al. 2003). Therefore, there is at present no evidence linking *CX43* mutations to **sensorineural** hearing loss.

More recently, 16 different missense mutations and one codon duplication were found in 17 families with oculo–dento–digital dysplasia (ODDD), an autosomal dominant syndrome that presents with skeletal anomalies, spastic paraplegia and neurodegeneration, associated in some patients with **conductive** hearing loss (Paznekas et al. 2003). Cx43 has been shown to be crucial for bone differentiation (Lecanda et al. 2000), and is expressed in the adult otic capsule (Cohen-Salmon et al. 2004; Fig. 6.2). Consequently, Cx43 may play a role in the development and physiology of the temporal

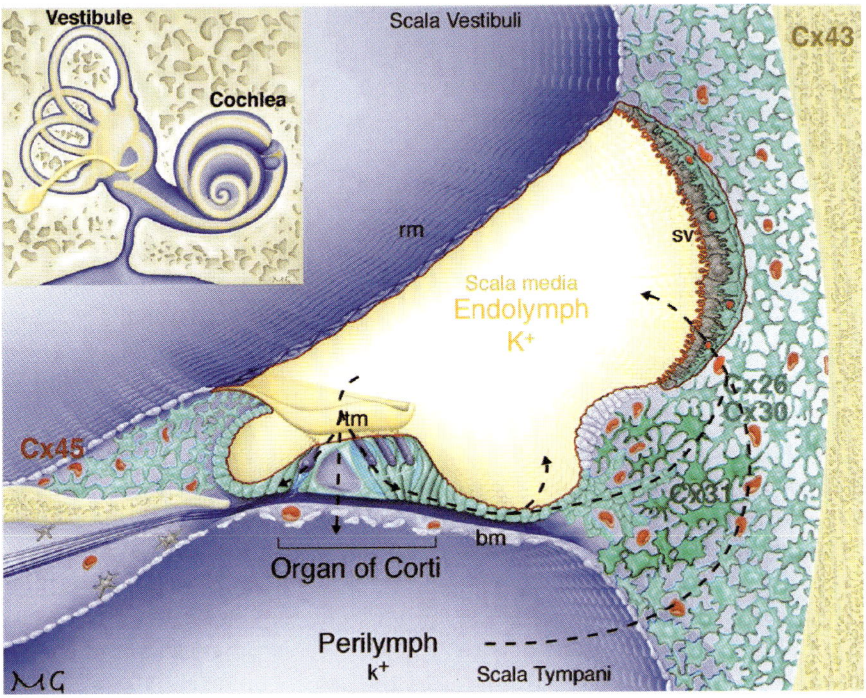

Fig. 6.2. Schematic transverse section of the cochlear duct. *Inset* Schematic representation of the inner ear. Endolymphatic (*yellow*) and strial (*sv*) compartments are delimited by a tight junction network (*red line*). Perilymph (*blue*) bathes the rest of the cochlea. *bm* Basilar membrane, *rm* Reissner's membrane. Cells expressing Cx26 and Cx30 are in *light green*. They form two independent cellular networks: the epithelial network comprising supporting cells and adjacent epithelial cells, and the connective tissue network comprising fibrocytes (*f*) and basal and intermediate cells of the stria vascularis (*sv*). Capillaries (*red*) express Cx45. The bone (*dark yellow*) forming the otic capsule expresses Cx43. Type II fibrocytes (*dark green*) express Cx31. Sensory transduction depends on a K^+ current driven by the endocochlear potential generated at the level of the stria vascularis. This current flows through the sensory hair cells (*dark blue*). Endolymphatic K^+ is thought to be subsequently recycled (*black arrows*)

bone. Computed-tomography scan analysis of temporal bones would be of particular interest in ODDD patients, as it may reveal morphological anomalies of the otic capsule.

CX32 (*GJB1*)

Mutations in *CX32* are known to cause X-linked, demyelinating Charcot-Marie-Tooth disease, a generalized motor and sensory neuropathy that is sometimes accompanied by mild hearing loss (reviewed in Suter and

Scherer 2003). Since there is no evidence for expression of Cx32 within the adult or developing inner ear, including the myelinated portions of the eighth cranial nerve within the cochlea (Forge et al. 2003), hearing impairment should be caused by neuropathy which impairs nerve function somewhere farther up the auditory pathway.

6.3 Inner Ear and Gap Junctions

6.3.1 Inner Ear Architecture

The inner ear of mammals consists of a series of interconnecting fluid-filled tubes present in the cavity of the temporal bone (Fig. 6.2), collectively named the bony labyrinth. It contains a membranous duct immersed in perilymph, a high-sodium solution (Na^+ concentration 140 mM, K^+ concentration 3.5 mM), and filled with endolymph, which is unusually rich in K^+ (about 150 mM) and almost devoid of sodium (1 mM). This membranous duct contains six sensory areas; five of them (the saccule, the utricle, and the three cristae ampulares of the semicircular canals) form the vestibule, the organ responsible for balance perception. The sixth sensory area is the organ of Corti within the cochlea, the hearing organ.

The cochlea is divided into three chambers, namely, the scala vestibuli, the scala tympani, and the scala media (Fig. 6.2). The scalae vestibuli and tympani are both filled with perilymph, whereas the scala media is filled with endolymph. At the base of the cochlea, the scala media communicates with the saccule by the ductus reuniens. The scalae vestibuli and tympani are connected through an opening at the apex of the cochlea, called the helicotrema. The scala vestibuli communicates with the perilymphatic space of the vestibule. The scala tympani is connected by the cochlear aqueduct to the subarachnoidal space of the cranial cavity which is filled with cerebrospinal fluid.

Strikingly, endolymph in the cochlea exhibits an electric potential of about +80 mV, called the endocochlear potential (EP) that is essential for sound transduction. The EP is generated by the stria vascularis, a bilayered, highly vascularized epithelium on the lateral side of the cochlear duct which is also responsible for the secretion of K^+ into the endolymph (Wangemann et al. 1995; Figs. 6.2, 6.3, see Sect. 6.3.3). Maintenance of EP is dependent on a tight junction network formed by all epithelial cells facing the scala media, as well as the basal cells of the stria vascularis. This network constitutes an electrochemical barrier isolating the stria vascularis and the endolymphatic space (see Figs. 6.2, 6.3). The lateral wall of the cochlea (spiral ligament)

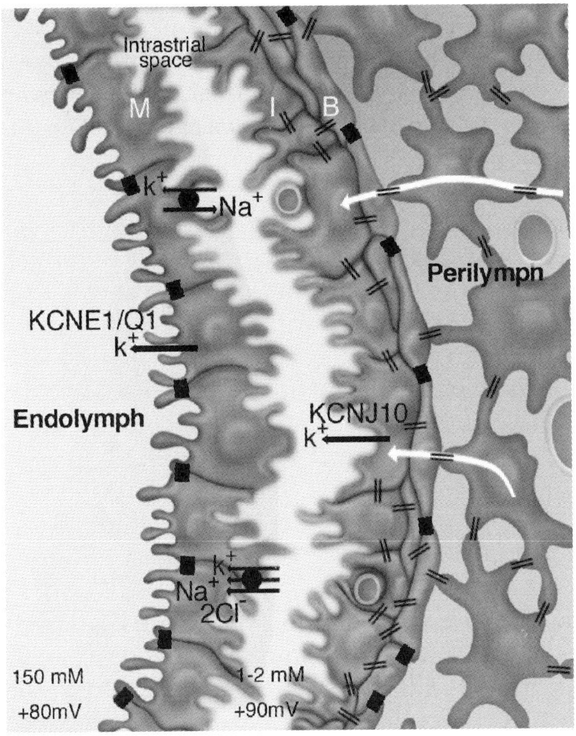

Fig. 6.3. Schematic transverse section of stria vascularis. The stria vascularis is a bilayered, highly vascularized epithelium responsible for the secretion of K^+ into the endolymph and generation of the endocochlear potential (EP). *Black boxes* represent tight junctions which form an electrochemical barrier isolating the stria vascularis. Gap junctions (*short double lines*) couple fibrocytes of the spiral ligament, basal (*B*) and intermediate (*I*) cells. The EP is a K^+ diffusion potential generated across the KCNJ10 K^+ channel by a low K^+ concentration in the intrastrial space and a high K^+ concentration in the cytosol of intermediate cells. Na^+/K^+-ATPase pumps and Na^+-K^+-$2Cl^-$ cotransporters in the marginal cell basal membrane ensure a low K^+ concentration in the intrastrial space. *White arrows* indicate hypotheses for the passage of K^+ through the connective tissue gap junction network

is composed of connective tissue containing blood vessels, extracellular matrix material, and fibrocytes of four different types (Spicer and Schulte 1991, 1996).

The organ of Corti, lying on the basilar membrane, is composed of supporting cells and sensory hair cells (Fig. 6.2). Hair cells (HC) owe their name to a characteristic, highly organized bundle of microvilli, which houses the mechanotransduction machinery and is located on their apical surface, facing the scala media. There are two types of HCs: inner hair cells (IHC), the genuine sensory cells, which receive up to 95% of the afferent innervations of the eighth cranial nerve; and outer hair cells (OHC), which are mechanical effectors that enhance the frequency selectivity and sensitivity in the output of IHCs, through voltage-driven length changes upon sound stimulation (the so-called cochlear amplifier). The basal portion of each HC rests on top of a supporting cell which partially envelops it. The lateral membranes of HCs, particularly in the OHCs, are exposed to large intercellular spaces, filled by perilymph, allowing a free circulation of fluid around the cell bodies; on the other hand, their apical surfaces are in contact with endolymph. Upon sound stimulation, endolymphatic K^+, which carries most of the mechanotransduction current (Hudspeth et al. 2000), flows through the hair cells into the perilymph bathing the organ of Corti. Depolarization of each IHC triggers neurotransmitter glutamate release to connected afferent nerve fibers, which in turn send action potentials to the brain.

6.3.2 Connexin Expression in the Cochlea

The first evidence for gap junctions in the inner ear comes from ultrastructural observations and electrophysiological experiments (Jahnke 1975; Reale et al. 1975; Kikuchi et al. 1994, 1995; Santos-Sacchi 2000). Investigators observed that injecting current into one supporting cell produced voltage drops in adjacent cells. Voltage-gated conductance was found to be sensitive to intracellular pH, Ca^{2+}, temperature, membrane tension, and several uncoupling agents (reviewed in Santos-Sacchi 2000). In addition, asymmetrical voltage gating in double voltage clamp experiments, as well as non-uniform dye spread, were observed in the organ of Corti, which may reflect the variety of gap junction types in this organ (Santos-Sacchi 1986; Zhao and Santos-Sacchi 2000).

Within the mammalian inner ear, gap junctions appear to be very numerous, and some of the plaques are unusually large, as shown by studies combining freeze-fracture analysis, confocal microscopy and immunogold labeling (Forge et al. 2003). Almost all cells are extensively coupled to their neighbors, which suggests that the cochlea may be spanned by large functional syncytia. No evidence of gap junctions in either hair cells of the organ of Corti, or marginal cells in the stria vascularis has ever been found (Kikuchi et al. 1994; Forge et al. 2003).

Several members of the connexin family are expressed in the inner ear. Among them, **Cx26** and **Cx30** are the major gap junction molecules in the inner ear. They are expressed in the same structures (Lautermann et al. 1998), and co-localize. Recently, immunogold labeling experiments demonstrated that labeling for each connexin was always located on both sides of the junctions, a pattern which is consistent with individual connexons being heteromeric with both Cx26 and Cx30 (Forge et al. 2003). Further support for this observation comes from co-immunoprecipitation of both connexins in cochlear tissues (Ahmad et al. 2003; Forge et al. 2003).

In the cochlea, there are two independent networks of cells connected by means of Cx26- and Cx30-containing gap junctions (Kikuchi et al. 1994, 1995; Fig. 6.2). The ***epithelial network*** connects supporting cells and adjacent epithelial cells (Kikuchi et al. 1994, 1995; Frenz and Van De Water 2000); in the mouse, it appears about embryonic day 16 (E16). The ***connective tissue network*** is composed of fibrocytes and mesenchymal cells, and also includes the basal and intermediate cells of the stria vascularis (Kikuchi et al. 1994, 1995; Xia et al. 1999); it starts to develop around birth in mouse.

Cx31 expression in mouse cochlear tissue has been substantiated by RT-PCR experiments (Xia et al. 1999). Cx31 immunolabeling patterns observed in different laboratories are difficult to reconcile. Nevertheless, they all agree on an expression in type II fibrocytes below the spiral prominence from postnatal day 12 (P12) in mouse (Liu et al. 2000; Forge et al. 2003; Fig. 6.2).

Immunostaining experiments revealed the presence of **Cx43** in the supporting cells of the organ of Corti and in the fibrocytes (Liu et al. 2001; Suzuki et al. 2003). However, this expression was found to be transient throughout development using X-gal labeling in a transgenic mouse model with targeted replacement of *CX43* with *lacZ*. Instead, expression of Cx43 was only found in the bone of the otic capsule from P8 onwards (Cohen-Salmon et al. 2004; Fig. 6.2). Forge et al. also reported Cx43 immunostaining in "tension" fibrocytes lining the otic capsule in adult cochlea (Forge et al. 2003).

Cx45 was found to be expressed in the mature inner ear vasculature, by using *CX45-lacZ* transgenic mice (Cohen-Salmon et al. 2004; Fig. 6.2).

As suggested by RT-PCR and cDNA microarray experiments (Ahmad et al. 2003; Forge et al. 2003), other connexins are expressed in the cochlea, such as **Cx29** and **Cx50**, although further work is required to identify their precise sites of expression.

6.3.3 Gap Junctions and Inner Ear Homeostasis

Auditory transduction depends on the production of endocochlear potential (EP) and maintenance of the ionic composition of the endolymph and perilymph. In particular, the concentration of K^+, the major component of the hair cell mechanoelectrical transduction current, has to be tightly regulated. The ionic composition of the inner ear extracellular fluids, endolymph and perilymph, is controlled by local transports. Perilymphatic K^+ partly originates from filtration of blood circulating in spiral ligament and spiral limbus capillaries, which are devoid of fenestrae, as in the blood–brain barrier (Kimura et al. 1974). Endolymphatic K^+, which is actively secreted by the stria vascularis, originates from perilymph, and is thought to cycle in the cochlea (Konishi and Kelsey 1973; Wada et al. 1979). EP is generated in the stria vascularis by KCNJ10, a K^+ channel located in the apical membrane of intermediate cells facing the intrastrial space (Marcus et al. 2002; Fig. 6.3). Ionic pumps and ion cotransporters expressed by marginal cells ensure a very low K^+ concentration (1–2 mM) in the intrastrial space, while the cytosolic K^+ concentration in intermediate cells remains high. Ultimately, the current created is maintained through the marginal cell barrier by KCNQ1/KCNE1 K^+ channels which secrete K^+ into endolymph (Wangemann et al. 1995). EP generation and maintenance require electrochemical barriers formed by tight junction networks on both sides of the stria vascularis as well as between epithelial cells lining the scala media (Gow et al. 2004; Kitajiri et al. 2004; see Figs. 6.2, 6.3). Due to the existence of such barriers, it has been suggested that endolymphatic K^+ might be conveyed from perilymph to the stria vascularis by means of intracellular transport. On the basis of the localization of gap junctions in the inner ear, it has been proposed that intercellular communication may play a central role in these transport processes. At the lateral wall of the cochlea, K^+ might be taken up by type II fibrocytes through the combined action of Na^+/K^+-ATPase and Na^+-K^+-$2Cl^-$ cotransporter. Intracellular transport might then be performed along the cochlear wall by the connective-tissue gap junction network that links fibrocytes with basal and intermediate cells of the stria vascularis (Figs. 6.2, 6.3). To date, the role of Cx26 in this process has not yet been addressed, for instance by targeted inactivation of *Cx26* in the fibrocytes. Inactivation experiments in mice cast doubt on the role of Cx30 in the transport of K^+ to the stria vascularis. Indeed, in the absence of Cx30, mutant mice retain the typical high K^+ concentration in the endolymph (before neuroepithelium degenerates, see below). However, they never develop EP, indicating therefore that Cx30 is dispensable for thetransport of K^+ to the endolymph, but not for the EP generation. To date, no histological abnormality that could explain the absence of EP in $Cx30^{-/-}$ mutants has been found. In addition, expression of genes known to be involved in EP generation is unchanged in $Cx30^{-/-}$ cochlea (Cohen-Salmon, unpublished). Therefore, the role of Cx30 in the EP generation remains to be determined. In conjunction with Cx30, Cx26 might also be required for the maintenance of EP. Indeed, $Cx26^{+/-}/Cx30^{+/-}$ double heterozygous mice showed a decreased EP compared to single heterozygotes (Michel et al. 2003).

Sound stimulation in the cochlea results in an increased flow of K^+ from endolymph through the hair cells into perilymph. Electrophysiological measurements have suggested that K^+ released by hair cells could be recycled back to endolymph by following two distinct pathways (see Fig. 6.2). In the first pathway, K^+ would directly flow through the basilar membrane into scala tympani perilymph (Zidanic and Brownell 1990), and then follow the connective tissue pathway described above.

The second pathway involves the uptake of K^+ by supporting cells. Since these cells are all coupled by gap junctions, they could provide a way to rapidly remove excess K^+ from the vicinity of sensory cells upon sound stimulation (Fig. 6.2). Deiters' cells facing OHCs express KCC4, a K^+-Cl^- cotransporter that could be one of the transporters required for K^+ uptake (Boettger et al. 2002). Finally, targeted deletion of *Cx26* in the epithelial network (Cohen-Salmon et al. 2002), ubiquitous inactivation of *Cx30* (Teubner et al. 2003), expression of dominant-negative R75W mutant Cx26 (Kudo et al. 2000), and ubiquitous inactivation of KCC4 (Boettger et al. 2002), all result in degeneration of the organ of Corti by cell death. The fact that degeneration starts, in all cases, at about the time of hearing onset, suggests that cellular coupling through Cx26 and/or Cx30 is crucial for the survival of the neurosensory epithelium upon sound stimulation, and that cell death likely results from the lack of K^+ spatial buffering. Furthermore, we observed that cell death caused by the absence of Cx26 in supporting cells starts in IHC supporting cells. In this case, the inability of these cells to rapidly remove K^+ secreted by IHC could also reverse the functioning mode of the glutamate transporter GLAST, present in IHC supporting cells. Therefore, not only K^+ would accumulate around IHCs, but also glutamate, provoking cell death (Cohen-Salmon et al. 2002).

Another mechanism by which epithelial gap junctions could control K^+ homeostasis has been recently proposed by Beltramello et al. (2005). They demonstrated the role of epithelial network gap junctions in the regulation of IP_3 diffusion. IP_3 is a second messenger involved in the mobilization of Ca^{2+} intracellular stores and the spreading of Ca^{2+} waves. Delivery of IP_3 via a patch pipette to a single supporting cell (Hensen's cell) in an organotypic culture of rat organ of Corti resulted in a rapid elevation of intracellular Ca^{2+} in the patched cell, as well as Ca^{2+} oscillations in the surrounding supporting cells. Such coordinated activity is likely required for the cochlear supporting cells to accomplish K^+ spatial buffering. Indeed it has been suggested that the activation of Ca^{2+}-dependent K-Cl co-transport systems in the supporting cells may be essential in maintaining the ionic balance of the cochlear fluids, with alteration in K^+ levels leading to excitotoxic death of hair cells (Lagostena et al. 2001).

Given the role of supporting cells in the uptake of K^+ secreted by hair cells upon sound stimulation, several authors have proposed that K^+ would be funneled all through the epithelial gap junction network and secreted into the extracellular spaces of the spiral ligament. Once there, K^+ ions would be actively absorbed by type II fibrocytes, transported via the connective tissue gap junction network to the stria vascularis as discussed previously (Kikuchi et al. 1995; Spicer and Schulte 1996), and the stria vascularis would secrete them again into the endolymph. Although this hypothesis has been largely broadcast in the literature, there is for the moment no experimental evidence to support it.

6.4 Future Prospects

The role of gap junctions in the inner ear is still far from being understood. Nevertheless, recent years have witnessed several major advances in the field, each of them emphasizing the intricacies of inner ear gap junction biology, and pinpointing unresolved issues. First, a precise atlas of connexin expression in the inner ear is essential. Then, with animal models on hand, it will be possible to precisely dissect the role of each connexin in inner ear physiology. Progress in genetic characterization of patients should uncover the involvement of other connexin genes in deafness, as well as genes acting as modifiers of connexin phenotypes. The products of some of those modifier genes may actually be connexin ligands that will provide novel entry points into further aspects of connexin cellular biology.

Acknowledgements. We thank F. Mammano for discussion, R. Bruzzone, and I. del Castillo for critical reading of the manuscript. F.J.d.C. was a recipient of a Marie Curie Post-doctoral Fellowship.

References

Abe S, Usami S, Shinkawa H, Kerlley PM, Kimberling WJ (2000) Prevalent connexin 26 gene mutations in Japanese. J Med Genet 37:41–43

Ahmad S, Chen S, Lin X (2003) Cellular distribution and molecular assembly of connexins in the inner ear of mice. ARO abstract:442

Azaiez H, Chamberlin GP, Fischer SM, Welp CL, Prasad SD, Taggart RT, del Castillo I, Van Camp G, Smith RJH (2004) GJB2: the spectrum of deafness-causing allele variants and their phenotype. Hum Mut 24:305–311

Beltramello M, Piazza V, Bukauskas FF, Pozzan T, Mammano F (2005) Impaired permeability to IP3 in a mutant connexin underlies recessive hereditary deafness. Nat Cell Biol 7: 63–69

Boettger T, Hubner CA, Maier H, Rust MB, Beck FX, Jentsch TJ (2002) Deafness and renal tubular acidosis in mice lacking the K-Cl co-transporter Kcc4. Nature 416:874–878

Brandner JM, Houdek P, Husing B, Kaiser C, Moll I (2004) Connexins 26, 30, and 43: differences among spontaneous, chronic, and accelerated human wound healing. J Invest Dermatol 122:IX-X

Brobby GW, Muller-Myhsok B, Horstmann RD (1998) Connexin 26 R143 W mutation associated with recessive nonsyndromic sensorineural deafness in Africa. New Engl J Med 338:548–550

Bruzzone R, Veronesi V, Gomes D, Bicego M, Duval N, Marlin S, Petit C, D'Andrea P, White TW (2003) Loss-of-function and residual channel activity of connexin26 mutations associated with non-syndromic deafness. FEBS Lett 533:79–88

Cohen-Salmon M, Ott T, Michel V, Hardelin J, Perfettini I, Eybalin M, Wu T, Marcus D, Wangemann P, Willecke K, Petit C (2002) Targeted ablation of connexin26 in the inner ear epithelial gap junction network causes hearing impairment and cell death. Current Biol 12:1106–1111

Cohen-Salmon M, Maxeiner S, Krüger O, Theis M, Willecke K, Petit C (2004) Expression of connexin43- and connexin45- encoding genes in the developing and mature mouse inner ear. Cell Tissue Res 316:15–22

Common JEA, Di W-L, Davies D, Kelsell DP (2004) Further evidence for heterozygote advantage of GJB2 deafness mutations: a link with cell survival. J Med Genet 41:573–575

Cryns K, Orzan E, Murgia A, Huygen PLM, Moreno F, delCastillo I, Pearker Chamberlin G, Azaiez H, Prasad S, Cucci RA, Leonardi E, Snoechx RL, Govaerts PJ, Van de Heyning PH, Van de Heyning CM, Smith RJH, Van Camp G (2004) A genotype-phenotype correlation for GJB2 (Connexin26) deafness. J Med Genet 41:147–154

D'Andrea P, Veronesi V, Bicego M, Melchionda S, Zelante L, Di Iorio E, Bruzzone R, Gasparini P (2000) Hearing loss: frequency and functional studies of the most common connexin26 alleles. Biochem Biophys Res Commun 296:685–691

del Castillo FJ, Rodríguez-Ballesteros M, Álvarez A, Hutchin T, Leonardi E, de Oliveira CA, Azaiez H, Brownstein Z, Avenarius MR, Marlin S, Pandya A, Shahin H, Siemering KR, Weil D, Wuyts W, Aguirre LA, Martín Y, Moreno-Pelayo MA, Villamar M, Avraham KB, Dahl HHM, Kanaan M, Nance WE, Petit C, Smith RJH, Van Camp G, Sartorato EL, Murgia A, Moreno F, del Castillo I (2005) A novel deletion involving the connexin-30 gene, del(GJB6-D13S1854), found in *trans* with mutations in the GJB2 gene (connexin-26) in subjects with autosomal recessive nonsyndromic hearing impairment. J Med Genet (in press)

del Castillo I, Villamar M, Moreno-Pelayo MA, delCasillo FJ, Alvarez A, Telleria D, Menendez I, Moreno F (2002) A deletion involving Connexin30 gene in nonsyndromic hearting impairment. N Engl J Med 346:243–249

del Castillo I, Moreno-Pelayo MA, del Castillo FJ, Brownstein Z, Marlin S, Adina Q, Cockburn DJ, Pandya A, Siemering KR, Chamberlin GP, Ballana E, Wuyt sW, Maciel-Guerra AT, Álvarez A, Villamar M, Shohat M, Abeliovich D, Dahl HHM, Estivill X, Gasparini P, Hutchin T, Nance WE, Sartorato EL, Smith RJH, Van Camp G, Avraham KB, Petit C, Moreno F (2003) Prevalence and evolutionary origins of the del(GJB6-D13S1830) mutation in the DFNB1 locus in hearing-impaired subjects: a multicenter study. Am J Hum Genet 73:1452–1458

Denoyelle F, Lina-Granade G, Plauchu H, Bruzzone R, Chaïb H, Levi-Acobas F, Weil D, Petit C (1998) Connexin26 gene linked to a dominant deafness. Nature 393:319–320

Denoyelle F, Marlin S, Weil D, Moatti L, Chauvin P, Garabédian E-N, Petit C (1999) Clinical features of the prevalent form of childhood deafness, *DFNB1*, due to a connexin26 gene defect: implications for genetic counselling. Lancet 353:1298–1303

Di WL, Monypenny J, Common JEA, Kennedy CT, Holland KA, Leigh IM, Rugg EL, Zicha D, Kelsell DP (2002) Defective trafficking and cell death is characteristic of skin disease-associated connexin 31 mutations. Hum Mol Genet 11:2005–2014

Estivill X, Fortina P, Surrey S, Rabionet R, Melchionda S, D'Agruma L, Mansfield E, Rappaport E, Govea N, Milà M, Zelante L, Gasparini P (1998) Connexin-26 mutations in sporadic and inherited sensorineural deafness. Lancet 351:394–398

Feldmann D, Denoyelle F, Loundon N, Weil D, Garabedian EN, Couderc R, Joannard A, Schmerber S, Delobel B, Leman J, Journel H, Catros H, Ferrec C, Drouin-Garraud V, Obstoy MF, Moati L, Petit C, Marlin S (2004) Clinical evidence of the nonpathogenic nature of the M34T variant in the connexin 26 gene. Eur J Hum Genet 12:279–284

Forge A, Becker DL, Casalotti SO, Edwards J, Marziano NK, Nevill G (2003) Gap junctions in the inner ear: Comparison of distribution patterns in different vertebrates and assessment of connexin composition in mammals. J Comp Neurol 467:207–231

Frenz CM, Van De Water T (2000) Immunolocalization of connexin26 in the developing mouse cochlea. Brain Res Rev 32:172–180

Fuse Y, Doi K, Hasegawa T, Sugii A, Hibino H, Kubo T (1999) Three novel connexin26 gene mutations in autosomal recessive non-syndromic deafness. Neuroreport 10:1853–1857

Gabriel HD, Jung D, Butzler C, Temme A, Traub O, Winterhager E, Willecke K (1998) Transplacental uptake of glucose is decreased in embryonic lethal connexin26-deficient mice. J Cell Biol 140:1453–1461

Gasparini P, Rabionet R, Barbujani G, Melchionda S, Petersen M, Brøndum-Nielsen K, Metspalu A, Oitmaa E, Pisano M, Fortina P, Zelante L, Estivill X, 35delG TGACoG (2000) High carrier frequency of the 35delG deafness mutation in European populations. Eur J Hum Genet 8:19–23

Gow A, Davies C, Southwood CM, Frolenkov G, Chrustowski M, Ng L, Yamauchi D, Marcus DC, Kachar B (2004) Deafness in Claudin 11-null mice reveals the critical contribution of basal cell tight junctions to stria vascularis function. J Neurosci 24:7051–7062

Grifa A, Wagner CA, D'Ambrosio L, Melchionda S, Bernardi F, Lopez-Bigas N, Rabionet R, Arbones M, Monica MD, Estivill X, Zelante L, Lang F, Gasparini P (1999) Mutations in *GJB6* cause nonsyndromic autosomal dominant deafness at DFNA3 locus. Nature Genet 23:16–18

Hamelmann C, Amedofu GK, Albrecht K, Muntau B, Gelhaus A, Brobby GW, Hortsmann RD (2001) Pattern of Connexin26 (GJB3) mutations causing sensorineural hearing impairment in Ghana. Hum Mut 18:84–85

Heathcote K, Syrris P, Carter ND, Patton MA (2000) A connexin 26 mutation causes a syndrome of sensorineural hearing loss and palmoplantar hyperkeratosis (MIM 148350). J Med Genet 37:50–51

Hudspeth AJ, Choe Y, Mehta AD, Martin P (2000) Putting ion channels to work: mechanoelectrical transduction, adaptation, and amplification by hair cells. Proc Natl Acad Sci USA 97:11765–11772

Jahnke K (1975) The fine structure of the freeze-fractured intercellular junctions in the guinea pig inner ear. Acto Otolaryngol (Stockh) 336:1–40

Jan AY, Amin S, Ratajczak P, Richard G, Sybert VP (2004) Genetic heterogeneity of KID syndrome: identification of a Cx30 gene (GJB6) mutation in a patient with KID syndrome and congenital atrichia. J Invest Dermatol 122:1108–1113

Kelley PM, Abe S, Askew JW, Smith SD, Usami S, Kimberling WJ (1999) Human connexin 30 (GJB6), a candidate gene for nonsyndromic hearing loss: molecular cloning, tissue-specific expression, and assignment to chromosome 13q12. Genomics 62:172–176

Kelsell DP, Dunlop J, Stevens HP, Lench NJ, Liang JN, Parry G, Mueller RF, Leigh IM (1997) Connexin 26 mutations in hereditary non-syndromic sensorineural deafness. Nature 387:80–83

Kenneson A, Van Naarden Braun K, Boyle C (2002) GJB2 (connexin 26) variants and non-syndromic sensorineural hearing loss: a HuGE review. Genet Med 4:258–274

Kikuchi T, Adams LC, Paul DL, Kimura RS (1994) Gap junction systems in the rat vestibular labyrinth: Immunohistochemical and ultrastructural analysis. Acta Otolaryngol 114:520–528

Kikuchi T, Kimura RS, Paul DL, Adams JC (1995) Gap junctions in the rat cochlea: immunohistochemical and ultrastructural analysis. Anat Embryol 191:101–118

Kimura RS, Schuknecht HF, Ota CY (1974) Blockage of the cochlear aqueduct. Acta Otolaryngol 77:1–12

Kitajiri SI, Furuse M, Morita K, Saishin-Kiuchi Y, Kido H, Ito J, Tsukita S (2004) Expression patterns of claudins, tight junction adhesion molecules, in the inner ear. Hearing Res 187:25–34

Konishi T, Kelsey E (1973) Effect of potassium deficiency on cochlear potentials and cation contents of the endolymph. Acta Otolaryngol (Stockh) 76:410–418

Kudo T, Ikeda K, Kure S, Matsubara Y, Oshima T, Watanabe K, Kawase T, Narisawa K, T (2000) Novel mutations in the connexin 26 gene (GJB2) responsible for childhood deafness in the Japanese population. J Med Genet 90:141–145

Kudo T, Kure S, Ikeda K, Xia AP, Katori Y, Suzuki M, Kojima K, Ichinohe A, Suzuki Y, Aoki Y, Kobayashi T, Matsubara Y (2003) Transgenic expression of a dominant-negative connexin26 causes degeneration of the organ of Corti and non-syndromic deafness. Hum Mol Genet 12:995–1004

Lagostena L, Ashmore JF, Kachar B, Mammano F (2001) Purinergic control of intercellular communication between Hensen's cells of the guinea-pig cochlea. J Physiol 531:693–706

Lamartine J, Munhoz Essenfelder G, Kibar Z, Lanneluc I, Callouet E, Laoudj D, Lemaitre G, Hand C, Hayflick SJ, Zonana J, Antonarakis S, Radhakrishna U, Kelsell DP, Christianson AL, Pitaval A, Der Kaloustian V, Fraser C, Blanchet-Bardon C, Rouleau G, Waksman G (2000) Mutations in GJB6 cause hidrotic ectodermal dysplasia. Nature Genet 26:142–144

Lautermann J, ten Cate WJ, Altenhoff P, Grummer R, Traub O, Frank H, Jahnke K, Winterhager E (1998) Expression of the gap-junction connexins 26 and 30 in the rat cochlea. Cell Tissue Res 294:415–420

Lecanda F, Warlow P, Sheikh S, Furlan F, Steinberg T, Civitelli R (2000) Connexin43 deficiency causes delayed ossification, carniofacial abnormalities, and osteoblast dysfunction. J Cell Biol 151:931–943

Lerer I, Sagi M, Ben-Neriah Z, Wang T, Levi H, Abeliovich D (2001) A deletion mutation in GJB6 cooperating with a GJB2 mutation in *trans* in non-syndromic deafness: a novel founder mutation in Ashkenazi Jews. Hum Mut 18:460–465

Liu XZ, Xia XJ, Xu LR, Pandya A, Liang CY, Blanton SH, Brown SD, Steel KP, Nance WE (2000) Mutations in connexin31 underlie recessive as well as dominant non-syndromic hearing loss. Hum Mol Genet 9:63–67

Liu XZ, Xia XJ, Adams J, Chen ZY, Welch KO, Tekin M, Ouyang XM, Kristiansen A, Pandya A, Balkany T, Arnos KS, Nance WE (2001) Mutations in GJA1 (connexin 43) are associated with non-syndromic autosomal recessive deafness. Hum Mol Genet 25:2945–2951

Liu XZ, Xia XJ, Ke XM, Ouyang XM, Du LL, Liu YH, Angeli S, Telisch FF, Nance WE, Balkany T, Xu LR (2002) The prevalence of connexin 26 (GJB2) mutations in the Chinese population. Hum Genet 111:394–397

Lopez-Bigas N, Olive M, Rabionet R, Ben-David O, Martinez-Matos JA, Bravo O, Banchs I, Volpini V, Gasparini P, Avraham KB, Ferrer I, Arbones ML, Estivill X (2001) Connexin 31 (GJB3) is expressed in the peripheral and auditory nerves and causes neuropathy and hearing impairment. Hum Mol Genet 10:947–952

Maestrini E, Korge BP, Ocana-Sierra J, Calzolari E, Cambiaghi S, Scuder PM, Hovnanian A, Monaco AP, Munro CS (1999) A missense mutation in connexin 26, D66H, causes mutilating keratoderma with sensorineural deafness (Vohwinkel's syndrome) in three unrelated families. Hum Mol Genet 8:1237–1243

Maheshwari M, Vijaya R, Ghosh M, Shastri S, Kabra M, Menon PSN (2003) Screening of families with autosomal recessive non-syndromic hearing impairment (ARNSHI) for mutations in GJB2 gene: Indian scenario. Am J Med Genet 120A:180–184

Marcus DC, Wu T, Wangemann P, Kofuji P (2002) KCNJ10 (Kir4.1) potassium channel knockout abolishes endocochlear potential. Am J Physiol Cell Physiol 282:C403–407

Martin PE, Coleman SL, Casalotti SO, Forge A, Evans WH (1999) Properties of connexin26 gap junctional proteins derived from mutations associated with non-syndromal hereditary deafness. Hum Mol Genet 8:2369–2376

Marziano NK, Casalotti SO, Portelli AE, Becker DL, Forge A (2003) Mutations in the gene for connexin26 (GJB2) that cause hearing loss have a dominant negative effect on connexin30. Hum Mol Genet 12:805–812

Meyer CG, Amedofu GK, Brandner JM, Pohland D, Timmann C, Horstmann RD (2002) Selection for deafness? Nature Med 8:1332–1333

Michel V, Hardelin J, Petit C (2003) Molecular mechanism of a frequent genetic form of deafness. New Engl J Med 349:716–717

Minarik G, Ferak V, Ferakova E, Ficek A, Polakova H, Kadasi L (2003) High frequency of GJB2 mutation W24X among Slovak Romany (Gypsy) patients with non-syndromic hearing loss (NSHL). Gen Physiol Biophys 22:549–556

Morell RJ, Kim HJ, Hood LJ, Goforth L, Friderici K, Fisher R, Van Camp G, Berlin CI, Oddoux C, Ostrer H, Keats B, Friedman TB (1998) Mutations in the connexin 26 gene (GJB2) among Ashkenazi Jews with nonsyndromic recessive deafness. New Engl J Med 339:1500–1505

Morle L, Bozon M, Alloisio N, Latour P, Vandenberghe A, Plauchu H, Collet L, Edery P, Godet J, Lina-Granade G (2000) A novel C202F mutation in the connexin26 gene (GJB2) associated with autosomal dominant isolated hearing loss. J Med Genet 37:368–370

Ohtsuka A, Yuge I, Kimura RS, Namba A, Abe S, Van Laer L, Van Camp G, Usami S (2003) GJB2 deafness gene shows a specific spectrum of mutations in Japan, including a frequent founder mutation. Hum Genet 112:329–333

Oliveira CA, Maciel-Guerra AT, Sartorato EL (2002) Deafness resulting from mutations in the GJB2 (connexin 26) gene in Brazilian patients. Clin Genet 61:354–358

Pallares-Ruiz N, P. B, Mondain M, Claustres M, Roux AF (2002) A large deletion including most of GJB6 in recessive non syndromic deafness: a digenic effect? Eur J Hum Genet 10:72–76

Park HJ, Hahn SH, Chun YM, Park K, Kim HN (2000) Connexin26 mutations associated with nonsyndromic hearing loss. Laryngoscope 110:1535–1538

Paznekas W, Boyadjiev S, Shapiro R, Daniels O, Wollnik B, Keegan C, Innis J, Dinulos M, Christian C, Hannibal M, Jabs E (2003) Connexin 43 (GJA1) mutations cause the pleiotropic phenotype of oculodentodigital dysplasia. Am J Hum Genet 72:408–418

Petit C, Levilliers J, Hardelin J-P (2001) Molecular genetics of hearing loss. Annu Rev Genet 35:589–646

Plum A, Winterhager E, Pesch J, Lautermann J, Hallas G, Rosentreter B, Traub O, Herberhold C, Willecke K (2001) Connexin31-deficiency in mice causes transient placental dysmorphogenesis but does not impair hearing and skin differentiation. Dev Biol 231:334–347

RamShankar M, Girirajan S, Dagan O, Ravi Shankar HM, Jalvi R, Rangasayee R, Avraham KB, Anand A (2003) Contribution of connexin26 (GJB2) mutations and founder effect to non-syndromic hearing loss in India. J Med Genet 40:68

Reale E, Luciano L, Franke K, Pannese E, Wermbter G, Iurato S (1975) Intercellular junctions in the stria vascularis and spiral ligament. J Ultrastruct Res 53:284–297

Richard G, White TW, Smith LE, Bailey RE, Compton JG, Paul DL, Bale SJ (1998) Functional defects of Cx26 resulting from a heterozygous missense mutation in a family with dominant deaf-mutism and palmoplantar keratoderma. Hum Genet 103:393–399

Richard G, Rouan F, Willoughby CE, Brown N, Chung P, Ryynanen M, Jabs EW, Bale SJ, DiGiovanna JJ, Uitto J, Russell L (2002) Missense mutations in GJB2 encoding connexin-26 cause the ectodermal dysplasia keratitis-ichthyosis-deafness syndrome. Am J Hum Genet 70:1341–1348

Richard G, Brown N, Ishida-Yamamoto A, Krol A (2004) Expanding the phenotypic spectrum of Cx26 disorders: Bart-Pumphrey syndrome is caused by a novel missense mutation in GJB2. J Invest Dermatol 123:856–863

Rothrock CR, Murgia A, Sartorato EL, Leonardi E, Wei S, Lebeis SL, Yu LE, Elfenbein JL, Fisher RA, Friderici KH (2003) Connexin 26 35delG does not represent a mutational hotspot. Hum Genet 113:18–23

Rouan F, White TW, Brown N, Taylor AM, Lucke TW, Paul DL, Munro CS, Uitto J, Hodgins MB, Richard G (2001) Trans-dominant inhibition of connexin43 by mutant connexin26: implications for dominant connexin disorders affecting epidermal differentiation. J Cell Sci 114:2105–2113

Santos-Sacchi J (1986) Dye coupling in the organ of Corti. Cell Tissue Res 245:525–529

Santos-Sacchi J (2000) Cell coupling in Corti's organ. Brain Research reviews 32:167–171

Skerrett IM, Di WL, Kasperek EM, Kelsell DP, Nicholson BJ (2004) Aberrant gating, but a normal expression pattern, underlies the recessive phenotype of the deafness mutant Connexin26M34T. FASEB J 18:860–862

Smith FJ, Morley SM, McLean WH (2002) A novel connexin 30 mutation in Clouston syndrome. J Invest Dermatol 118:530–532

Sobe T, Erlich P, Berry A, Korostichevsky M, Vreugde S, Avraham KB, Bonné-Tamir B, Shohat M (1999) High frequency of the deafness-associated 167delT mutation in the connexin 26 (*GJB2*) gene in Israeli Ashkenazim. Am J Med Genet 86:499–500

Sobe T, S. V, Shahin H, Berlin M, Davis N, Kanaan M, Yaron Y, Orr-Urtreger A, Frydman M, Shohat M, Avraham KB (2000) The prevalence and expression of inherited connexin 26 mutations associated with nonsyndromic hearing loss in the Israeli population. Hum Genet 106:50–57

Spicer S, Schulte BA (1991) Differentiation of inner ear fibrocytes according to their ion transport related activity. Hearing Res 56:53–64

Spicer SS, Schulte BA (1996) The fine structure of spiral ligament cells relates to ion return to the stria and varies with place-frequency. Hearing Res 100:80–100

Suter U, Scherer SS (2003) Disease mechanisms in inherited neuropathies. Nature Rev Neurosci 4:714–726

Suzuki T, Takamatsu T, Oyamada M (2003) Expression of gap junction protein Connexin43 in the adult rat cochlea: comparison with Connexin26. J Histochem Cytochem 51:903–912

Teubner B, Michel V, Pesch J, Lautermann J, Cohen-salmon M, Söhl G, Janhke K, Winterhager E, Heberhold C, Hardelin J, Petit C, Willecke K (2003) Connexin30 (GjB6)-deficiency causes severe hearing impairment and lack of endocochlear potential. Hum Mol Genet 12:13–21

Thonnissen E, Rabionet R, Arbones M, Estivill X, Willecke K, Ott T (2002) Human connexin26 (GJB2) deafness mutations affect the function of gap junction channels at different levels of protein expression. Hum Genet 111:190–197

Van Laer L, Coucke P, Mueller RF, Caethoven G, Flothmann K, Prasad SD, Chamberlin GP, Houseman M, Taylor GR, Van de Heyning CM, Fransen E, Rowland J, Cucci RA, Smith RJH, Van Camp G (2001) A common founder for the 35delG GJB2 gene mutation in connexin 26 hearing impairment. J Med Genet 38:515–518

Wada J, Kambayashi J, Marcus DC, Thalmann R (1979) Vascular perfusion of the cochlea: effect of potassium-free and rubidium-substituted media. Arch Otorhinolaryngol 225:79–81

Wang HL, Chang WT, Li AH, Yeh TH, Wu CY, Chen MS, Huang PC (2003) Functional analysis of connexin-26 mutants associated with hereditary recessive deafness. J Neurochem 84:735–742

Wang YC, Kung CY, Su MC, Su CC, Hsu HM, Tsai CC, Lin CC, Li SY (2002) Mutations of Cx26 gene (GJB2) for prelingual deafness in Taiwan. Eur J Hum Genet 10:495–498

Wangemann P, Liu J, Marcus DC (1995) Ion transport mechanisms responsible for K^+ secretion and the transepithelial voltage across marginal cells of stria vascularis in vitro. Hearing Res 84:19–29

White TW, Deans MR, Kelsell DP, Paul DL (1998) Connexin mutations in deafness. Nature 394:630–631

Xia A-P, Kikuchi T, Hozawa K, Katori Y, Takasaka T (1999) Expression of connexin26 and Na, K-ATPase in the developing mouse cochlear lateral wall: functional implications. Brain Res 846:106–111

Xia JH, Liu CY, Tang BS, Pan Q, Huang L, Dai HP, Zhang BR, Xie W, Hu DX, Zheng D, Shi XL, Wang DA, Xia K, Yu KP, Liao XD, Feng Y, Yang YF, Xiao JY, Xie DH, Huang JZ (1998) Mutations in the gene encoding gap junction protein beta-3 associated with autosomal dominant hearing impairment. Nat Genet 20(4):3:370–373

Yan D, Park HJ, Ouyang XM, Pandya A, Doi K, Erdenetungalag R, Du LL, Matsushiro N, Nance WE, Griffith AJ, Liu XZ (2003) Evidence of a founder effect for the 235delC mutation of GJB2 (connexin 26) in east Asians. Hum Genet 114:44–50

Zhao HB, Santos-Sacchi J (2000) Voltage gating gap junctions in cochlear supporting cells: evidence for nonhomotypic channels. J Membr Biol 175:17–24

Zidanic M, Brownell WE (1990) Fine structure of the intracochlear potential field. I. The silent current. Biophys J 57:1253–1268

7 Human Connexins in Skin Development and Skin Disorders

Gabriele Richard[1]

7.1 Introduction

The skin is our barrier to the terrestrial environment. Its uppermost layer, the epidermis, is a highly specialized, keratinizing stratified epithelium designed to protect the organism from water loss and physical, chemical, and mechanical insults. In order to establish and constantly maintain this barrier, epidermal keratinocytes undergo a complex, highly organized and tightly controlled process of terminal differentiation, during which they transform from undifferentiated basal cells to cornified cells while migrating to the surface. Under normal conditions, cell proliferation and desquamation are in equilibrium. Any external or internal disturbance of this balance, however, is bound to impair the barrier function of the skin eventually resulting in disease. Disorders of cornification are characterized by scaling and thickening of the epidermis and cornified layer (hyperkeratosis) due to an increased cell proliferation, a delay in cell shedding, or as in most circumstances, a combination of both mechanisms. In the last decade, genetic tools have been successfully applied to uncover the molecular basis of a large number of disorders of cornification. This knowledge substantially advanced our understanding of disease pathobiology as well as the role of structural proteins (keratins, desmosomal proteins, proteins of the cornified cell envelope), enzymes involved in lipid, protein, and amino acid metabolism, and regulatory molecules (calcium, gap junction proteins) in the process of keratinocyte differentiation. In fact, the recent identification and analysis of a growing number of genetic connexin gene defects have directly linked connexin function and gap junctional intercellular communication with the control and coordination of epidermal development, proliferation, and differentiation.

[1]Department of Dermatology and Cutaneous Biology, Thomas Jefferson University, 833 S. 10th Street, Philadelphia, PA and GeneDx, Inc., 207 Perry Parkway, Gaithersburg, MD, USA, E-mail: gabi@genedx.com

Gap Junctions in Development and Disease
(ed. by E. Winterhager)

7.2 The Gap Junction System of Human Skin

7.2.1 Changing Patterns of Connexin Expression During Human Skin Development

In the human embryo, skin development starts at around 4 weeks of estimated gestational age (EGA) with the proliferation of the single-layered ectoderm into a two-layered epidermis. The outer layer is the periderm and the inner layer is formed by proliferating basal cells adhering to an immature basement membrane (Holbrook and Odland 1975; Holbrook 1979). Around the 11th week of EGA an intermediate layer is established. By 21 weeks, the epidermis has further stratified into spinous, granular, and cornified layers and the periderm is gradually shed into the amniotic fluid (Holbrook 1979). At birth, a dramatic change in environmental conditions takes place, which requires the epidermis to mature and adapt to a terrestrial milieu by forming a competent skin barrier (Williams et al. 1998).

The different stages of morphogenesis of the epidermis are accompanied by changes in gap junction distribution and connexin expression. In fetal mouse skin, Cx26 and Cx43 were found to be uniformly expressed in all epidermal layers during the early phase. In the later stages of development, Cx43 expression is limited to the intermediate layer and Cx26 expression is restricted to the outermost differentiated epidermal layer (Choudhry et al. 1997). In human fetal skin, Cx26 appears to be expressed during the early stages of fetal epidermal development and Cx43 during the stratification and differentiation (Figs. 7.1, 7.2). As early as 7 weeks EAG, Cx26 can be detected in the basal layer and the periderm. As the epidermis further stratifies, Cx26 expression becomes limited to the periderm while basal expression declines and ceases. After 16 weeks EAG, no Cx26 immunostaining has been detected in the four-layered epidermis (Arita et al. 2002). In neonatal and adult interfollicular epidermis, Cx26 is usually found only in a patchy distribution between basal keratinocytes in the skin of palms and soles. However, there is a high expression of Cx26 in hair follicles and eccrine sweat glands, including acroinfundibulum and sweat duct epithelium (Salomon et al. 1994; Lucke et al. 1999). Cx43 expression has been detected at the surface of basal cells after 8 weeks EAG (Hentula et al. 2001), and later also in the intermediate layer around 12–13 weeks EAG (Arita et al. 2002). Nevertheless, the gap-junction plaques formed before 16 weeks EAG appear immature on electron microscopic examination. Contrary to Cx26, Cx43 expression increases throughout the differentia-

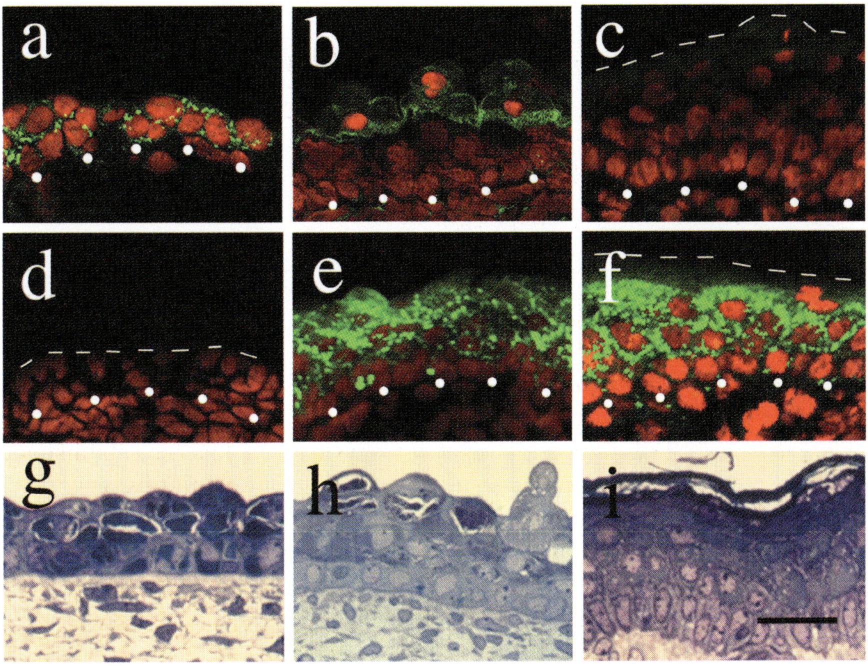

Fig. 7.1a–i. Connexin 26 is expressed in the early stages of human fetal epidermal development and Cx43 in the later stages. **a** Expression of Cx26 within the periderm and basal cells in the two-layered stage of fetal epidermis (49 days EGA). **b** Cx26 is restricted to the periderm at the four- or more layered stage (96 days EGA). **c** Lack of epidermal Cx26 during the stage of interfollicular keratinization (163 days EGA). **d** Conversely, there is no expression of Cx43 in the epidermis at 49 days EGA. **e** Cx43 expression is detectable within the periderm and the intermediate cells of the four or more layered epidermis (88 days EGA). **f** Cx43 is visible between all intermediate cells, but there is only weak expression at basal cell borders during the stage of interfollicular keratinization (163 days EGA). **a–f** Fluorescent micrographs. *Green* represents FITC staining of connexins; *red* is nuclear PI staining. **g–i** Light micrographs depict representative morphological features of developing human epidermis (toluidine blue). **g** The two-layered stage. **h** The four- or more layered stage. **i** The stage of interfollicular keratinization. *White dots* represent the basement membrane zone, *dashed line* the top of the epidermis (top of the periderm or cornified layer). *Bar* 100 µm. (Reproduced with permission from Arita et al. 2002)

tion process of fetal epidermis. In neonatal epidermis, Cx43 shows a typical punctate plasma membrane staining, which is strongest in the granular and spinous cell layers and less intense and focal in the basal layer (Hentula et al. 2001; Arita et al. 2002). After 17–21 weeks EAG, ultrastructural analyses revealed morphologically mature and intact gap junction plaques with typical pentalaminar structures, predominantly between cells of the periderm and less frequently between cells of the intermediate layers. Basal

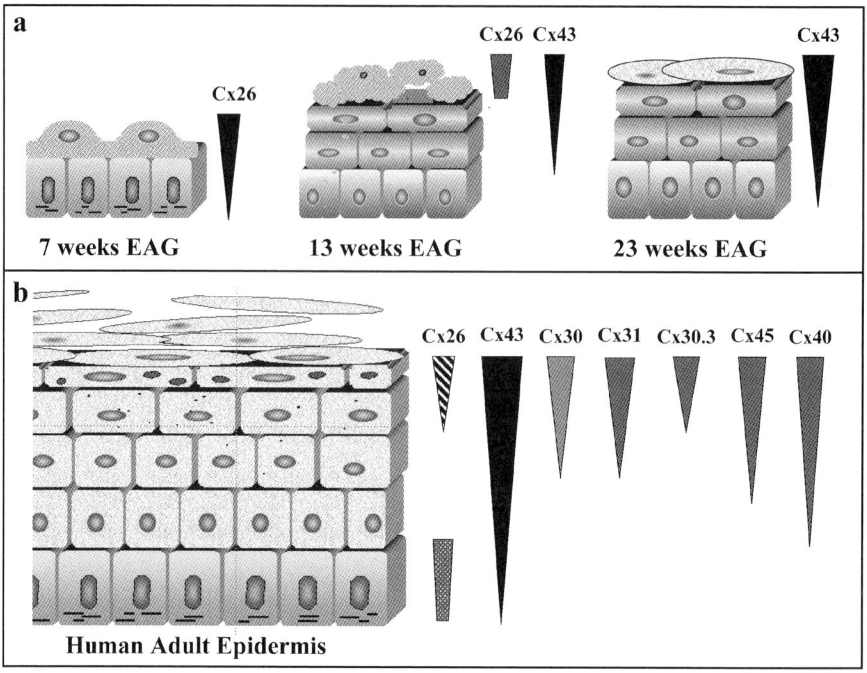

Fig. 7.2. Schematic of the connexin expression patterns in fetal and mature epidermis

cells showed no obvious gap junctions at this stage, but developed a small number of gap junctions later on. Overall, the number of gap junction plaques increases significantly with gestational age, especially in the upper differentiated layers (Arita et al. 2002). It still remains to be determined, if and which other connexin proteins are required during fetal skin development.

It is also difficult to determine the actual level and pattern of intercellular communication that is established by this intricate network of different gap junctions in vivo. In developing mouse epidermis, the cytoplasmic microinjection of a fluorescent tracer into single cells revealed a rapid and extensive vertical dye spread between the basal and all differentiating layers. Surprisingly, the obvious switch in connexin expression during maturation and differentiation of the epidermis did not result in a change in the pattern of dye transfer, perhaps suggesting that more sensitive methods have to be developed to detect all biological changes in intercellular communication (Choudhry et al. 1997).

7.2.2 Connexin Expression in the Mature Human Epidermis

Similar to other epithelia of ectodermal origin, the epidermis and its appendages are exceptionally well coupled by gap junctions. During terminal differentiation, keratinocytes utilize at least nine different connexin proteins, thus producing a very complex and redundant system (Di et al. 2001). Keratinocytes of the basal layer encompass predominantly proliferating, transiently amplifying cells and a small number of slow-cycling stem cells (Watt 2002). The only connexin consistently expressed within this layer in interfollicular epidermis is Cx43, although spotty Cx26 immunostaining has been observed in palmoplantar basal epidermis. With initiation and progress of keratinocyte differentiation, cells gradually expand their gap junction network and express different connexins, including Cx30, Cx30.3, Cx31, Cx40, and Cx45 (Fig. 7.2). The expression patterns of Cx43 and Cx26 in adult epidermis do not significantly deviate from those of neonatal skin. Cx43 is expressed throughout the interfollicular epidermis preferentially in differentiated keratinocytes (Fig. 7.3). It is also abundant in the sebaceous glands and hair follicles, especially in the cortex, inner and outer root sheets

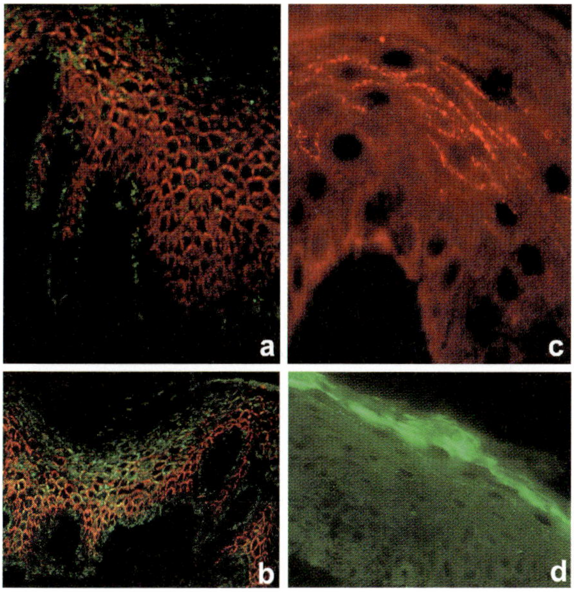

Fig. 7.3a–d. Expression pattern of selected epidermal connexin isoforms. Immunolocalization of Cx26 (*green*) and Cx43 (*red*). **a** In normal plantar epidermis and **b** in a patient with PPK and SNHL due to mutation E42del in Cx26. **c** Immunolocalization of Cx30 (*red*) and **d** Cx31 (*green*) in normal truncal epidermis. (Reproduced with permission from Richard 2003)

(Salomon et al. 1994; Tada and Hashimoto 1997; Lucke et al. 1999; Iguchi et al. 2003). Moreover, Cx43 is the most prominent gap-junction protein of fibroblasts. Strong Cx43 expression in the hair papilla and the surrounding connective tissue led to speculation that these structures might communicate via Cx43 gap junctions, thus synchronizing the dynamic structural changes during hair cycling (Iguchi et al. 2003). On the contrary, no intercellular coupling has been detected between the epidermis or follicular epithelium and the underlying mesenchyme in mouse whiskers hair, suggesting that gap junctions do not participate in direct epidermal–dermal interactions (Salomon et al. 1988; Choudhry et al. 1997). Cx26 is usually limited to palmoplantar epidermis, but shows strong presence in hair follicles and eccrine sweat glands and ducts, where it co-localizes with Cx43 (Fig. 7.3; Salomon et al. 1994; Lucke et al. 1999; Rouan et al. 2001). Cx30 mirrors the prominent adnexal expression of Cx26 and both connexins widely co-localize. Similar to Cx30.3 and Cx31, Cx30 is also expressed during the late stages of epidermal differentiation in the granular and upper spinous layers of interfollicular epidermis (Fig. 7.3; Macari et al. 2000; Di et al. 2001). Like Cx43, Cx31 is also found in sebocytes and the outer root sheath of hair follicles. The reported immunolocalization of Cx40 widely overlaps with Cx43, while Cx45 appears be distributed at the cell membranes between the upper epidermal layers (Di et al. 2001). Cx32, usually not detectable in keratinocytes, is expressed by melanocytes (Salomon et al. 1994; Masuda et al. 2001). In addition, transcripts of Cx31.1 and Cx37 have been detected by RT-PCR from RNA of human adult epidermis (Di et al. 2001) and localized by immunocytochemistry in neonatal foreskin to the granular and upper spinous cell layers (Richard et al. 1997a). A similarly complex pattern of connexin expression also exists in rodent skin, which has been shown to include Cx26, Cx43 Cx45, Cx40, Cx37, Cx31.1, and Cx31 (Butterweck et al. 1994; Goliger and Paul 1994).

7.2.3 Changes in Connexin Expression During Epidermal Differentiation

Any change in the proliferation and differentiation program of keratinocytes apparently coincides with a dynamically regulated switch of connexin expression. Compared to normal epidermis, stratified nonkeratinizing mucosal epithelia show significant expression of Cx26 in the basal and suprabasal layers, widely overlapping with Cx43 (Lucke et al. 1999). Hyperplasia of the epidermis, as found in psoriasis, viral warts, or constitutively in palmoplantar skin, is usually associated with an increase in expression of all epidermal connexins. In addition, an impressive upregulation of Cx26 is triggered as determined by cDNA differential display, Northern blot analysis or immunohistochemistry (Rivas et al. 1997; Labarthe et al. 1998). Similar observations have also been made after tape stripping of human skin. Mechanical irritation initially induces a patchy Cx26 expression in the basal cell layer, which subsequently spreads throughout the basal and suprabasal layers within 48 h (Lucke et al. 1999). In psoriatic plaques, Cx26 is abundantly expressed throughout all layers of the epidermis and localizes both to the cell membranes and the cytoplasm, suggesting a disturbance of Cx26 biogenesis in psoriatic skin (Labarthe et al. 1998). Nevertheless, Cx26 positive cells are negative for the proliferation marker Ki67 and apparently do not proliferate, implying that Cx26 is associated with differentiation rather than proliferation.

Epidermal wound healing, which requires cell proliferation, migration, and differentiation of keratinocytes, appears to be tightly controlled by changes in connexin expression and gap junction communication (Goliger and Paul 1995; Coutinho et al. 2003; Kretz et al. 2003; Qiu et al. 2003; Brandner et al. 2004). For instance, wounding of the adult rat-tail is associated with a strong upregulation of Cx26 expression in differentiated cells proximal to the wound edge, while simultaneously the expression of Cx31.1 and Cx43 declines, and in vivo dye transfer patterns change (Goliger and Paul 1995). Coutinho et al. (2003) described the lack of Cx26, Cx30, Cx43 and Cx31.1 at the wound edges in mouse skin 6 h after wounding. Similarly, in human skin, initial wound healing is associated with a drastic loss of Cx43 staining at the wound margins as early as 5 h after wounding, followed by a subsequent induction of Cx26 and Cx30 near the margins after 18 h (Coutinho et al. 2003; Brandner et al. 2004). Therefore, it has been suggested that the downregulation or change in properties of Cx43 is a prerequisite for proliferation, migration of keratinocytes into the wound bed, and effective wound closure (Kretz et al. 2003). The complete absence of gap junction proteins in keratinocytes at the wound margins might be required to isolate these cells and to allow them to freely migrate into the wounded area. These findings are further supported by the observation that reducing expression of Cx43 in mouse skin (using antisense oligonucleotides or conditional gene knockout) can accelerate closure of incisional wounds (Kretz et al. 2003; Qiu et al. 2003) and that Cx43 is strongly expressed at the edges of nonhealing skin ulcers (Brandner et al. 2004). Later stages of epithelial regeneration apparently necessitate expression of Cx26 and Cx30, predominantly in keratinocytes at or near the wound edges (Coutinho et al. 2003; Qiu et al. 2003; Brandner et al. 2004). These recent advances in our understanding of the role of connexin during wound healing provide exciting new possibilities to modulate cell behavior and hasten wound healing by targeted suppression or induction of specific connexins.

In summary, it has become clear that multiple connexins control epidermal homeostasis during periods of rapid epidermal growth and differentiation. Dependent on their compatibility code, they likely also interact with each other, forming mixed hemichannels or mixed gap junction channels with unique permeability properties. These possibilities further increase the diversity and complexity of this epidermal signaling system and probably critically determine tissue- and differentiation-specific functions of epidermal connexins (Goldberg et al. 1999; Niessen et al. 2000; Forge et al. 2003). However, connexins not only facilitate intercellular communication through gap-junction channels. There is mounting evidence that connexins can form functional hemichannels in the plasma membranes. Such hemichannels may regulate diverse mechanisms crucial for cell survival and differentiation, such as extracellular calcium ion concentration, cell volume control (Quist et al. 2000), or modulation of paracrine signaling through the release of ATP (for review, see Goodenough and Paul 2003).

7.3 Connexin Disorders of the Skin

The functional importance of epidermal gap junctions manifests itself in a series of inherited connexin disorders with cutaneous involvement (Table 7.1). To date, germline mutations in five different connexin genes have been identified, including Cx26, Cx30, Cx30.3, Cx31 and Cx43, and are outlined in the following. While some of these genetic defects are limited to the skin, others involve several epithelia of ectodermal origin, thus producing hearing loss, hair and nail abnormalities, keratitis, or neuropathy (Table 7.1). Even within one gene, distinct mutations can cause a wide spectrum of clinical presentations as exemplified by the allelic series of cutaneous Cx26 disorders (Fig. 7.4, Table 7.1). In contrast to the inner ear, where autosomal recessive null alleles of Cx26, Cx30 or Cx31 result in nonsyndromic sensorineural hearing loss (SNHL; Nance 2003), the mere loss of expression or compromised function of one connexins is usually not sufficient to produce an overt skin disorder. These observations, confirmed in transgenic mice deficient for individual connexins (Willecke et al. 2002), reflect a high level of functional redundancy in the cutaneous gap junction system. The majority of pathogenic connexin gene mutations with cutaneous manifestations is autosomal dominant in nature and has been shown to dominantly interfere with the formation and function of gap junctions in which they are incorporated. The specific composition of the gap junction system in subpopulations of keratinocytes or other epithelial cells in which the mutant connexin protein is expressed thus might be a determining factor for the pleiotropic effects of individual mutations and the genetic diversity of connexin disorders, which result in intriguing genotype-phenotype correlations.

Table 7.1. Cutaneous connexin disorders (*continued on next page*)

Disorder	OMIM no.	Inheritance	Mutated Gene(s)	Locus	Affected protein(s)	Clinical features	References
Erythrokeratodermia variabilis (EKV)	133200	AD	*GJB3*, *GJB4*	1q35.1 1q35.1	Connexin 31, Connexin 30.3	Transient erythematous patches and generalized hyperkeratosis or well demarcated hyperkeratotic plaques; PPK in about 50% of patients	Richard et al. (1998a); Wilgoss et al. (1999); Macari et al. (2000); Richard et al. (2003); Morley et al. (2004)
		or rarely AR					Gottfried et al. (2002); Terrinoni et al. (2004)
Diffuse palmoplantar keratoderma with sensorineural hearing loss	124500	AD	*GJB2*	13q11-q12	Connexin 26	Congenital, severe to profound SNHL; diffuse or plaque-like PPK	Richard et al. (1998b); Heathcote et al. (2000); Loffeld and Kelsell (2000); Rouan et al. (2001); Uyguner et al. (2002)
Vohwinkel syndrome (VW syndrome)	148350	AD	*GJB2*	13q11-q12	Connexin 26	Mild-moderate SNHL; diffuse PPK with honeycomb-like surface; starfish-shaped hyperkeratotic plaques on distal extremities; circular digital constriction bands (pseudoainhum)	Maestrini et al. (1999); Kelsell et al. (2000)
Bart-Pumphrey syndrome (BPS)	149200	AD	*GJB2*	13q11-q12	Connexin 26	Severe-profound SNHL, variable degree of diffuse PPK; leukonychia; knuckle pads and hyperkertotic plaques over interdigital joints	Richard et al. (2004)

Table 7.1. (*continued*)

Disorder	OMIM no.	Inheritance	Mutated Gene(s)	Locus	Affected protein(s)	Clinical features	References
Keratitis–ichthyosis–deafness syndrome (KIDS)	148210	AD, mostly sporadic	*GJB2*	13q11-q12	Connexin 26	Severe-profound SNHL; PPK with honeycomb-like surface; erythematous, hyperkeratotic plaques on face and extremities and/or follicular hyperkeratoses; keratitis and corneal neovascularization; increased susceptibility to mucocutaneous infections and squamous cell carcinoma; variable nail dystrophy, alopecia or hypotrichosis	Richard et al. (2002); van Geel et al. (2002); van Steensel et al. (2002); Alvarez et al. (2003); Wasserman et al. (2003); Yotsumoto et al. (2003); Montgomery et al. (2004)
			GJB6	13q11-q12	Connexin 30		Jan et al. (2004)
Hidrotic ectodermal dysplasia (Clouston syndrome)	129500	AD	*GJB6*	13q11-q12	Connexin 30	Mild, diffuse PPK; hypo-or hypertrophic nail dystrophy; sparse or absent hair; normal sweating; usually normal hearing	Lamartine et al. (2000); Smith et al. (2002); van Steensel et al. (2003); Zhang et al. (2003)
Oculo–dento–digital dysplasia (ODDD)	1642000	AD	*GJA1*	6q22-q24	Connexin 43	PPK, curly, sparse hair, nail dystrophy, craniofacial and limbal dysmorphism, ophthalmological, neurological and cardiac anomalies	Paznekas et al. (2003); Kjaer et al. (2004); Richardson et al. (2004); van Steensel et al. (2004); Kellermayer et al. (2004)

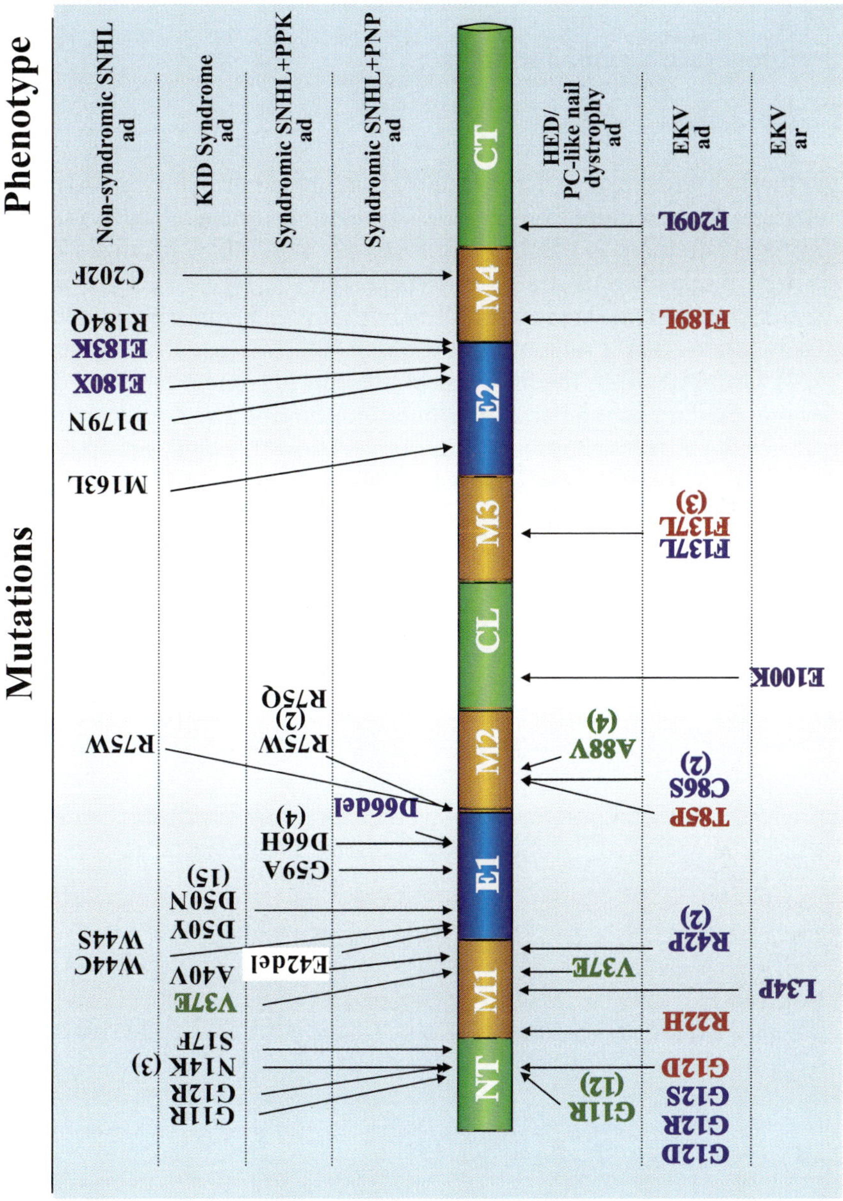

Fig. 7.4. Topographic distribution of connexin mutations in Cx26 (*black*), Cx30 (*green*), Cx31 (*blue*) and Cx30.3 (*red*) across a generic β-connexin molecule and their clinical phenotypes. Due to their large number, mutations associated with ODDD and autosomal recessive sensorineural hearing loss are not depicted. The structural motifs of connexins are *NT* amino-terminus, *M1–M4* transmembrane domains 1–4, *E1*/*E2* extracellular domains 1 and 2, *CL* cytoplasmic loop, *CT* carboxy-terminus, *SNHL* sensorineural hearing loss, *KID* keratitis-ichthyosis-deafness syndrome, *PPK* palmoplantar keratoderma, *PNP* peripheral neuropathy, *HED* hidrotic ectodermal dysplasia (Clouston syndrome), *EKV* erythrokeratodermia variabilis, *ad* autosomal dominant, *ar* autosomal recessive. (Modified, with permission, from Richard 2003)

7.3.1 Erythrokeratodermia Variabilis

Clinical Features

Erythrokeratodermia variabilis (EKV) is a rare genetic skin disorder characterized by a combination of two distinct morphological features, short-lasting erythematous patches and relatively persistent, localized or generalized thickening of the skin (hyperkeratosis; Fig. 7.5). Both features demonstrate a remarkable variability in shape, size, duration and location of skin lesions that is reflected in the name of the disease (Mendes da Costa 1925). Usually, within the first year of life, sharply demarcated red patches develop that are shaped and distributed randomly. They only last for several hours to days, resulting in the impression that they are 'migrating' over the body surface. The red areas, sometimes surrounded by an anemic halo, can coalesce into large, figurate patches or have a circinate, targetoid appearance. Simultaneously or subsequently, sharply defined plaques of rough, thickened skin develop in a strikingly symmetric distribution with predilection of extremities, buttocks and lateral trunk. Sometimes, the hyperkeratosis can be generalized and so severe that the built-up stratum corneum is several centimeters thick (so-called ichthyosis hystrix-like appearance). The skin on palms and soles is thickened in about half of all patients (palmoplantar keratoderma-PPK; Richard et al. 1998a). Both

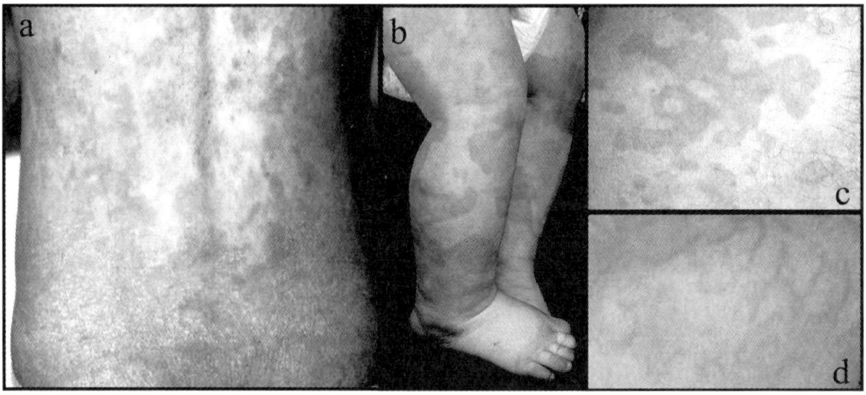

Fig. 7.5a–d. Erythrokeratodermia variabilis due to autosomal dominant mutations of *GJB3* (Cx31) and *GJB4* (Cx30.3). **a** Characteristic features of EKV with coexistence of transient, sharply demarcated erythematous patches and generalized, brown hyperkeratosis, mutation G12R in Cx31. **b** Erythema and symmetrical hyperkeratotic plaques, mutation F137L in Cx30.3. **c** Sharply outlined, figurate erythematous patches, mutation G12R in Cx31. **d** Circinate erythematous patches, mutation F137L in Cx30.3. (Reproduced with permission from Richard, in press)

erythema and hyperkeratosis can be triggered or exaggerated by sudden temperature changes, mechanical friction, and other endogenous or environmental factors. Histopathologic and ultrastructural abnormalities of EKV are not unique. Typically, they are a prominent thickening of the stratum corneum with normal basket-weaved appearance (orthokeratotic hyperkeratosis), acanthosis, church spire-like papillomatosis, dilatation of superficial capillaries and a very mild perivascular inflammation (Vandersteen and Muller 1971; Rappaport et al. 1986).

Erythrokeratodermia variabilis (EKV) broadly overlaps with progressive symmetric erythrokeratoderma (PSEK), an ill-defined group of disorders with persistent and progressive hyperkeratosis, PPK and underlying erythema. PSEK lacks the occurrence of independent 'migrating' red patches, the hallmark of EKV. The considerable phenotypic variability of each disorder and the observation of both phenotypes within a single family suggested that both are allelic disorders (MacFarlane et al. 1991; van Steensel 2004). Nevertheless, the lack of connexin gene mutations in all cases of PSEK tested to date supports the view that EKV and PSEK are distinct disease entities of different etiology (Richard 2005).

Molecular Basis

Erythrokeratodermia variabilis is predominantly inherited as an autosomal dominant trait with nearly complete penetrance, but considerable intra- and interfamilial variability (Richard et al. 2003). However, two families with autosomal recessive transmission have been documented recently (Gottfried et al. 2002; Terrinoni et al. 2004). In several large multigeneration families of different origins EKV has been mapped to chromosome 1p34-p35.1, a region harboring a cluster of four connexin genes (van der Schroeff et al. 1984; Richard et al. 1997b; Macari et al. 2000). Nevertheless, EKV is genetically heterogeneous and may be caused by mutations in at least two different connexin genes (Table 7.1; Fig. 7.4). In 1998, Richard et al. identified disease-causing mutations in the *GJB3* gene encoding Cx31, which made EKV the first human connexin disorder of the skin to be discovered. Subsequently, several new missense mutations of *GJB3* (Wilgoss et al. 1999; Richard et al. 2000) and pathogenic mutations in the neighboring connexin gene *GJB4* encoding Cx30.3 were identified (Macari et al. 2000; Richard et al. 2003). Despite these findings, a small number of EKV patients do not harbor pathogenic connexin gene mutations and the molecular basis of the disorder in these patients still remains elusive (Ishida-Yamamoto et al. 2000; Arita et al. 2003; Richard et al. 2003). Both Cx31 and Cx30.3 belong to the group of β-type connexins and are preferentially expressed in the upper, differentiated keratinocytes of human epidermis, suggesting they play a crucial role during epidermal differentiation (Macari et al. 2000;

Richard et al. 2003). While Cx30.3 has a very limited tissue distribution, including placenta and thymus (Fonseca et al. 2004), Cx31 is further expressed in the inner ear, ciliary body, CNS, testis, kidney, placenta, human preimplantation embryos and blastocysts and, as suggested in one report, in Schwann cells of the peripheral nervous system (Xia et al. 2000; Di et al. 2001; Gabriel et al. 2001; Lopez-Bigas et al. 2001a; Coffey et al. 2002; Jungbluth et al. 2002; Forge et al. 2003; Juneja 2003). Hence, it is not surprising that Cx31 mutations have been associated with other disorders, specifically autosomal dominant or recessive sensorineural hearing loss and peripheral neuropathy, albeit without any apparent overlap with EKV (Xia et al. 1998; Liu et al. 2000; Lopez-Bigas et al. 2001a).

To date, pathogenic connexin mutations have been reported in 18 unrelated EKV patients and families of predominantly Northern European origin, including 7 distinct missense mutations in the Cx31 gene *GJB3* and 5 missense mutations in the Cx30.3 gene *GJB4* (Fig. 7.4). The clinical features of mutations are in general indistinguishable. In three out of seven EKV families, however, Cx30.3 gene mutations were associated with the occurrence of transient erythematous patches with peculiar, circinate or gyrate borders reminiscent of erythema gyratum repens (Macari et al. 2000; Richard et al. 2003). The nature and location of mutations in both genes are very similar. All mutations result in replacement of amino acid residues highly conserved among β-connexins and are predicted to exert a dominant negative effect on the function of co-expressed wild-type connexins. With the exception of one recurrent mutation that lies in the cytoplasmic carboxy-terminal tail of Cx31 (L209F; Morley et al. 2005, Feldmeyer et al. 2005), all other mutations cluster either in the cytoplasmic amino-terminus (G12D, G12R) or the membrane-crossing α-helices of Cx31 (R42P, C86S, F137L) and Cx30.3 (R22H, T85P, F137L, F189L; Fig. 7.4; Richard et al. 1998a, 2000, 2003; Wilgoss et al. 1999; Macari et al. 2000). Mutations in the transmembrane domains have been predicted to hinder regulation of voltage gating or the kinetics of channel closure. The remainders substitute a small, conserved glycine residue at position 12 of the cytoplasmic amino-terminus of Cx31 or Cx30.3 with a positively (G12R) or negatively charged residue (G12D). These mutations are assumed to interfere with the flexibility of this domain, connexin selectivity, or gating polarity of gap junction channels (Lagree et al. 2003).

In addition to the dominant connexin gene mutations described above, two recent genetic studies have confirmed the existence of an autosomal recessive variant of EKV (Gottfried et al. 2002; Terrinoni et al. 2004). In each EKV family, the affected individuals were homozygous for a missense mutation in Cx31, L34P in the first transmembrane domain or E100 K in the cytoplasmic loop (Fig. 7.4). The unaffected parents were heterozygous mutation carriers. Expression studies of L34P-Cx31 in human keratinocytes revealed an abnormal cytoplasmic accumulation of the mutant Cx31 protein and lack of gap junction plaques, although gap junction-mediated cell coupling or the possibility of functional hemichannels was not tested (Gottfried et al. 2002).

Several sequence variants of the coding regions of *GJB3* and *GJB4* exist, which have not been associated with a disease state. Missense mutation R32W in Cx31, for example, has a relatively high allele frequency (up to 17%) in the general population (Lopez-Bigas et al. 2000). The biological relevance of this amino acid substitution remains disputed as there is evidence supporting R32W to be a nonconsequential polymorphism vs. an autosomal recessive allele with subnormal function (Lopez-Bigas et al. 2001b; Di et al. 2002; Rouan et al. 2003a). Another striking finding is a common four-base pair deletion (154del4) in the coding sequence of *GJB4* with an allele frequency ranging from 3.3 to 4.5% (Lopez-Bigas et al. 2002; Richard et al. 2003). This deletion leads to a frameshift and premature termination of protein translation. Since RNA analysis in a person homozygous for this deletion has excluded nonsense-mediated RNA decay (van Geel et al. 2002b), the deletion likely results in synthesis of a shortened protein, which is rapidly degraded, thus resulting in a null allele. Heterozygous carriers of 154del4 from different population groups have had no skin abnormalities (Lopez-Bigas et al. 2002; Richard et al. 2003). Two homozygous individuals have been identified who are in essence human Cx30.3 'knock-outs'. One person had early onset hearing loss, but the Cx30.3 deletion did not cosegregate with hearing loss in several other families (Lopez-Bigas et al. 2002). The other person had normal hearing, but skin features consistent with progressive symmetric erythrokeratoderma and also carried a heterozygous sequence variant (R32W) in Cx31. Nevertheless, an unaffected sibling shared the same genotype, excluding the possibility that 154del4 in Cx30.3 or R32W in Cx31 were sufficient to cause the patient's skin disorder (van Geel et al. 2002b). These observations are consistent with data on transgenic mice deficient for Cx31, Cx30 or Cx43, which also do not develop an obvious skin phenotype (Willecke et al. 2002).

Functional Implications of GJB3 (Cx31) and GJB4 (Cx30.3) Mutations

In vitro expression studies recently revealed that amino acid substitutions in Cx31 and Cx30.3 disturb the intracellular processing and trafficking of the gap junction proteins to the plasma membrane (Fig. 7.6). It has been proposed that the mutant proteins are incorrectly folded, and thus are recognized and tagged for degradation by intracellular quality control processes (VanSlyke et al. 2000), leading to paucity or loss of visible gap junction plaques and impaired intercellular communication (Di et al. 2002; Plantard et al. 2003; Rouan et al. 2003a). In addition, the expression of

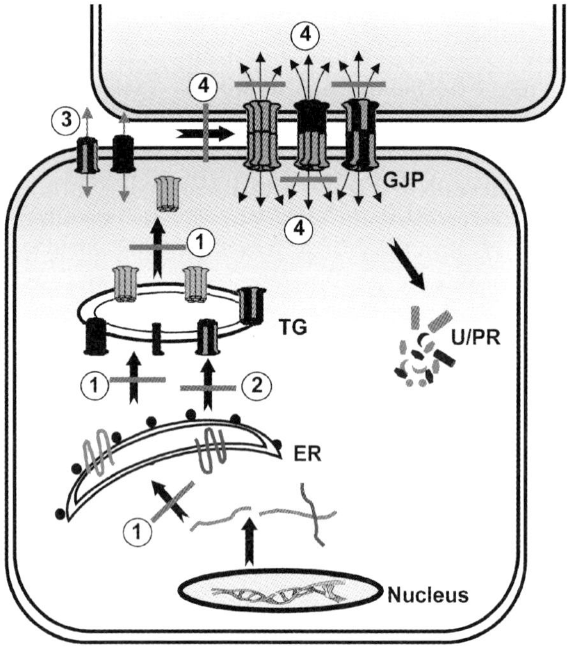

Fig. 7.6. Principal consequences of skin disease-associated connexin mutations on connexin biogenesis and function. *ER* Endoplasmic reticulum, *TG trans*-Golgi network, *GJP* gap-junction plaque, *U/PR* ubiquitination/proteasomal degradation. 1 Missense mutations interfere with protein folding and/or oligomerization and stabilization of homomeric hemichannels. The mutant protein is not transported to the plasma membrane and trapped in the cytoplasm. Examples: R42P, C86S in Cx31; G11R, A88 V in Cx30 (Common et al. 2002; Di et al. 2002). 2 Dominant missense mutations alter the formation and/or function of mixed connexons or potentially change the compatibility code of the mutant protein (change hetero-oligomerization with other isoforms). Example: R142W in Cx32, perhaps G12D/G12R in Cx26 or Cx31 or G11R in Cx30 (Ressot and Bruzzone 2000; Lagree et al. 2003). 3 Gain of function, formation of functional hemichannels in the plasma membrane. Examples: G11R and A88V in Cx30 (Essenfelder et al. 2004). 4 Dominant mutations, especially in the extracellular loops, prevent docking of connexons or proper channel formation and thus gap junction coupling. Examples: D66del in Cx31; W44C, E42del, R75W in Cx26 (Martin et al. 1999; Rouan et al. 2001; Common et al. 2002; Di et al. 2002; Oshima et al. 2003). (Modified, with permission, from Richard 2003)

several different Cx31 mutations, including G12R (Diestel et al. 2002), G12D, G12R, R42P, C86S (Di et al. 2002; Common et al. 2003) and F137L (Rouan et al. 2003b), has been described to induced cell death. If and how this mechanism could be of biological relevance for the epidermis in EKV, however, remains to be determined. Nevertheless, cell death induced by one of these mutations (G12R-Cx31) was due to a gain of function, stemming from a doubling of the dye transfer capacity of transfected HeLa cells

(Diestel et al. 2002). In contrast to Cx31 mutations, reduced cell survival was not observed for the Cx30.3 mutation F137L, although the function of co-expressed wild-type Cx31 was inhibited consistent with a dominant negative mechanism (Plantard et al. 2003; Rouan et al. 2003b). A study by Plantard et al. (2004) revealed for the first time direct evidence for a co-oligomerization of Cx31 and Cx30.3 into mixed connexin hemichannels, which are more stable and functionally active compared to homomeric Cx31 or Cx30.3 channels (Plantard et al. 2003). This important finding is crucial for understanding how mutations affecting either Cx31 or Cx30.3 result in a similar clinical EKV phenotype.

7.3.2 Palmoplantar Keratodermas

Hereditary palmoplantar keratodermas (PPKs) are a very large, clinically, genetically and etiologically heterogeneous group of disorders characterized by thickening of the skin ('keratoderma') of palms and soles due to disturbed epidermal keratinization. In addition, PPK is a common feature of

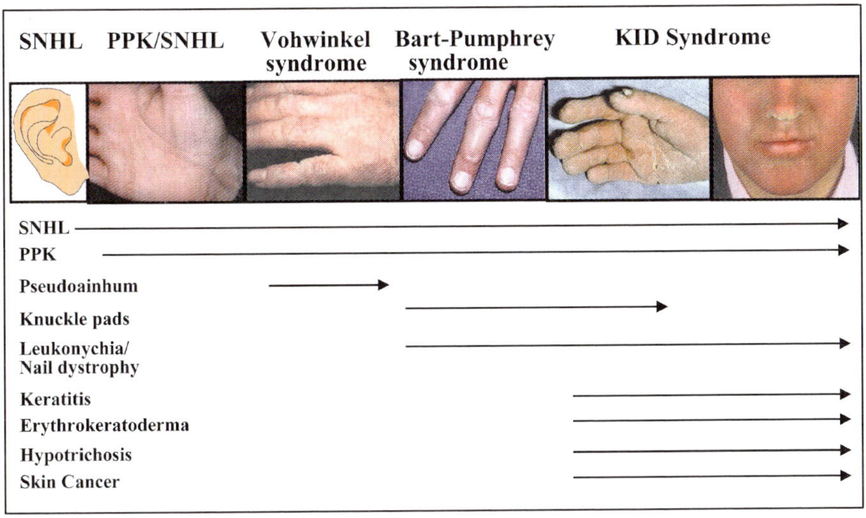

Fig. 7.7. The spectrum of genetic disorders due to autosomal dominant mutations of *GJB2* (Cx26) and their major clinical features. *SNHL* Sensorineural hearing loss, *PPK* palmoplantar keratoderma, *KID* keratitis–ichthyosis–deafness syndrome. Diffuse PPK and deafness, mutation E42del. Circular constriction band and verruciform hyperkeratoses on dorsum of palm in Vohwinkel syndrome, mutation D66H. Leukonychia and knuckle pads in Bart-Pumphrey syndrome, mutation N54 K. Palmoplantar keratoderma with rough surface. Symmetrical erythematous, hyperkeratotic plaques of the face in KID syndrome, mutation D50N

many generalized skin disorders, in particular ichthyoses and related disorders of cornification (Richard 2003). On microscopic examination, there is epidermal hyperplasia and papillomatosis with a tremendous thickening of the corneal layer called hyperkeratosis. Although formally 'hyperkeratosis' is a histological feature, it is increasingly used as a clinical term to describe a thickening of the skin without apparent scaling. To date, mutations in at least 13 different genes have been reported in various subtypes of hereditary PPKs (Kimyai-Asadi et al. 2002). One of these subgroups is caused by autosomal dominant mutations in the connexin gene *GJB2* encoding connexin 26 (Cx26). In these Cx26 disorders, bilateral PPK is consistently associated with sensorineural hearing loss (SNHL) and sometimes various other features (Fig. 7.7).

7.3.2.1 Palmoplantar Keratoderma Associated with Sensorineural Hearing Loss

Clinical Features

This very rare autosomal dominant disorder usually manifests during infancy or early childhood with slowly progressive thickening of the skin of palms and soles, which may extend to the dorsal surface of hands and feet (Fig. 7.7). The dry and rough skin may crack easily causing deep and painful fissures. In some patients, the keratoderma may be diffuse, in others striate or confined to the skin over pressure points. All affected individuals also have congenital SNHL, which can vary in severity from bilateral prelingual deafness to slowly progressive, high frequency SNHL. Hair, nails, teeth, and mucous membranes are normal.

Molecular Basis

In a handful of families studied to date, PPK and SNHL consistently cosegregated with heterozygous mutations in *GJB2* on chromosome 13q11-q12.1. Three missense mutations (G59A, R75W, R75Q) and a codon deletion (E42del) cluster in or at the border of the first extracellular loop of Cx26 (Fig. 7.4; Richard et al. 1998b; Heathcote et al. 2000; Loffeld and Kelsell 2000; Rouan et al. 2001; Uyguner et al. 2002). One missense mutation replacing arginine 75 with tryptophan (R75W) has been found in several unrelated individuals with SNHL, but variable involvement of the skin (Richard et al. 1998b; Loffeld and Kelsell 2000), thus indicating that other genetic and environmental factors can influence the clinical expression of Cx26 mutations.

7.3.2.2
Vohwinkel Syndrome (Mutilating PPK)

Clinical Features

This rare, autosomal dominant type of PPK was first described by Vohwinkel in 1929 (Vohwinkel 1929). The keratoderma is diffuse and has a rough, patterned surface likened to a honeycomb. The hyperkeratosis usually extends on the dorsal surface of the digits, hands, feet, elbows and knees with knuckle pads and linear or starfish shaped plaques. Starting in adolescence or adulthood, small, hardened, circular constriction bands may form on the distal digits, which can lead to autoamputation and mutilation, also known as pseudoainhum (Fig. 7.7). The hearing loss in Vohwinkel syndrome, albeit congenital, is relatively mild or moderate (Vohwinkel 1929) and often can be treated with hearing aids. Nails or other ectodermal tissues are usually not involved.

Molecular Basis

Vohwinkel syndrome has been mapped to a cluster of connexin genes on chromosome 13q11-q12 and is due to a specific missense mutation in the *GJB2* gene encoding Cx26 (Maestrini et al. 1999). In four unrelated families from the UK, Spain and Italy, sequence analysis of *GJB2* revealed a heterozygous nucleotide transversion, which substitutes an invariably present aspartic acid residue at position 66 of Cx26 with histidine (D66H; Fig. 7.4; Maestrini et al. 1999; Kelsell et al. 2000). This mutation also lies within the evolutionary conserved first extracellular domain of Cx26 and obviously gives rise to a consistent phenotype. Nevertheless, genotype-phenotype correlations in a British family with Vohwinkel syndrome segregating additional sequence variants in Cx26 and Cx31 suggested that these variants perhaps modulate some phenotypic features of the disease, such as type and severity of hearing loss (Kelsell et al. 2000). Many sporadic cases of Vohwinkel syndrome have been reported and are likely due to spontaneous de novo mutations of *GJB2*.

Interestingly, similar cutaneous features are also seen in loricrin keratoderma (aka, the Camisa-type of PPK; Camisa and Rossana 1984). In contrast to Vohwinkel syndrome, the hearing is normal, there is more widespread hyperkeratosis or scaling of the skin and an abnormal intracellular distribution of loricrin, a major component of the cornified cell envelope, on histopathological examination. The disorder is caused by small insertion mutations in the loricrin gene (*LOR*) resulting in frameshift and translation of an abnormally long and aberrant carboxy-terminal tail of loricrin (Maestrini et al. 1996). This abnormal tail domain interferes with the proper assembly of the cornified cell envelope, a crucial step in the formation of the cornified layer and skin barrier (Ishida-Yamamoto 2003). The striking similarity of the skin phenotype in Vohwinkel syndrome and loricrin keratoderma suggests a close connection between Cx26-mediated intercellular signaling and terminal differentiation of keratinocytes.

7.3.2.3 Bart-Pumphrey Syndrome

Clinical Features

First described in 1967, this autosomal dominant PPK with a honeycomb-like surface is characterized by prominent hyperkeratotic knuckle pads and white nail plates (leukonychia; Fig. 7.7). BPS lacks digital constriction bands and starfish shaped hyperkeratoses, and the prelingual SNHL is usually more severe than in Vohwinkel syndrome. Nevertheless, BPS and Vohwinkel syndrome are two closely related entities, especially since the clinical features of BPS often vary with age and between different family members (Bart and Pumphrey 1967; Ramer et al. 1994).

Molecular Basis

In one multigeneration family with Bart-Pumphrey syndrome, a missense mutation of *GJB2* has been identified, replacing asparagine codon 54 with a negatively charged lysine (N54 K) in the first extracellular loop of Cx26 (Fig. 7.4; Richard et al. 2004).

Functional Implications of GJB2 (Cx26) Mutations

Apparently all dominant *GJB2* mutations producing PPK and SNHL cluster in or at the boundaries of the first extracellular loop of Cx26 (Fig. 7.4). The amino acid sequence of this protein domain is highly conserved throughout evolution and among all connexins. In general, the first extracellular loop is essential for gap gating properties and interaction of connexin hemichannels, assembly of intercellular channels and their voltage gating behavior (Goodenough et al. 1996; White and Bruzzone 1996; Oshima et al. 2003). Recently, systematic changes of the charge of residue D66 in Cx26 within the first extracellular loop have revealed that this sequence motif is critical for Cx26 transport to the cell surface as well as function of the resulting gap junction channels (Thomas et al. 2004). Therefore, it has been speculated that the detected mutations seriously interfere with those functions in a dominant negative fashion. In vivo immunohistochemical studies of the epidermis and sweat gland system of a patient with PPK/SNHL exclusively expressing the mutant allele E42del-Cx26 did not expose any abnormalities in protein expression levels, hexamer assembly or intracellular trafficking. However, dual voltage clamp studies of *Xenopus l.* oocytes expressing this and other Cx26 mutants with a cutaneous phenotype revealed a complete loss of intercellular conductance and inhibition of the function of co-expressed wild-type Cx26 as well as wild-type Cx43 (Rouan et al. 2001). Such a dominant and *trans*-dominant negative effect of autosomal dominant Cx26 mutations has also been confirmed in various mammalian cells, such as HeLa cells, fibroblasts or keratinocytes. For example, some dominant Cx26 mutants tested (E42del [SNHL+PPK], R75W [SNHL±PPK] and W44C/S [SNHL, no skin involvement]) were found to oligomerize into hemichannels and form gap junctional plaques at the plasma membranes albeit gap junction channels were nonfunctional (Fig. 7.6). These mutants also significantly reduced dye transfer of co-expressed wild-type Cx26 or Cx30 (Martin et al. 1999; Marziano et al. 2003; Thomas et al. 2003). Interestingly, in another study R75W-Cx26 was found to impede interactions between connexin subunits and thereby destabilized hemichannels and obliterated their function (Fig. 7.6; Oshima et al. 2003). Other Cx26 mutations, such as G59A (PPK/SNHL), D66H (VWS) and D50N (KIDS), predominantly resulted in perinuclear accumulation of the mutant protein within the brefeldin A-insensitive trans-Golgi network suggestive of protein misfolding or an intracellular trafficking defect and were unable to produce functional gap junctional channels (Fig. 7.6; Marziano et al. 2003; Shurman et al. 2004; Thomas et al. 2004). In one study, the faulty hexamer formation or membrane targeting of mutants G59A and D66H was rescued in the presence of wild-type Cx26 or Cx30, providing evidence that Cx26 and Cx30 form heteromeric connexons. Nevertheless, the incorporation of G59A-Cx26 selectively inhibited the function of coexpressed wild-type Cx30 (Marziano et al. 2003), Cx32 and Cx43 (Thomas et al. 2004), whereas D66H-Cx26 exerted a dominant negative effect on wild-type Cx26 and Cx43, but not on Cx32 (Marziano et al. 2003; Thomas et al. 2004). The KIDS mutation D50N-Cx26 had a dominant negative effect on both wild-type Cx26 and Cx30 (Shurman et al. 2004). Based on these observations, it has been proposed that the severity of the skin phenotype depends on the specific nature of each Cx26 mutation and its *trans*-dominant effects on other coexpressed connexins (Rouan et al. 2001; Thomas et al. 2004).

Recently, the first in vivo model of a dominant connexin disorder of the skin has been established and provides valuable new insights in disease pathogenesis. Bakirtzis et al. (2003) established and analyzed transgenic mice expressing mutant D66H-Cx26, which causes in humans Vohwinkel syndrome, using a bovine keratin 10 promoter to target suprabasal keratinocytes. Soon after birth, transgenic mice developed scaling and hyperkeratosis of the skin on the body, legs, and tail. The hyperkeratosis persisted on the legs and tail and led to autoamputation of the distal tail and lymphedema of the legs ensuing from hard, keratotic, circular constriction bands, mimicking the human disorder. Transgene expression was also associated with a loss of Cx26 and Cx30 positive gap-junction plaques at points of cell–cell contacts and increased epidermal TUNEL staining, albeit the formation of the epidermal water barrier during late embryonic development appeared normal (Bakirtzis et al. 2003). The K10-Cx26(D66H) mice will certainly provide a valuable new asset to study the role of gap-junctional intercellular communication in epidermal differentiation.

7.3.3 Ectodermal Dysplasias

7.3.3.1 Hypohidrotic Ectodermal Dysplasia (Clouston Syndrome)

Clinical Features

Clouston syndrome is characterized by hypotrichosis and alopecia with sparse, brittle, blond hair, PPK, and progressive nail dystrophy with thickened, discolored and shortened nails that may be easily shed, and occasionally discrete hyperpigmentations in skin folds. In contrast to KID Syndrome, cornea, hearing and sweating are usually normal (Rajagopalan and Tay 1977; Jan et al. 2004).

Molecular Basis

This autosomal dominant ectodermal dysplasia, which is particularly common among the French-Canadian population due to a founder effect, has been mapped to a connexin gene cluster on chromosome 13q11-q12.1 including *GJB2* and *GJB6* (Kibar et al. 1996; Lamartine et al. 2000). Subsequently, two common pathogenic missense mutations were identified in the Cx30 gene *GJB6*, accounting for Clouston syndrome in 12 families of French-Canadian and other backgrounds, including the Chinese Han (Lamartine et al. 2000; Zhang et al. 2003). Mutation G11R is located in the amino-terminus of Cx30, while A88V is predicted to lie in the second transmembrane domain of Cx30 (Fig. 7.4). Both mutations were also found in patients with very thickened, hypertrophic nails with and without plaque-like PPK (van Steensel et al. 2003) resembling pachyonychia congenita, a heritable nail disorder due to mutations in nail matrix keratin genes (Smith 2003).

Interestingly, a third Cx30 gene mutation has been associated with different phenotypes. Smith et al. (2002) first reported a mutation replacing valine 37 with glutamic acid (V37E) in a patient with congenital atrichia, dystrophy of all nails, PPK with a velvety surface, and papular hyperkeratotic lesion especially on the forearms (Smith et al. 2002). While in this case hearing and vision were normal, supporting a diagnosis of Clouston syndrome, a similar mutation was also detected in a patient exhibiting congenital atrichia as well as all characteristic features of KID Syndrome (Sect. 7.3.2.3; Jan et al. 2004). Finally, a fourth missense mutation (M5T) has been reported in a family with autosomal dominant SNHL without skin involvement (Grifa et al. 1999), while a partial deletion of the Cx30 gene was found to be a common cause of autosomal recessive SNHL (Lerer et al. 2001; del Castillo et al. 2002), which further underscores the clinical complexity of Cx30 defects.

Functional Implications of GJB6 (Cx30) Mutations

The clinical phenotype of Cx30 mutations involving the epidermis as well as hair and nails points to a specific role of Cx30 in adnexal epithelia, which apparently cannot be substituted by Cx26 or other connexins. Autosomal dominant Cx30 mutations with cutaneous involvement appear not to interfere with the normal distribution of Cx30 or other cutaneous connexins in affected palmoplantar epidermis (Essenfelder et al. 2004). Nevertheless, mutations G11R-Cx30, V37E-Cx30, and A88V-Cx30 were found to impair trafficking of the mutant protein to the cell membrane and lead to cytoplasmic accumulation when transiently expressed in cultured keratinocytes and HeLa cells, thus resulting in a complete loss of gap-junction function (Common et al. 2002). In a recent study, Essenfelder et al. (2004) have demonstrated that the presence of the wild-type Cx30 rescued the impaired trafficking of mutated Cx30 to the plasma membrane and allowed G11R-Cx30 and A88V-Cx30 to form functional intercellular channels, which slightly differed in their electrophysiological channel characteristics. Most importantly, the authors unveiled a gain of function of G11R-Cx30 and A88V-Cx30 (Fig. 7.6). Both mutant proteins formed not only typical gap-junction channels, but also functional hemichannels in the cell membrane, which permitted the release of ATP into the extracellular medium. A similar mechanism in vivo could alter the paracrine signaling of keratinocytes and thus lead to the HED phenotype (Essenfelder et al. 2004).

7.3.3.2
Keratitis–Ichthyosis–Deafness Syndrome

Clinical Features

Keratitis-Ichthyosis-Deafness Syndrome (KIDS)is a rare ectodermal dysplasia featuring a combination of PPK, erythrokeratoderma, and SNHL in addition to abnormalities of other ectodermal epithelia of the cornea, hair and nails (Caceres-Rios et al. 1996). More than 100 cases have been published to date, most of which (> 90%) occurred sporadically, although autosomal dominant inheritance has been observed (Tuppurainen et al.

1988). PPK is a consistent feature and typically has a grainy, pitted surface pattern (Fig. 7.7). Other cutaneous features vary, ranging from symmetrically distributed, sharply demarcated, erythematous and hyperkeratotic plaques on the face and extremities (Fig. 7.7), to a thickening of the skin with a coarse-grained appearance and furrowing or generalized filiform follicular hyperkeratoses without erythema. Hair and nail dystrophy, scarring alopecia, dental anomalies, and heat intolerance occur less frequently. KIDS is also associated with increased susceptibility to mucocutaneous infections, including multiple draining abscesses, and squamous cell carcinoma of the skin and oral cavity (Caceres-Rios et al. 1996). The SNHL is congenital, often but not always bilateral and severe to profound (Szymko-Bennett et al. 2002). A hallmark of KIDS is the involvement of the corneal epithelium. During childhood corneal epithelial defects and advancing neovascularization develop (Fig. 7.7), which may result in progressive visual decline and blindness, compounding the patients' sensorial and communication problems. A subgroup of patients with very severe, follicular hyperkeratoses, deep furrowing of the face, and unusual electron microscopic findings has been published under the name of Hystrix-like–Ichthyosis–Deafness syndrome (HIDS; Traupe 1989; van Geel et al. 2002a).

Molecular Basis

Keratitis–ichthyosis–deafness syndrome is genetically heterogeneous and may be caused by heterozygous, autosomal dominant mutations in the Cx26 gene (*GJB2*) or in the Cx30 gene (*GJB6*; Fig. 7.4), both of which cluster on chromosome 13q11-q12.1 and have been also implicated in the etiology of nonsyndromic SNHL (Richard et al. 2002; van Steensel et al. 2002; Jan et al. 2004). About 75% percent of KIDS patients (15 out of 20 cases) of various ethnic and geographic origins have been found to harbor a recurrent nucleotide transition in exon 2 of *GJB2*, replacing aspartic acid with asparagine in codon 50 (D50N) (Richard et al. 2002; van Geel et al. 2002a; van Steensel et al. 2002; Alvarez et al. 2003; Wasserman et al. 2003; Yotsumoto et al. 2003). The vertical transmission of mutation D50 N across subsequent generations in a family with KIDS confirmed the autosomal dominant inheritance of this disorder on a molecular level (Richard et al. 2002). The mutation site appears to be a mutational hotspot in *GJB2* due to the presence of a hypermutable CpG dinucleotide. With few exceptions, D50 is invariably present in the first extracellular loop of human connexins. The entire sequence motif is highly conserved and any change in composition and charge is likely to be detrimental for formation and function of gap junction channels, in particular for connexon–connexon interactions and voltage gating (Rubin et al. 1992; White et al. 1995). Mutation D50N has been also identified in a patient with HIDS, thus demonstrating that this disorder is allelic with KIDS and likely represents a clinical variant (van Geel et al. 2002a).

Other pathogenic KIDS mutations of *GJB2* cluster in the 5' quarter of the coding sequence of *GJB2* and alter residues in the cytoplasmic amino-terminus of Cx26 that are conserved among all β-type connexins across different species. To date, four distinct missense mutations have been identified, including G11R, G12R, N14K and S17F (Richard et al. 2002; Wasserman et al. 2003; van Steensel et al. 2004). One of these mutations, N14K, was detected in three unrelated patients with sporadic KIDS originating from the Netherlands (described as 'Clouston syndrome with deafness'), Switzerland and the US and, thus, might represent another recurrent mutation. Preliminary genotype–phenotype correlations suggest that mutation D50N usually results in 'classical' features of KIDS, while amino-terminal Cx26 mutations present with variable and deviant features, such as diffuse follicular hyperkeratoses (N14K), scattered erythematous papules and small scaly plaques with psoriasiform appearance (G11R, N14K), and lack of PPK or ophthalmological symptoms (Sundaram et al. 2003; Wasserman et al. 2003; van Steensel et al. 2004). Finally, another variant of KIDS with follicular hyperkeratoses, inflammatory dissecting folliculitis of the scalp, hidradenitis suppurativa, and cystic acne (also know as 'follicular occlusion triad') has been associated with the missense mutation A40V, which lies in the first transmembrane domain of Cx26 (Montgomery et al. 2004).

In addition to the KIDS mutations in *GJB2*, on one occasion a KIDS patient with SNHL, keratitis, corneal neovascularization, PPK, spiny hyperkeratoses reminiscent of HIDS, but also congenital absence of hair and nail abnormalities was found to harbor a missense mutation in the Cx30 gene *GJB6* (Jan et al. 2004). The identified mutation introduces a highly charged glutamic acid residue instead of valine at position 37 (V37E) in the first transmembrane helix of Cx30 (Fig. 7.4). A similar mutation had been previously implicated in Clouston syndrome without evidence for abnormal sweating, hearing, photophobia and keratitis (Smith et al. 2002). These findings illustrate an astonishing phenotypic variability of this *GJB6* mutation, which has also been noted for certain Cx26 or Cx31 mutations, pointing to the importance of genetic and epigenetic factors for the clinical expression of human connexin disorders. Moreover, this case illustrates again that mutations in more than one connexin gene can result in similar phenotypes, such as KIDS (*GJB2* and *GJB6* mutations), EKV (*GJB3* and *GJB4* mutations), nonsyndromic SNHL (*GJB2*, *GJB6*, and *GJB3* mutations) or cataract (*GJA3* and *GJA8* mutations).

To further broaden the phenotypic spectrum of inherited *GJB2* mutations, Brown et al. (2003) reported the missense mutation F142L in the third transmembrane domain of Cx26 in a child with unusual features outside the spectrum of KIDS. The patient had congenital SNHL, periorificial and mucosal erythrokeratoderma, dental lamina cysts, enamel defects, and a tendency to develop excessive granulation tissue (Brown et al. 2003). Screening of other connexin genes expressed in the skin and inner ear did not reveal additional sequence variants, excluding the possibility that a combination of subnormal connexin variants is responsible for this rare and atypical phenotype. In vitro expression of F142L-Cx26-EGFP in coupling-incompetent HeLa cells revealed a failure of the mutant protein to reach the plasma membrane and form gap junction plaques. In a dominant negative fashion, co-expressed wild-type Cx26-ECFP was also trapped in the cytoplasm. Moreover, the viability of cells expressing F142L-Cx26-EGFP was significantly reduced, changing the ratio of living:dead cells from 6:1 for wild-type Cx26 to 1.1:1 for the mutant Cx26 protein (Tang et al. 2004). Collectively, these data demonstrated that F142L is a functionally relevant mutation associated with an unusual clinical presentation.

Functional Implications of GJB2 (Cx26) Mutations in KIDS

The functional consequences of autosomal dominant Cx26 or Cx30 mutations in KIDS and the rationale for their pleiotropic effect remain largely unknown. In vitro expression of S17F-Cx26 and D50N-Cx26 in coupling-incompetent HeLa cells or primary corneal epithelial cells, respectively, showed cytoplasmic accumulation and a failure in forming functional gap-junction channels (Fig. 7.6; Richard et al. 2002; Shurman et al. 2004). Similar to other connexin mutants with a skin phenotype, expression of D50N-Cx26 dominantly inhibited the function of co-expressed wild-type Cx26 and Cx30 (Shurman et al. 2004) and significantly induced cell death (Common et al. 2004; Shurman et al. 2004). The latter observation could perhaps reflect a gain in hemichannel function and subsequent ATP leakage as demonstrated for certain Cx30 mutants in Clouston syndrome (Essenfelder et al. 2004).

If a similar mechanism applies to the KIDS mutation, V37E in Cx30 also remains to be investigated. In contrast to the discussed Cx26 mutations, this mutation lies in the first transmembrane helix of Cx30 (Fig. 7.4). Based on a three-dimensional map of truncated Cx43 (Unger et al. 1999) as well as SCAM (scanning cysteine accessibility mutagenesis) analysis of Cx32 (Zhou et al. 1997), this domain lines the cytoplasmic end of the channel pore and is part of a voltage sensing and transduction mechanism. Therefore, any change in sequence, charge, or side chain conservation of this domain has likely detrimental consequences for the channel properties and ion selectivity. Replacing a hydrophobic valine in Cx30 with a charged glutamate residue can be expected to disrupt the alpha-helical structure of this transmembrane domain and ultimately alter the assembly and function of Cx30 in our patient. This prediction is supported by the finding that

V37E-Cx30 failed to be transported to the cell membranes and could not produce functional gap junction channels when transiently expressed in HeLa cells and keratinocytes (Common et al. 2003).

The genetic heterogeneity of KIDS strongly supports the notion that Cx26 and Cx30 fulfill closely related functions and that mutations in both genes share a common pathomechanism. Cx26 and Cx30 are not only tightly related proteins with 76% amino acid identity, but also widely overlap in their tissue expression, especially in the human and murine epidermis, sweat glands, sweat ducts, hair follicles, cochlea and corneal epithelium (Richard et al. 2002; Coutinho et al. 2003). As recently demonstrated, the inner ear contains heteromeric connexons composed of Cx26 and Cx30 and deafness-related mutations in Cx26 may exert a dominant negative effect on Cx30 (Forge et al. 2003). These findings, together with the similar, but not identical phenotypes of Cx26 and Cx30 mutations in KIDS, strongly suggest the existence of such mixed Cx26/Cx30 connexons and/or heterotypic gap-junction channels in the skin and its appendages. Nevertheless, the complete absence of hair in patients with the Cx30 mutation V37E implies a hitherto unexplored role of Cx30 in the keratinization of hair follicles (Jan et al. 2004).

7.3.3.3 Oculo–Dento–Digital Dysplasia

Clinical Features

This is a rare autosomal dominant syndrome with a multitude of developmental abnormalities, yet only infrequent hair or skin findings. Characteristic features are craniofacial and limbal dysmorphism, including microcephaly, a narrow nose with small, anteverted nares, syndactyly, camptodactyly, and skeletal hyperostosis. Also common are various ophthalmological and neurological anomalies, including microphthalmia, microcornea, cataracts, ataxia and conductive hearing loss. Occasionally, hypotrichosis, curly hair, nail dystrophy and mild PPK have been reported (Gorlin et al. 2001; Paznekas et al. 2003; Kjaer et al. 2004; van Steensel et al. 2005).

Molecular Basis

Oculo–dento–digital dysplasia (ODDD) has been mapped to chromosome 6q22-q24. Paznekas et al. (2003) identified *GJA1* (Cx43) as the disease gene in ODDD and reported 16 pathogenic missense mutations and one codon duplication in their large cohort of 17 unrelated ODDD families and sporadic cases of various origins. A single patient with sporadic disease had also mild PPK, carrying mutation K134E in the cytoplasmic loop of Cx43. Since then, ten additional new missense mutations have been reported, one of which (V96M in the second transmembrane domain) was associated with curly hair and another (780_781delTG in the carboxy terminal tail domain) with PPK (Kjaer et al. 2004; Richardson et al. 2004; van Steensel et al. 2005). The latter mutation results in a frameshift and is translated into a shortened protein with an aberrant tail of 46 amino acids. Another patient has been identified with abortive features of ODDD, such as shortened middle phalanges of hands and feet and early tooth decay, who also had extensive redness and hyperkeratosis of the skin, PPK, alopecia and nail dystrophy reminiscent of KIDS or Clouston syndrome albeit hearing was normal. This individual harbored mutations in two different connexin genes, *GJB2* and *GJA1*, including the sporadic missense mutation V41L in the first transmembrane helix of Cx43 and the heterozygous sequence variant R172H in Cx26. These findings suggest that *GJA1* mutations can produce variable clinical phenotypes on the background of sequence variants in other connexins (Kellermayer et al. 2005). All reported autosomal dominant mutations in ODDD are scattered across the Cx43 gene, sparing only the 3' end of the coding region encoding the cytoplasmic tail of Cx43. Similar to the spectrum of mutations in Cx26 or Cx31 however, certain mutations appear to be inherited in an autosomal recessive fashion. Pizzutti et al. (2004) detected a homozygous missense mutation (R76H) in *GJA1* in a sporadic case with overlapping features to Hallermann-Streiff syndrome, such as short stature, brachycephaly, frontal bossing, maxillary and mandibulary hypoplasia with dental malocclusion, clinodactily and mild psychomotor retardation. Interestingly, the patient also had hypotrichosis.

Functional Implications of GJA1 (Cx43) Mutations

The pathologic mechanisms leading to ODDD are still unclear. Recently, Seki et al. (2004) have studied the functional consequences of three ODDD mutations within the cytoplasmic loop, I130T, K134E, and G138R, in communication-deficient murine neuroblastoma cells (N2A). These mutations were found to seriously diminish or abolish the ability of connexins to form functional channels, despite the presence of Cx43 gap junction plaques. In contrast to mutant G138R, occasional channel openings were observed for mutants I130T and K134E. However, while the unitary conductance of I130T-Cx43 was similar to that of the wild-type channels, it was reduced for mutant K134E-Cx43 (Seki et al. 2004). Considering the dominant and *trans*-dominant negative effects of autosomal dominant connexin mutations in a variety of other human connexin disorders, it is likely that similar mechanisms are at play in ODDD and may account for the pleiotropic expression of Cx43 mutations. Using whole mount in situ hybridization on mouse embryos at different stages of fetal development, Richardson et al. (2004) demonstrated strong expression of the murine Cx43 gene *Gja1* in the eye, limbs, maxillary and mandibulary arches, nasal region, fusing secondary palate, and hair follicles. This spatiotemporal expression pattern in the developing craniofacial complex and limbs correlates well with the clinical features in ODDD. Nevertheless, Cx43 is ubiquitously expressed and the most abundant gap junction protein in the heart, epidermis and hair follicles. Despite its widespread tissue expression, however, cardiac defects are very rare and cutaneous manifestations are limited to hairs and palmoplantar skin, suggesting that other connexins might be important for the embryonic development of heart and skin structures or can substitute for the function of Cx43 in humans.

7.4 Summary

Gap junctions play a crucial role in the development, proliferation and differentiation of the epidermis. Keratinocytes express at least nine members of the human connexin family in a temporal and differentiation-specific manner. The expression patterns of different connexins are distinct, but widely overlap. The majority of connexins are utilized during the late stages of terminal differentiation. The connexin expression undergoes dynamic changes to adapt to the specific needs of the epithelium during embryological skin morphogenesis, wound healing, or other physiological and pathological processes. Recent advances in molecular genetics disclosed a wide array of genetic disorders due to autosomal dominant mutations in five epidermal connexin genes. The cutaneous manifestations encompass erythrokeratodermia variabilis, palmoplantar keratoderma associated with sensorineural hearing loss (including Vohwinkel syndrome and Bart-Pumphrey syndrome) and ectodermal dysplasias with involvement of other ectodermal epithelia (Clouston syndrome, keratitis–ichthyosis–deafness syndrome and oculo–dento–digital dysplasia). Distinct mutations of one connexin gene, such as Cx26 or Cx30, may lead to several different skin disorders with involvement of multiple tissues. It is tempting to speculate that the complexity and redundancy of the epidermal gap junction system is responsible for these puzzling genotype–phenotype correlations. Most skin disease-associated connexin mutations exert a dominant negative effect and interfere with the functional properties of homotypic or heterotypic channels in which mutant connexin subunits are incorporated. They may inhibit the proper assembly and transport of connexon hemichannels leading to cytoplasmic accumulation and paucity or lack of gap junction plaques. Alternatively, they may interfere with gating properties of intercellular channels, or change the behavior of hemichannels in the plasma membrane, which might be responsible for the reduced cell survival rate observed for some skin disease-associated mutations in vitro. The discovery of genetic gap junction defects has significantly advanced our current understanding of the biological role of gap junction signalling in the skin and other ectoderm-derived epithelia and will certainly further rekindle the interest in biophysiology of gap junction channels.

Acknowledgements. I thank all my colleagues and collaborators for their referrals, continuous support, and stimulating discussions. The original work has been supported in part by the Dermatology Foundation, The Foundation for Ichthyosis and Related Skin Disorders, the American Skin Association, the National Foundation for Ectodermal Dysplasias, and NIH/NIAMS grants K08-AR02141 and P01-AR38923. With deepest gratitude and thanks, this paper is dedicated to my mentor and colleague, Dr. Jouni Uitto, for his guidance and unfaltering support. His enormous enthusiasm and passion for dermatological research have been truly inspiring.

References

Alvarez A, Del Castillo I, Pera A, Villamar M, Moreno-Pelayo MA, Moreno F, Moreno R, Tapia MC (2003) De novo mutation in the gene encoding connexin-26 (GJB2) in a sporadic case of keratitis–ichthyosis–deafness (KID) syndrome. Am J Med Genet 117A:89–91

Arita K, Akiyama M, Tsuji Y, McMillan JR, Eady RA, Shimizu H (2002) Changes in gap junction distribution and connexin expression pattern during human fetal skin development. J Histochem Cytochem 50:1493–1500

Arita K, Akiyama M, Tsuji Y, Onozuka T, Shimizu H (2003) Erythrokeratoderma variabilis without connexin 31 or connexin 30.3 gene mutation: immunohistological, ultrastructural and genetic studies. Acta Derm Venereol 83:266–270

Bakirtzis G, Choudhry R, Aasen T, Shore L, Brown K, Bryson S, Forrow S, Tetley L, Finbow M, Greenhalgh D, Hodgins M (2003) Targeted epidermal expression of mutant Connexin 26(D66H) mimics true Vohwinkel syndrome and provides a model for the pathogenesis of dominant connexin disorders. Hum Mol Genet 12:1737–1744

Bart RS, Pumphrey RE (1967) Knuckle pads, leukonychia and deafness. A dominantly inherited syndrome. N Engl J Med 276:202–207

Brandner JM, Houdek P, Husing B, Kaiser C, Moll I (2004) Connexins 26, 30, and 43: differences among spontaneous, chronic, and accelerated human wound healing. J Invest Dermatol 122:1310–1320

Brown CW, Levy ML, Flaitz CM, Reid BS, Manolidis S, Hebert AA, Bender MM, Heilstedt HA, Plunkett KS, Fang P, Roa BB, Chung P, Tang HY, Richard G, Alford RL (2003) A novel GJB2 (connexin 26) mutation, F142L, in a patient with unusual mucocutaneous findings and deafness. J Invest Dermatol 121:1221–1223

Butterweck A, Elfgang K, Willeke K, Traub O (1994) Differential expression of the gap junction proteins connexin45, -43, -40, -31, and –26 in mouse skin. Eur J Cell Biol 65:152–163

Caceres-Rios H, Tamayo-Sanchez L, Duran-Mckinster C, de la Luz Orozco M, Ruiz-Maldonado R (1996) Keratitis, ichthyosis, and deafness (KID syndrome): review of the literature and proposal of a new terminology. Pediatr Dermatol 13:105–113

Camisa C, Rossana C (1984) Variant of keratoderma hereditaria mutilans (Vohwinkel's syndrome). Treatment with orally administered isotretinoin. Arch Dermatol 120:1323–1328

Choudhry R, Pitts JD, Hodgins MB (1997) Changing patterns of gap junctional intercellular communication and connexin distribution in mouse epidermis and hair follicles during embryonic development. Dev Dyn 210:417–430

Coffey KL, Krushinsky A, Green CR, Donaldson PJ (2002) Molecular profiling and cellular localization of connexin isoforms in the rat ciliary epithelium. Exp Eye Res 75:9–21

Common JE, Becker D, Di WL, Leigh IM, O'Toole EA, Kelsell DP (2002) Functional studies of human skin disease- and deafness-associated connexin 30 mutations. Biochem Biophys Res Commun 298:651–656

Common JE, Di WL, Davies D, Galvin H, Leigh IM, O'Toole EA, Kelsell DP (2003) Cellular mechanisms of mutant connexins in skin disease and hearing loss. Cell Commun Adhes 10:347–351

Common JE, Di WL, Davies D, Kelsell DP (2004) Further evidence for heterozygote advantage of GJB2 deafness mutations: a link with cell survival. J Med Genet 41:573–575

Coutinho P, Qiu C, Frank S, Tamber K, Becker D (2003) Dynamic changes in connexin expression correlate with key events in the wound healing process. Cell Biol Int 27:525–541

del Castillo I, Villamar M, Moreno-Pelayo MA, del Castillo FJ, Alvarez A, Telleria D, Menendez I, Moreno F (2002) A deletion involving the connexin 30 gene in nonsyndromic hearing impairment. N Engl J Med 346:243–249.

Di WL, Rugg EL, Leigh IM, Kelsell DP (2001) Multiple epidermal connexins are expressed in different keratinocyte subpopulations including connexin 31. J Invest Dermatol 117:958–964

Di WL, Monypenny J, Common JE, Kennedy CT, Holland KA, Leigh IM, Rugg EL, Zicha D, Kelsell DP (2002) Defective trafficking and cell death is characteristic of skin disease-associated connexin 31 mutations. Hum Mol Genet 11:2005–2014

Diestel S, Richard G, Doring B, Traub O (2002) Expression of a connexin31 mutation causing erythrokeratodermia variabilis is lethal for HeLa cells. Biochem Biophys Res Commun 296:721–728

Essenfelder GM, Bruzzone R, Lamartine J, Charollais A, Blanchet-Bardon C, Barbe MT, Meda P, Waksman G (2004) Connexin30 mutations responsible for hidrotic ectodermal dysplasia cause abnormal hemichannel activity. Hum Mol Genet 13:1703–1714

Feldmeyer L, Plantard L, Mevorah B, Huber M, Hohl D (2005) Novel mutation of connexin31 causing erythrokeratodermia variabilis. Br J Dermatol 152:1072–1074

Fonseca PC, Nihei OK, Urban-Maldonado M, Abreu S, De Carvalho AC, Spray DC, Savino W, Alves LA (2004) Characterization of connexin 30.3 and 43 in thymocytes. Immunol Lett 94:65–75

Forge A, Marziano NK, Casalotti SO, Becker DL, Jagger D (2003b) The inner ear contains heteromeric channels composed of cx26 and cx30 and deafness-related mutations in cx26 have a dominant negative effect on cx30. Cell Commun Adhes 10:341–346

Gabriel HD, Strobl B, Hellmann P, Buettner R, Winterhager E (2001) Organization and regulation of the rat Cx31 gene. Implication for a crucial role of the intron region. Eur J Biochem 268:1749–1759

Goldberg GS, Lampe PD, Nicholson BJ (1999) Selective transfer of endogenous metabolites through gap junctions composed of different connexins. Nat Cell Biol 1:457–459

Goliger JA, Paul DL (1994) Expression of gap junction proteins cx26, cx31.1, cx37, and cx43 in developing and mature rat epidermis. Dev Dyn 200:1–13

Goliger JA, Paul DL (1995) Wounding alters epidermal connexin expression and gap junction-mediated intercellular communication. Mol Biol Cell 6:1491–1501

Goodenough DA, Paul DL (2003) Beyond the gap: functions of unpaired connexon channels. Nat Rev Mol Cell Biol 4:285–294

Goodenough DA, Goliger JA, Paul DL (1996) Connexins, connexons, and intercellular communication. Annu Rev Biochem 65:475–502

Gorlin RJ, Cohen MMJ, Hennekam RCM (2001) Oculodentoosseous dysplasia (oculodentodigital syndrome). Syndromes of the head and neck. Oxford University Press, New York, pp 234–235

Gottfried I, Landau M, Glaser F, Di WL, Ophir J, Mevorah B, Ben-Tal N, Kelsell DP, Avraham KB (2002) A mutation in GJB3 is associated with recessive erythrokeratodermia variabilis (EKV) and leads to defective trafficking of the connexin 31 protein. Hum Mol Genet 11:1311–1316

Grifa A, Wagner CA, D'Ambrosio L, Melchionda S, Bernardi F, Lopez-Bigas N, Rabionet R, Arbones M, Monica MD, Estivill X, Zelante L, Lang F, Gasparini P (1999) Mutations in GJB6 cause nonsyndromic autosomal dominant deafness at DFNA3 locus. Nat Genet 23:16–18

Heathcote K, Syrris P, Carter ND, Patton MA (2000) A connexin 26 mutation causes a syndrome of sensorineural hearing loss and palmoplantar hyperkeratosis (MIM 148350). J Med Genet 37:50–51

Hentula M, Peltonen J, Peltonen S (2001) Expression profiles of cell-cell and cell-matrix junction proteins in developing human epidermis. Arch Dermatol Res 293:259–267

Holbrook KA (1979) Human epidermal embryogenesis. Int J Dermatol 18:329–356

Holbrook KA, Odland GF (1975) The fine structure of developing human epidermis: light, scanning, and transmission electron microscopy of the periderm. J Invest Dermatol 65:16–38

Iguchi M, Hara M, Manome H, Kobayasi H, Tagami H, Aiba S (2003) Communication network in the follicular papilla and connective tissue sheath through gap junctions in human hair follicles. Exp Dermatol 12:283–288

Ishida-Yamamoto A (2003) Loricrin keratoderma: a novel disease entity characterized by nuclear accumulation of mutant loricrin. J Dermatol Sci 31:3–8

Ishida-Yamamoto A, Kelsell D, Common J, Houseman MJ, Hashimoto M, Shibaki H, Asano K, Takahashi H, Hashimoto Y, Senshu T, Leigh IM, Iizuka H (2000) A case of erythrokeratoderma variabilis without mutations in connexin 31. Br J Dermatol 143:1283–1287

Jan AY, Amin S, Ratajczak P, Richard G, Sybert VP (2004) Genetic heterogeneity of KID syndrome: identification of a Cx30 gene (GJB6) mutation in a patient with KID syndrome and congenital atrichia. J Invest Dermatol 122:1108–1113

Juneja SC (2003) mRNA expression pattern of multiple members of connexin gene family in normal and abnormal fetal gonads in mouse. Indian J Physiol Pharmacol 47:147–156

Jungbluth S, Willecke K, Champagnat J (2002) Segment-specific expression of connexin31 in the embryonic hindbrain is regulated by Krox20. Dev Dyn 223:544–551

Kellermayer R, Keller M, Ratajczak P, Richardson PE, Harangi F, Mérei E, Melegh B, Kosztolányi G, Richard G (2005) Bigenic connexin mutations in a patient with hidrotic ectodermal dysplasia. Eur J Dermatol 15:75–79

Kelsell DP, Wilgoss AL, Richard G, Stevens HP, Munro CS, Leigh IM (2000) Connexin mutations associated with palmoplantar keratoderma and profound deafness in a single family. Eur J Hum Genet 8:141–144

Kibar Z, Der Kaloustian VM, Brais B, Hani V, Fraser FC, Rouleau GA (1996) The gene responsible for Clouston hidrotic ectodermal dysplasia maps to the pericentromeric region of chromosome 13q. Hum Mol Genet 5:543–547

Kimyai-Asadi A, Kotcher LB, Jih MH (2002) The molecular basis of hereditary palmoplantar keratodermas. J Am Acad Dermatol 47:327–343
Kjaer KW, Hansen L, Eiberg H, Leicht P, Opitz JM, Tommerup N (2004) Novel Connexin 43 (GJA1) mutation causes oculo-dento-digital dysplasia with curly hair. Am J Med Genet 127A:152–157
Kretz M, Euwens C, Hombach S, Eckardt D, Teubner B, Traub O, Willecke K, Ott T (2003) Altered connexin expression and wound healing in the epidermis of connexin-deficient mice. J Cell Sci 116:3443–3452
Labarthe MP, Bosco D, Saurat JH, Meda P, Salomon D (1998) Upregulation of connexin 26 between keratinocytes of psoriatic lesions. J Invest Dermatol 111:72–76
Lagree V, Brunschwig K, Lopez P, Gilula NB, Richard G, Falk MM (2003) Specific amino-acid residues in the N-terminus and TM3 implicated in channel function and oligomerization compatibility of connexin43. J Cell Sci 116:3189–3201
Lamartine J, Munhoz Essenfelder G, Kibar Z et al. (2000) Mutations in GJB6 cause hidrotic ectodermal dysplasia. Nat Genet 26:142–144
Lerer I, Sagi M, Ben-Neriah Z, Wang T, Levi H, Abeliovich D (2001) A deletion mutation in GJB6 cooperating with a GJB2 mutation in trans in non-syndromic deafness: A novel founder mutation in Ashkenazi Jews. Hum Mutat 18:460
Liu XZ, Xia XJ, Xu LR, Pandya A, Liang CY, Blanton SH, Brown SD, Steel KP, Nance WE (2000) Mutations in connexin31 underlie recessive as well as dominant non-syndromic hearing loss. Hum Mol Genet 9:63–67
Loffeld A, Kelsell DP (2000) Palmoplantar kerato-derma and sensorineural deafness in an 8-year-old boy: a case report. Br J Dermatol 143:A38
Lopez-Bigas N, Rabionet R, Martinez E, Banchs I, Volpini V, Vance JM, Arbones ML, Estivill X (2000) Identification of seven novel SNPS (five nucleotide and two amino acid substitutions) in the connexin31 (GJB3) gene. Hum Mutat 15:481–482
Lopez-Bigas N, Olive M, Rabionet R, Ben-David O, Martinez-Matos JA, Bravo O, Banchs I, Volpini V, Gasparini P, Avraham KB, Ferrer I, Arbones ML, Estivill X (2001a) Connexin 31 (GJB3) is expressed in the peripheral and auditory nerves and causes neuropathy and hearing impairment. Hum Mol Genet 10:947–952
Lopez-Bigas N, Rabionet R, Arbones ML, Estivill X (2001b) R32 W variant in Connexin 31: mutation or polymorphism for deafness and skin disease? Eur J Hum Genet 9:70
Lopez-Bigas N, Melchionda S, Gasparini P, Borragan A, Arbones ML, Estivill X (2002) A common frameshift mutation and other variants in GJB4 (connexin 30.3): Analysis of hearing impairment families. Hum Mutat 19:458
Lucke T, Choudhry R, Thom R, Selmer IS, Burden AD, Hodgins MB (1999) Upregulation of connexin 26 is a feature of keratinocyte differentiation in hyperproliferative epidermis, vaginal epithelium, and buccal epithelium. J Invest Dermatol 112:354–361
Macari F, Landau M, Cousin P, Mevorah B, Brenner S, Panizzon R, Schorderet DF, Hohl D, Huber M (2000) Mutation in the gene for connexin 30.3 in a family with erythrokeratodermia variabilis. Am J Hum Genet 67:1296–1301
MacFarlane AW, Chapman SJ, Verbov JL (1991) Is erythrokeratoderma one disorder? A clinical and ultrastructural study of two siblings. Br J Dermatol 124:487–491
Maestrini E, Korge BP, Ocana-Sierra J, Calzolari E, Cambiaghi S, Scudder PM, Hovnanian A, Monaco AP, Munro CS (1999) A missense mutation in connexin26, D66H, causes mutilating keratoderma with sensorineural deafness (Vohwinkel's syndrome) in three unrelated families. Hum Mol Genet 8:1237–1243
Maestrini E, Monaco AP, McGrath JA, Ishida-Yamamoto A, Camisa C, Hovnanian A, Weeks DE, Lathrop M, Uitto J, Christiano AM (1996) A molecular defect in loricrin, the major component of the cornified cell envelope, underlies Vohwinkel's syndrome. Nat Genet 13:70–77

Martin PE, Coleman LS, Casalotti SO, Forge A, Evans WH (1999) Properties of connexin26 gap junctional proteins derived from mutations associated with non-syndromal hereditary deafness. Hum Mol Genet 8:2369–2376

Marziano NK, Casalotti SO, Portelli AE, Becker DL, Forge A (2003) Mutations in the gene for connexin 26 (GJB2) that cause hearing loss have a dominant negative effect on connexin 30. Hum Mol Genet 12:805–812

Masuda M, Usami S, Yamazaki K, Takumi Y, Shinkawa H, Kurashima K, Kunihiro T, Kanzaki J (2001) Connexin 26 distribution in gap junctions between melanocytes in the human vestibular dark cell area. Anat Rec 262:137–146

Mendes da Costa S (1925) Erythro- et keratodermia variabilis in a mother and a daughter. Acta Derm Venerol 6:255–261

Montgomery JR, White TW, Martin BL, Turner ML, Holland SM (2004) A novel connexin 26 gene mutation associated with features of the keratitis-ichthyosis-deafness syndrome and the follicular occlusion triad. J Am Acad Dermatol 51:377–382

Morley SM, White MI, Rogers M, Wasserman D, Ratajczak P, McLean WHI, Richard G (2005) A new, recurrent mutation of GJB3 (Cx31) in erythrokeratodermia variabilis. Br J Dermatol 152:1143–1148

Nance WE (2003) The genetics of deafness. Ment Retard Dev Disabil Res Rev 9:109–119

Niessen H, Harz H, Bedner P, Kramer K, Willecke K (2000) Selective permeability of different connexin channels to the second messenger inositol 1,4,5-trisphosphate. J Cell Sci 113:1365–1372

Oshima A, Doi T, Mitsuoka K, Maeda S, Fujiyoshi Y (2003) Roles of M34, C64, and R75 in the assembly of human connexin 26: Implication for key amino acid residues for channel formation and function. J Biol Chem 278:1807–1816.

Paznekas WA, Boyadjiev SA, Shapiro RE, Daniels O, Wollnik B, Keegan CE, Innis JW, Dinulos MB, Christian C, Hannibal MC, Jabs EW (2003) Connexin 43 (GJA1) mutations cause the pleiotropic phenotype of oculodentodigital dysplasia. Am J Hum Genet 72:408–418

Pizzuti A, Flex E, Mingarelli R, Salpietro C, Zelante L, Dallapiccola B (2004) A homozygous GJA1 gene mutation causes a Hallermann-Streiff/ODDD spectrum phenotype. Hum Mutat 23:286

Plantard L, Huber M, Macari F, Meda P, Hohl D (2003) Molecular interaction of connexin 30.3 and connexin 31 suggests a dominant-negative mechanism associated with erythrokeratodermia variabilis. Hum Mol Genet 12:3287–3294

Qiu C, Coutinho P, Frank S, Franke S, Law LY, Martin P, Green CR, Becker DL (2003) Targeting connexin43 expression accelerates the rate of wound repair. Curr Biol 13:1697–1703

Quist AP, Rhee SK, Lin H, Lal R (2000) Physiological role of gap-junctional hemichannels. Extracellular calcium-dependent isosmotic volume regulation. J Cell Biol 148:1063–1074

Rajagopalan K, Tay CH (1977) Hidrotic ectodermal dysplasia: study of a large Chinese pedigree. Arch Dermatol 113:481–485

Ramer JC, Vasily DB, Ladda RL (1994) Familial leuconychia, knuckle pads, hearing loss, and palmoplantar hyperkeratosis: an additional family with Bart-Pumphrey syndrome. J Med Genet 31:68–71

Rappaport IP, Goldes JA, Goltz RW (1986) Erythrokeratodermia variabilis treated with isotretinoin. A clinical, histologic, and ultrastructural study. Arch Dermatol 122:441–445

Ressot C, Bruzzone R (2000) Connexin channels in Schwann cells and the development of the X-linked form of Charcot-Marie-Tooth disease. Brain Res Brain Res Rev 32:192–202

Richard G (2003) Ichthyosis and disorders of keratinization. In: Schachner LA, Hansen RC (eds) Pediatric dermatology. Mosby, New York, pp 385–445

Richard G (2005) Erythrokeratodermia variabilis: an update on clinics and genetics. G Ital Dermatol Venereol, in press

Richard G, Andreoli JM, Compton JG (1997a) Expression of epidermal connexins in erythrokeratodermia variabilis (EKV) and normal skin. J Invest Dermatol 108:587
Richard G, Lin JP, Smith L, Whyte YM, Itin P, Wollina U, Epstein E, Jr., Hohl D, Giroux JM, Charnas L, Bale SJ, DiGiovanna JJ (1997b) Linkage studies in erythrokeratodermias: fine mapping, genetic heterogeneity and analysis of candidate genes. J Invest Dermatol 109:666–671
Richard G, Smith LE, Bailey RA, Itin P, Hohl D, Epstein J, E.H., DiGiovanna JJ, Compton JG, Bale SJ (1998a) Mutations in the human connexin gene *GJB3* cause erythrokeratodermia variabilis. Nat Genet 20:366–369
Richard G, White TW, Smith LE, Bailey RA, Compton JG, Paul DL, Bale SJ (1998b) Functional defects of Cx26 resulting from a heterozygous missense mutation in a family with dominant deaf-mutism and palmoplantar keratoderma. Hum Genet 103:393–399
Richard G, Brown N, Smith LE, Terrinoni A, Melino G, Mackie RM, Bale SJ, Uitto J (2000) The spectrum of mutations in erythrokeratodermias – novel and de novo mutations in GJB3. Hum Genet 106:321–329
Richard G, Rouan F, Willoughby CE, Brown N, Chung P, Ryynanen M, Jabs EW, Bale SJ, DiGiovanna JJ, Uitto J, Russell L (2002) Missense mutations in GJB2 encoding connexin-26 cause the ectodermal dysplasia keratitis-ichthyosis-deafness syndrome. Am J Hum Genet 70:1341–1348
Richard G, Brown N, Rouan F, Van der Schroeff JG, Bijlsma E, Eichenfield LF, Sybert VP, Greer KE, Hogan P, Campanelli C, Compton JG, Bale SJ, DiGiovanna JJ, Uitto J (2003) Genetic heterogeneity in erythrokeratodermia variabilis: novel mutations in the connexin gene GJB4 (Cx30.3) and genotype-phenotype correlations. J Invest Dermatol 120:601–609
Richard G, Brown N, Ishida-Yamamoto A, Krol A (2004) Expanding the phenotypic spectrum of Cx26 disorders: Bart-Pumphrey syndrome is caused by a novel missense mutation in GJB2. J Invest Dermatol 123:856–863
Richardson R, Donnai D, Meire F, Dixon MJ (2004) Expression of Gja1 correlates with the phenotype observed in oculodentodigital syndrome/type III syndactyly. J Med Genet 41:60–67
Rivas MV, Jarvis ED, Morisaki S, Carbonaro H, Gottlieb AB, Krueger JG (1997) Identification of aberrantly regulated genes in diseased skin using the cDNA differential display technique. J Invest Dermatol 108:188–194
Rouan F, White TW, Brown N, Taylor AM, Lucke TW, Paul DL, Munro CS, Uitto J, Hodgins MB, Richard G (2001) trans-dominant inhibition of connexin-43 by mutant connexin-26: implications for dominant connexin disorders affecting epidermal differentiation. J Cell Sci 114:2105–2113
Rouan F, Lo CW, Fertala A, Wahl M, Jost M, Rodeck U, Uitto J, Richard G (2003a) Divergent effects of two sequence variants of GJB3 (G12D and R32 W) on the function of connexin 31 in vitro. Exp Dermatol 12:191–197
Rouan F, Yi LS, Uitto J, Richard G (2003b) Pathogenic mutations affecting Cx31and Cx30.3 impair gap junction function and induce cell death in vitro. J Invest Dermatol 121:A603
Rubin JB, Verselis VK, Bennett MV, Bargiello TA (1992) Molecular analysis of voltage dependence of heterotypic gap junctions formed by connexins 26 and 32. Biophys J 62:183–193; discussion 193–185
Salomon D, Saurat JH, Meda P (1988) Cell-to-cell communication within intact human skin. J Clin Invest 82:248–254
Salomon D, Masgrau E, Vischer S, Ullrich S, Dupont E, Sappino P, Saurat JH, Meda P (1994) Topography of mammalian connexins in human skin. J Invest Dermatol 103:240–247
Seki A, Coombs W, Taffet SM, Delmar M (2004) Loss of electrical communication, but not plaque formation, after mutations in the cytoplasmic loop of Cx43. Heart Rhythm 1:227–233

Shurman DL, Glazewski L, Gumpert A, Zieske JD, Richard G (2005) In vivo and in vitro expression of connexins in the human corneal epithelium. IOVS, in press
Smith F (2003) The molecular genetics of keratin disorders. Am J Clin Dermatol 4:347–364
Smith FJ, Morley SM, McLean WH (2002) A novel connexin 30 mutation in Clouston syndrome. J Invest Dermatol 118:530–532
Sundaram S, Willoughby C, Itin P, Ryynaenen M, Lewanda A, P. SV, Richard G (2003) The clinical spectrum of keratitis–ichthyosis–deafness syndrome. J Invest Dermatol 121:A660
Szymko-Bennett YM, Russell LJ, Bale SJ, Griffith AJ (2002) Auditory manifestations of keratitis–ichthyosis–deafness (KID) syndrome. Laryngoscope 112:272–280
Tada J, Hashimoto K (1997) Ultrastructural localization of gap junction protein connexin 43 in normal human skin, basal cell carcinoma, and squamous cell carcinoma. J Cutan Pathol 24:628–635
Tang H-Y, Brown CW, Richard G, Alford RL (2004) In vitro functional analysis of a Connexin 26 protein bearing an F142L amino acid substitution. 5th Meeting for Molecular Biology of Hearing and Deafness. Bethesda, Maryland
Terrinoni A, Leta A, Pedicelli C, Candi E, Ranalli M, Puddu P, Paradis M, Angelo C, Baggetta G, Melino G (2004) A novel recessive connexin 31 (GJB3) mutation in a case of erythrokeratodermia variabilis. J Invest Dermatol 122:837–839
Thomas T, Aasen T, Hodgins M, Laird DW (2003) Transport and function of cx26 mutants involved in skin and deafness disorders. Cell Commun Adhes 10:353–358
Thomas T, Telford D, Laird DW (2004) Functional domain mapping and selective trans-dominant effects exhibited by Cx26 disease-causing mutations. J Biol Chem 279:19157–19168
Tuppurainen K, Fraki J, Karjalainen S, Paljarvi L, Suhonen R, Ryynanen M (1988) The KID-syndrome in Finland. A report of four cases. Acta Ophthalmol (Copenh) 66:692–698
Unger VM, Kumar NM, Gilula NB, Yeager M (1999) Three-dimensional structure of a recombinant gap junction membrane channel. Science 283:1176–1180
Uyguner O, Tukel T, Baykal C, Eris H, Emiroglu M, Hafiz G, Ghanbari A, Baserer N, Yuksel-Apak M, Wollnik B (2002) The novel R75Q mutation in the *GJB2* gene causes autosomal dominant hearing loss and palmoplantar keratoderma in a Turkish family. Clin Genet 62:306–309
van der Schroeff JG, Nijenhuis LE, Meera Khan P, Bernini LF, Schreuder GM, van Loghem E, Volkers WS, Went LN (1984) Genetic linkage between erythrokeratodermia variabilis and Rh locus. Hum Genet 68:165–168
van Geel M, van Steensel MA, Kuster W, Hennies HC, Happle R, Steijlen PM, Konig A (2002a) HID and KID syndromes are associated with the same connexin 26 mutation. Br J Dermatol 146:938–942
van Geel M, van Steensel MA, Steijlen PM (2002b) Connexin 30.3 (GJB4) is not required for normal skin function in humans. Br J Dermatol 147:1275–1277
van Steensel M (2004) Does progressive symmetric erythrokeratoderma exist? Br J Dermatol 150:1043–1045
van Steensel MA, van Geel M, Nahuys M, Smitt JH, Steijlen PM (2002) A novel connexin 26 mutation in a patient diagnosed with keratitis-ichthyosis-deafness syndrome. J Invest Dermatol 118:724–727
van Steensel MA, Jonkman MF, van Geel M, Steijlen PM, McLean WH, Smith FJ (2003) Clouston syndrome can mimic pachyonychia congenita. J Invest Dermatol 121:1035–1038
van Steensel MA, Steijlen PM, Bladergroen RS, Hoefsloot EH, van Ravenswaaij-Arts CM, van Geel M (2004) A phenotype resembling the Clouston syndrome with deafness is associated with a novel missense GJB2 mutation. J Invest Dermatol 123:291–293

van Steensel MA, Spruijt L, van der Burgt I, Bladergroen RS, Vermeer M, Steijlen PM, van Geel M (2005) A 2-bp deletion in the GJA1 gene is associated with oculo-dento-digital dysplasia with palmoplantar keratoderma. Am J Med Genet 132:171–174

Vandersteen PR, Muller SA (1971) Erythrokeratodermia variabilis. An enzyme histochemical and ultrastructural study. Arch Dermatol 103:362–370

VanSlyke JK, Deschenes SM, Musil LS (2000) Intracellular transport, assembly, and degradation of wild-type and disease-linked mutant gap junction proteins. Mol Biol Cell 11:1933–1946

Vohwinkel KH (1929) Keratoma hereditarium mutilans. Arch Derm Syph 158:354–364

Wasserman D, Richard G, Sundaram S, Kozic H, Itin P, Lewanda A, Antaya R, Willoughby CE, Uitto J (2003) Genotype-phenotype correlations in KID syndrome: Novel GJB2 mutations and association with squamous cell carcinoma. J Invest Dermatol 121:A666

Watt FM (2002) The stem cell compartment in human interfollicular epidermis. J Dermatol Sci 28:173–180

White TW, Bruzzone R (1996) Multiple connexin proteins in single intercellular channels: connexin compatibility and functional consequences. J Bioenerg Biomembr 28:339–350

White TW, Paul DL, Goodenough DA, Bruzzone R (1995) Functional analysis of selective interactions among rodent connexins. Mol Biol Cell 6:459–470

Wilgoss A, Leigh IM, Barnes MR, Dopping-Hepenstal P, Eady RA, Walter JM, Kennedy CT, Kelsell DP (1999) Identification of a novel mutation R42P in the gap junction protein beta-3 associated with autosomal dominant erythrokeratoderma variabilis. J Invest Dermatol 113:1119–1122

Willecke K, Eiberger J, Degen J, Eckardt D, Romualdi A, Guldenagel M, Deutsch U, Sohl G (2002) Structural and functional diversity of connexin genes in the mouse and human genome. Biol Chem 383:725–737

Williams ML, Hanley K, Elias PM, Feingold KR (1998) Ontogeny of the epidermal permeability barrier. J Investig Dermatol Symp Proc 3:75–79

Xia AP, Ikeda K, Katori Y, Oshima T, Kikuchi T, Takasaka T (2000) Expression of connexin 31 in the developing mouse cochlea. Neuroreport 11:2449–2453

Xia JH, Liu CY, Tang BS et al. (1998) Mutations in the gene encoding gap junction protein beta-3 associated with autosomal dominant hearing impairment. Nat Genet 20:370–373

Yotsumoto S, Hashiguchi T, Chen X, Ohtake N, Tomitaka A, Akamatsu H, Matsunaga K, Shiraishi S, Miura H, Adachi J, Kanzaki T (2003) Novel mutations in GJB2 encoding connexin-26 in Japanese patients with keratitis-ichthyosis-deafness syndrome. Br J Dermatol 148:649–653

Zhang XJ, Chen JJ, Yang S, Cui Y, Xiong XY, He PP, Dong PL, Xu SJ, Li YB, Zhou Q, Wang Y, Huang W (2003) A mutation in the connexin 30 gene in Chinese Han patients with hidrotic ectodermal dysplasia. J Dermatol Sci 32:11–17

Zhou XW, Pfahnl A, Werner R, Hudder A, Llanes A, Luebke A, Dahl G (1997) Identification of a pore lining segment in gap junction hemichannels. Biophys J 72:1946–1953

8 Intercellular Communication in Lens Development and Disease

Adam M. DeRosa, Francisco J. Martinez-Wittinghan, Richard T. Mathias, Thomas W. White[1]

8.1 Introduction

Gap junctions provide an important pathway for the intercellular transport of molecules needed for lens growth and homeostasis (Goodenough 1992; White and Bruzzone 2000). In the lens, three different connexin subunits, connexin43 (Cx43), connexin46 (Cx46), and connexin50 (Cx50), contribute to the intercellular channels that comprise gap junctions (Gerido and White 2004). Each of these connexins has unique gating and permeation properties (Srinivas et al. 1999; Hopperstad et al. 2000; Valiunas et al. 2002), suggesting that the lens requires functionally distinct types of intercellular communication. Gap junction channels are particularly important in the lens, as they provide a major route for the exchange of essential molecules between the metabolically active epithelium and the quiescent fiber cell core (Mathias et al. 1997; Donaldson et al. 2001).

The lens is an avascular spherical organ suspended between the aqueous and vitreous humors of the eye. The anterior surface consists of a single layer of epithelial cells, while the remaining organ mass consists of highly differentiated fiber cells (Fig. 8.1). The lens first forms as a hollow ball of epithelial cells that invaginates from the surface ectoderm. The posterior cells then elongate toward the anterior surface, producing a solid sphere. Thereafter, epithelial cells proliferate, migrate to the equator and elongate until they stretch from the anterior to posterior poles, forming new lens fibers (Menko 2002). Morphological differentiation of fiber cells is accompanied by changes in protein synthesis, including the upregulation of a distinct set of lens membrane channels and a family of cytoplasmic proteins, the crystallins (Piatigorsky 1981; Donaldson et al. 2001). Differentiated fiber cells undergo further development as they are transformed into the mature fibers found deeper in the core of the lens. These changes include the proteolytic cleavage of the c-terminal of several membrane proteins including the connexins (Kistler et al. 1990; Lin et al. 1997). While the precise rele-

[1]Program in Genetics, Department of Physiology and Biophysics, T5-147 Basic Science Tower, State University of New York, Stony Brook, NY 11794-8661, USA, E-mail: thwhite@notes.cc.sunsyb.edu

Gap Junctions in Development and Disease
(ed. by E. Winterhager)

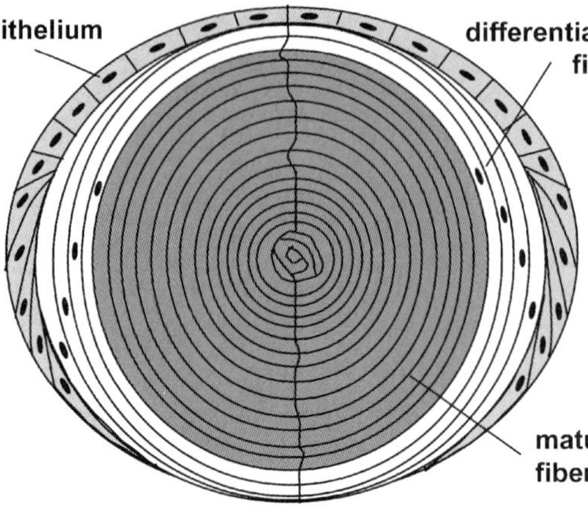

Fig. 8.1. A sketch of the cellular organization of the vertebrate lens cut in cross section along the anterior-posterior axis. There are three physiologically distinct zones. The anterior surface is covered by a simple epithelium (*light gray*). Below the epithelium, the peripheral 20% of the lens is made up of a shell of differentiating fibers (*DF*, *white*). The central 80% of the lens contains the mature fibers lacking organelles (*MF*, *dark gray*)

vance of this endogenous truncation remains illusive, c-terminal cleavage coincides with gap junctional plaque reorganization, changes in the pH gating of the channels, and the stabilization of these *trans*-membrane proteins (Gruijters et al. 1987; Mathias et al. 1991; Zampighi et al. 1992). The first lens fiber cells enter the final stages of differentiation on day 12 of mouse embryonic development (Bassnett 2002). However, the processes of fiber differentiation persist as the lens continues to accumulate new cells throughout life.

One function of the lens is accommodation, a process allowing the eye to focus on images at different focal distances. To provide accommodation, the lens requires several unique physiological properties. First, high levels of crystallin synthesis must occur to increase the refractive index of the fiber cell cytoplasm. Second, it must be devoid of light scattering elements like cytoplasmic organelles which are degraded during the final stages of fiber differentiation. Finally, the extracellular spaces between fiber cells must be very thin, narrower in fact than the wavelength of visible light, to maintain a continuous high refractive index throughout the organ. Once fiber cells become crystallin enriched and are depleted of all light scattering organelles, they gain the capacity to provide the lens with the clarity and high refractive index necessary to properly focus light onto the retina (Piatigorsky 1981; Goodenough 1992; Bassnett 2002).

The extensive specialization of fibers required to provide transparency presents unique challenges for these cells. With no cytosolic organelles, such as mitochondria normally essential for ATP production, nuclei and the endoplasmic reticulum critical for protein synthesis, or blood vessels for the removal of harmful cellular waste, the lens must create its own unique mode of intercellular transport for exchanging metabolites and ions with the surface epithelium. To address these transport challenges, the lens has developed its own unique circulatory system driven by a standing flow of ionic current generated via the differences in electromotive membrane potentials between surface and core fiber cells (Mathias et al. 1997).

Gap junctions play a critical role in the postnatal growth and differentiation of the lens. They are also a vital component of the circulating current that ensures lens homeostasis and prevents crystallin precipitation and cataract. While these roles have previously been postulated for the intercellular communication provided by lens gap junctions (Goodenough 1992), the need for three different connexin subunits Cx43, Cx46, and Cx50 has only recently been elucidated. This chapter will describe advances in our understanding of the specific requirements for each connexin in the processes of cell proliferation, differentiation, and cataractogenesis.

8.2 Connexin Specialization in the Lens

The process of lens fiber differentiation includes a change in the expression of cell-specific connexin proteins (Beyer et al. 1989; Goodenough 1992; White and Bruzzone 2000). In the mammalian lens, Cx43 is expressed at low levels in the lens epithelium, but not fibers, while Cx46 is absent from the epithelial cells, but becomes highly expressed during fiber cell differentiation (Paul et al. 1991; Gong et al. 1997). In contrast to the segregated expression of Cx43 and Cx46, Cx50 is produced in all cells of the lens, first being synthesized in the epithelium and persisting at high levels into the differentiating fibers (White et al. 1992; Dahm et al. 1999; Rong et al. 2002). Thus, three connexin family members are used in distinct, yet overlapping, expression patterns to maintain gap junctional communication between the metabolically active epithelium and the highly specialized fiber cells.

By using more than one connexin protein to form intercellular channels, the lens can generate functional diversity in the junctional communication between cells. This diversity arises from both the intrinsic differences in functional properties of the three connexins and their ability to selectively form mixed channels containing more than one subunit. Connexins first oligomerize to form single-membrane channels called connexons or hemichannels, which then align in the extracellular space between two cells to complete the intercellular channels found in gap junctions (White and Bruzzone 1996; Evans and Martin 2002). Each hemichannel is made of six subunits (Sosinsky 1996; Unger et al. 1999), and can contain either a single type of connexin forming a homomeric connexon, or more than one connexin creating a heteromeric connexon. Adjacent cells can contribute different types of connexons, forming homotypic (association of two identical connexons), heterotypic (combination of two homomeric connexins each of different connexin origin), or heteromeric channels (consisting of two distinct heteromeric hemichannels). In vivo, clear evidence for the ability of Cx46 and Cx50 to co-assemble into single connexons and intercellular channels has been obtained (Konig and Zampighi 1995; Jiang and Goodenough 1996), but the percentage of the total number of gap junction channels that are mixtures of these two proteins in the lens is not known. To date, heteromeric channels containing Cx43 with either Cx46 or Cx50 have not been identified in vivo.

The distinct gating properties of the lens connexins have been well characterized in vitro. Homotypic channels composed of Cx43, Cx46, or Cx50 were differentially regulated by transjunctional voltage, with Cx43 being the least voltage sensitive and Cx50 exhibiting the greatest sensitivity (Ebihara and Steiner 1993; White et al. 1994). The formation of heterotypic channels among lens connexins was a selective property requiring the presence of compatible connexins in adjacent cells. Cx50 did not form heterotypic channels with Cx43, while Cx46 was compatible with both Cx43 and Cx50. In addition, heterotypic channels displayed novel responses to transjunctional voltage that were not predicted by the properties of the corresponding homotypic channels (White et al. 1994). The unitary conductance of single channels were also very different: Cx43 had the smallest at 60–90 pS (Fishman et al. 1991; Moreno et al. 1994), Cx46 was intermediate at 140 pS (Hopperstad et al. 2000), and Cx50 had the largest at 220 pS (Srinivas et al. 1999; all measured in CsCl). Heterotypic and heteromeric channels formed from Cx46 and Cx50 displayed a range of unitary conductance that never exceeded the homotypic Cx50 value of 220 pS (Hopperstad et al. 2000).

As described above, epithelial to fiber cell differentiation coincides with a change in connexin expression from Cx43 and Cx50 to Cx46 and Cx50. As predicted by the in vitro studies, this change in gap junctional composition is accompanied by a change in junctional voltage dependence. Cultured epithelial cells are less voltage dependent than the gap junctions found in freshly dispersed fiber cells of the lens (Donaldson et al. 1995). Furthermore, lens cell differentiation is also associated with an increase in unitary gap junctional conductance (Donaldson et al. 1994). Although voltage gating is the best characterized functional property of the three lens connexins, it is likely that other functional differences in gating and/or permeability also play important roles in lens physiology. While much remains to be learned about the functional differences in gap junctional communication present in different lens regions, a great deal of information regarding the importance of the connexin diversity underlying these functional differences has been revealed by studies conducted with genetically engineered mice.

8.3 Genetic Manipulation of Lens Connexins

Knockout of lens connexins in mice has confirmed the established hypothesis that gap junctions are essential to proper lens development and function (Goodenough 1992). The long-term effects of knocking out Cx43 cannot be characterized in the lens, as deletion of this gene results in neonatal lethality (Reaume et al. 1995); however, prenatal ocular development has been successfully examined in Cx43 knockout mice (Gao and Spray 1998; White et al. 2001). In contrast, deletion of either Cx46 or Cx50 does not compromise mouse viability and the ocular phenotypes of these two animal models have been well studied (Gong et al. 1997, 1998, 1999; White et al. 1998; Baruch et al. 2001; Baldo et al. 2001; Rong et al. 2002; Gerido et al. 2003; Gao et al. 2004; Martinez-Wittinghan et al. 2004; Sellitto et al. 2004). Finally, functionally replacing Cx50 with Cx46 by genetic knockin has provided additional insight into the unique roles played by these two connexins in lens physiology (White 2002; Martinez-Wittinghan et al. 2003; Sellitto et al. 2004). Taken together, these studies have identified critical roles for two of the three connexins in lens physiology, and have underscored the importance of connexin diversity in vivo (Fig. 8.2).

8.3.1 Knockout of Cx43

Cx43 is not only expressed in the lens epithelium, but is also abundant in the heart and many other tissues. Deletion of the Cx43 gene in mice results in cardiac malformation and neonatal death (Reaume et al. 1995), thus limiting studies of Cx43's function in the lens to prenatal development. Prior to birth, one study of Cx43 knockout embryos showed development of histologically normal eyes and lenses, and expression of high levels of four different markers of lens fiber differentiation: aquaporin0 (previously known as MIP), α-crystallin, β-crystallin and γ-crystallin. Gap junctional communication in embryonic lenses was assayed by microinjection and subsequent intercellular transfer of neurobiotin. Cx43 knockout lenses showed a reduction in epithelial–epithelial and epithelial–fiber gap junctional communication, while fiber–fiber communication was unchanged

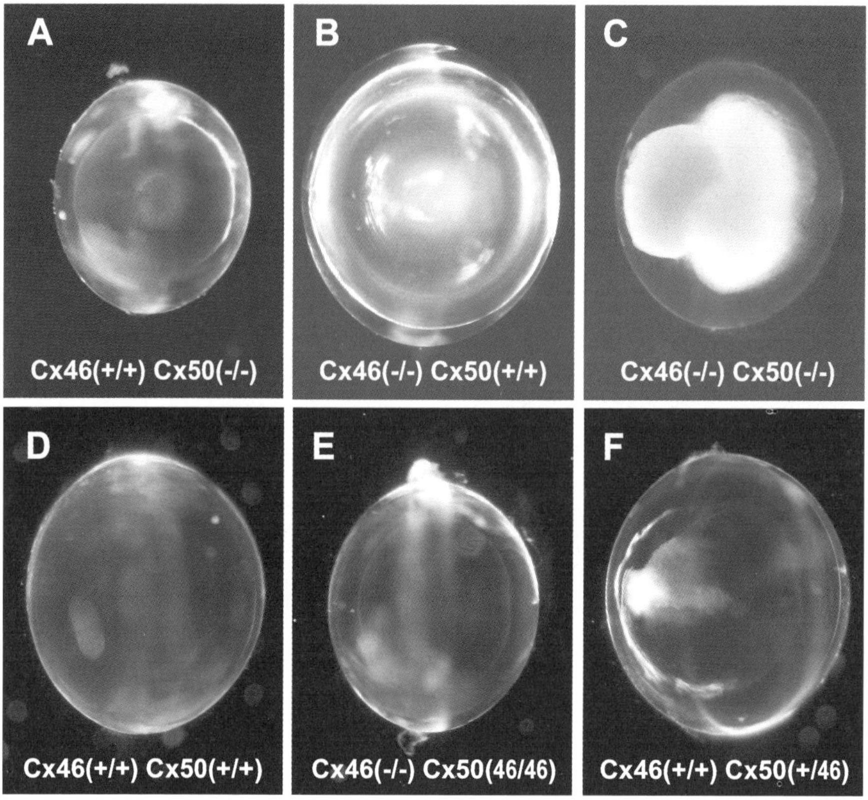

Fig. 8.2A–F. Cataracts form in connexin knockout lenses. Views looking down on the equatorial edge of Cx50 knockout (**A**), Cx46 knockout (**B**) and a Cx46/Cx50 double knockout (**C**) are shown. Deletion of Cx50 produced a small lens with a mild cataract. Knockout of Cx46 resulted in a severe cataract, but normal lens size. Disruption of both connexin genes resulted in dense cataract and small size. Wild-type lenses (**D**) maintained transparency, as did knockover lenses (**E**), which were smaller than normal, but clear. Thus, Cx46 alone could maintain lens clarity when it was expressed from the Cx50 locus (compare **E** with **A** and **B**). Heterozygous knockin lenses (**F**) developed dense cataracts in the central anterior region of the lens, in addition to an irregular diffraction under the epithelium. Therefore, incongruous mixing of Cx46 and Cx50 in heterozygous knockins caused cataract

from that observed in wild-type control lenses (White et al. 2001). These data suggested that Cx43 was not required for normal differentiation of the lens fibers, consistent with an earlier in vitro study showing that fiber differentiation could occur in the presence of pharmacological blockers of junctional communication (Le and Musil 1998).

In contrast, an earlier study of Cx43 knockouts detected lens structural abnormalities not reported by White et al. (2001). In this analysis, grossly dilated extracellular spaces and large intracellular vacuoles were reported in newborn Cx43 knockout lenses (Gao and Spray 1998). Abnormal cell-to-cell appositions were not reported in the later study, despite extensive examination by both light and electron microscopy throughout the late embryonic developmental period (White et al. 2001). Because the Cx43 null mice die at birth from anoxia, it is possible that the postnatally recovered Cx43 knockout lenses were affected by oxygen deprivation, or that the structural abnormalities observed by Gao and Spray (1998) resulted from trauma endured during birth. Although the two studies differed in the observation of lens structural abnormalities resulting from Cx43 deletion, their other findings were in general agreement. In particular, proper lens fiber differentiation was observed in both studies, including the normal expression of aquaporin0 and crystallins. Thus, Cx43-mediated gap junctional communication appeared to be dispensable with regard to embryonic lens fiber development; however, it could be involved in later processes of postnatal growth and maintenance of homeostasis.

8.3.2
Knockout of Cx46

Unlike Cx43, deletion of the Cx46 gene does not result in neonatal lethality, allowing both lens development and homeostasis in the absence of this connexin to be studied throughout life. Knockout of Cx46 resulted in severe cataracts associated with precipitation of several crystallin proteins and a specific proteolysis of γ-crystallin. Cataracts developed several weeks after birth and progressively worsened as the animals aged (Gong et al. 1997). The severity of cataracts in Cx46 knockouts was found to be influenced by the genetic background of the animals and cataract severity correlated well with the amount of cleaved γ-crystallin (Gong et al. 1999). Measurements of gap junctional coupling in knockout lenses showed that the total conductance was reduced by ∼50% in the differentiating fibers, while the mature fibers were completely uncoupled (Gong et al. 1998). Despite the development of severe cataracts, the Cx46 knockouts showed normal fiber differentiation and postnatal lens growth (Gong et al. 1997). These data argued that Cx46 played a central role in maintaining lens clarity, but had no effect on lens growth or development.

The specific cleavage of γ-crystallin had been proposed as the initiating factor in the loss of crystallin solubility and subsequent cataract formation in Cx46 knockout lenses. This cleavage is thought to be initiated by the calcium-dependent lens calpain, Lp82. When lenses were dissected from Cx46 knockout animals prior to the appearance of cataracts and maintained in tissue culture in the presence of a specific inhibitor of Lp82 (E-64), or greatly reduced extracellular calcium, both γ-crystallin cleavage and cataracts were prevented. Knockout lenses that were cultured without E-64 and in the presence of normal calcium levels went on to develop cataracts that were similar to those that occurred in vivo (Baruch et al. 2001). These results strongly suggested that Cx46 was essential for maintaining crystallin solubility and lens clarity, through a mechanism involving calcium homeostasis in the lens. More recent insights into this mechanism will be further discussed in Sect. 8.4 below.

8.3.3 Knockout of Cx50

Cx50 appears to be expressed only in the lens (Kistler et al. 1985; White et al. 1992; White et al. 2001; Rong et al. 2002), and like Cx46, deletion of its gene produces viable mice that can be studied throughout life (White et al. 1998). Genetic knockout of Cx50 resulted in several phenotypes, including the development of a mild nuclear cataract and a more pronounced growth defect of both the lens and eye (White et al. 1998). In addition, deletion of Cx50 caused a delay in fiber maturation characterized by a retardation of the denucleation process that marked the transition to mature fibers (Rong et al. 2002). These distinct phenotypes manifested during the first postnatal week of life, an earlier stage of development than the severe cataract phenotype found in Cx46 knockout animals (White et al. 1998; Rong et al. 2002). Like the Cx46 knockout, the severity of the mild cataract was dependent on the animal's genetic background, while the prominent growth defect persisted regardless of genetic background (Gerido et al. 2003). Coupling measurements in Cx50 knockout lenses showed that in the differentiating fibers conductance was again reduced by ~50%. However, in contrast to Cx46 knockouts, a significant level of coupling remained in the mature fibers following deletion of Cx50 (Baldo et al. 2001). These results argued that Cx46 and Cx50 had unique functions in lens physiology, and that the primary roles of Cx50 were to promote normal ocular growth and fiber differentiation.

The ocular growth defect in Cx50 knockouts was characterized by ~33% reduction in eye mass and ~50% decrease in lens mass compared to control animals, and resulted from a transient inhibition of the lens growth rate during the first postnatal week (White et al. 1998; Sellitto et al. 2004). The lens growth rate before and after this critical period was not significantly different from controls. The retardation in growth affected only the formation of secondary lens fibers, a process that continued throughout life, but was much more active during the early postnatal period (Brewitt and Clark 1988). Measurements of fiber diameter verified that the reduction of lens size was the result of fewer lens fibers, rather than smaller cells (White et al. 1998). This acute and transient decrease in lens growth suggested a specific requirement for Cx50 in epithelial cell proliferation and/or fiber differentiation. The role of Cx50 in lens cell proliferation will be discussed in greater detail in Sect. 8.5 below.

Recently, double knockout mice lacking both the Cx46 and Cx50 genes have been generated (Xiaohua Gong, pers. comm.). These animals developed a dense cataract that was far more extensive than that observed in either the Cx46 or Cx50 single knockout mice. Presumably, the increased density of the opacities in these mice resulted from the total absence of gap junction proteins in the lens fiber cells. Double knockout mice also displayed a significant reduction in lens growth comparable to that present in Cx50 knockout animals. The presence of both the cataract and growth defect phenotypes in the double knockout mice further confirms the general importance of connexins in both lens growth and in maintaining lens homeostasis.

8.3.4 Knockin of Cx46 into the Cx50 Gene

The diverse phenotypes that occurred in the Cx46 and Cx50 knockout mice could have resulted from either the unique intrinsic functional properties of these two connexins (Ebihara and Steiner 1993; White et al. 1994; Srinivas et al. 1999; Hopperstad et al. 2000), or their differential contributions to the magnitude and spatial arrangement of intercellular communication in the lens (Gong et al. 1998; Baldo et al. 2001; Rong et al. 2002). This question was directly addressed by engineering a genetic replacement of the Cx50 coding region with that of Cx46 by homologous recombination (knockin). Knockin mice exhibited the same total body weight when compared to wild type; however, their eyes and lenses were 25 and 33% smaller, respectively. Furthermore, knockin lenses lacked the nuclear cataracts exhibited by Cx50 knockout lenses (White 2002). Thus, in knockin mice, reduced lens growth lead to microphthalmia, a very similar phenotype to the growth defect associated with Cx50 deletion. However, lens fibers were able to maintain homeostasis and avoid crystallin precipitation, hence they lacked the cataract found after knockout of either Cx46 or Cx50. The simplest interpretation of these data was that the lens had segregated the contributions of connexin-specific gap-junctional communication to the control of normal growth and the maintenance of clarity, where Cx50 played an indispensable role in normal growth and Cx46 was required to prevent cataracts.

The knockin data established that when Cx46 was expressed from the Cx50 gene locus, it was as good at maintaining lens clarity as native Cx50. Evidence that Cx46 was even better than Cx50 at preventing cataract was provided by mating knockin and Cx46 knockout mice to produce animals where native Cx46 was completely deleted and homozygously knocked into the Cx50 locus (knockover, i.e., Cx46$^{(-/-)}$ Cx50$^{(46/46)}$). Knockover lenses maintained lens clarity although they had a significant growth defect equivalent to that of the Cx50 knockout (Martinez-Wittinghan et al. 2003). In Cx46 knockout mice, Cx50 expression from its native locus was unable to maintain lens clarity, while Cx46 expression from the Cx50 locus in knockover animals prevented cataract formation in the absence of native Cx46. These data clearly demonstrated that Cx46 was far more efficient than Cx50 at preventing the cataract phenotype when expressed from the same genetic locus.

If all else were equal, Cx46 should provide a lower level of ionic coupling than Cx50 when expressed from the same promoter, due to the smaller unitary conductance of its channels (140 pS vs. 220 pS). Moreover, the knockover lenses should have less coupling than wild type due to an overall reduction in channel number caused by the loss of native Cx46. The knockover data strongly suggest that the coupling provided by Cx46, when exclusively expressed from the Cx50 locus, was far more efficient than that provided by Cx50 in maintaining lens clarity. Since the knockout of Cx50 produced cataract (White et al. 1998; Rong et al. 2002), these results also suggested that connexin location and identity were more important than connexin quantity, as another significant difference between the Cx50 knockout and the Cx46 knockover was the locus-dependent pattern of expression of Cx46.

Further insight into lens connexin diversity was provided by analysis of heterozygous knockin mice. In wild-type animals, Cx46 and Cx50 are normally co-expressed in lens fibers, but not in the epithelium. In addition, they can co-oligomerize into heteromeric channels within the fiber cells (Konig and Zampighi 1995; Jiang and Goodenough 1996), and heteromeric channels are functional in vitro (White et al. 1994; Ebihara et al. 1999; Hopperstad et al. 2000). In knockin mice, immunocytochemical data showed that Cx46 was expressed in the epithelium where it could mix with Cx50 in heterozygous knockin animals. Surprisingly, this novel mixing of Cx46 and Cx50 in lens epithelial cells produced dense cataracts that were distinct from those resulting from the knockout of either Cx46 or Cx50. These unique cataracts consisted of a densely opaque core located in the central anterior region of the lens, as well as a larger irregular diffraction more closely underlying the epithelium. This phenotype occurred in all heterozygous knockin mice and was independent of total lens connexin quantity (Martinez-Wittinghan et al. 2003). Furthermore, there were no differences in the growth rates of wild-type and heterozygous knockin lenses throughout postnatal development. Thus, forced mixing of Cx46 and Cx50 in lens epithelial cells produced a dominant cataract phenotype without adversely affecting lens growth.

Analysis of junctional coupling in the different knockin-derived lenses supported the idea that the molecular composition of gap junction channels was more important than the magnitude of coupling they provided. Frequency domain impedance methods (Mathias et al. 1991), were used to directly measure the magnitude of coupling in the knockin animals and compare them to wild type, Cx46 knockout, and Cx50 knockout mice (Gong et al. 1998; Baldo et al. 2001; Martinez-Wittinghan et al. 2004). In wild-type lenses, the coupling conductance (G_j) was ~1 S/cm^2 of cell-to-cell contact in the outer shell of differentiating fibers (DF), whereas it was about half that value in the core of mature fibers (MF). In the DF, levels of ionic coupling appeared to be primarily determined by the respective unitary conductance of Cx46 and Cx50. Knockout of one Cx50 allele reduced G_j in DF to 66% of normal, whereas heterozygous knockin of Cx46 restored DF G_j to 82% of normal. Homozygous knockin of Cx46 produced a G_j ~60% of the wild-type value found in DF, or roughly equivalent to the heterozygous Cx50 knockout (Fig. 8.3A). Thus, the magnitude of coupling was dependent on which connexin was expressed on the Cx50 gene locus, and to a first approximation, correlated well with the fact that the unitary conductance of Cx46 is approximately half that of Cx50 (Srinivas et al. 1999;Hopperstad et al. 2000), suggesting that genetic knockin of the Cx46 gene resulted in an equal number of functional channels in the DF. In contrast to the DF results, knockin of Cx46 greatly increased coupling in the MF. Previous studies have shown that deletion of Cx46 eliminated coupling, while knockout of Cx50 had only marginal effects on MF coupling. This suggested that the observed 50% drop in coupling at the DF/MF transition in wild-type

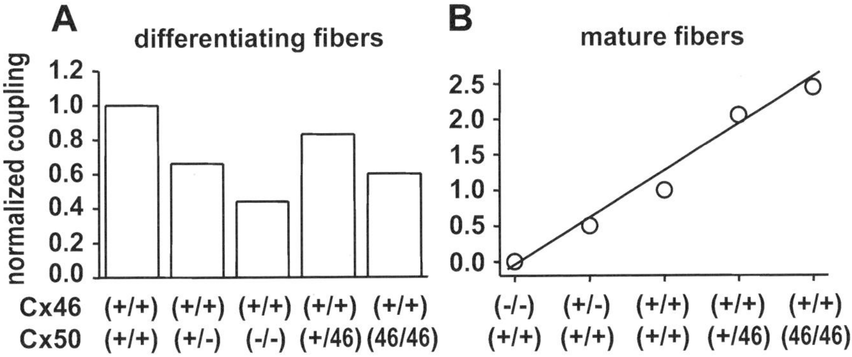

Fig. 8.3A, B. Measurement of gap junctional coupling in engineered mouse lenses. In differentiating fibers (**A**), genetic replacement of Cx50 with Cx46 resulted in a decrease in coupling conductance consistent with the smaller unitary conductance of Cx46 compared to Cx50. In the mature fibers (**B**), knockout of Cx46 reduced, while replacement of Cx50 with Cx46 by knockin increased coupling

lenses was due to the normal loss of functional Cx50 channels (Baldo et al. 2001). Heterozygous knockin lenses had approximately twice the MF coupling of wild-type and homozygous lenses showed a further increase in conductance. A plot of MF conductance vs. the number of active Cx46 alleles (native plus knockin) showed that MF coupling increased in a nearly linear fashion with each added Cx46 coding region (Fig. 8.3B). Taken together, the impedance data suggested that there are spatial differences in the contributions of Cx46 and Cx50 to lens fiber coupling. Both connexins contribute equally to DF coupling, whereas in the MF, Cx46 is critical for coupling and Cx50 plays no role. These impedance data also explained why knockover lenses lacked the cataract that normally resulted from Cx46 knockout. In the Cx46 knockout, Cx50 expression from its native locus was unable to maintain clarity. However, Cx46 expression from the Cx50 locus in knockover mice prevented cataract development, a finding consistent with the idea that native Cx50 is nonfunctional in the MF, while knockin Cx46 remains functional in the MF of knockover animals.

The heterozygous knockin lenses had high levels of electrical coupling, yet they still developed cataracts, suggesting that the simple magnitude of gap junctional coupling was not the limiting factor. Gap junctional communication can be simply divided into two forms, ionic coupling (the exchange of major cytoplasmic ions like K^+) and biochemical communication (the intercellular sharing of larger solutes like cAMP). While all connexins mediate ionic coupling without much selectivity, the spectrum of larger biochemical solutes that can be exchanged between coupled cells is highly dependent on the connexin composition of the channels (Bevans et al. 1998; Goldberg et al. 1999; Valiunas et al. 2002). Fluorescent dyes allow patterns of biochemical coupling to be visualized within the lens (Miller and Goodenough 1986; Rae et al. 1996; White et al. 1998, 2001). The extent of epithelial dye coupling in knockin lenses was examined by documenting the intercellular passage of the dye 4',6-diamidino-2-phenylindole dihydrochoride (DAPI). Injection of a single epithelial cell in wild-type lenses resulted in the intercellular passage of DAPI to a large number of neighboring cells. In contrast, DAPI was transferred to only a few neighboring epithelial cells in heterozygous knockin lenses. Thus, DAPI was able to pass through gap junctions in all lenses, however, there was a large qualitative difference in the extent of dye spread. Combined with the impedance results, these dye data suggested that the heterozygous knockin cataract was not a consequence of reduced levels of ionic gap junctional coupling, but instead correlated with alterations in dye coupling, and the incongruous mixing of Cx46 and Cx50 in epithelial cells altered the spatial patterns of biochemical coupling necessary to sustain lens clarity.

8.4 How Does Cx46 Prevent Cataract?

The experiments described above have identified a critical role for Cx46 in maintaining lens clarity, particularly in the highly specialized mature fiber cells which are dependent on Cx46 mediated gap junctional coupling for communication with the metabolically active surface epithelium. Furthermore, calcium-dependent proteolytic activity of Lp82 has been identified as an important trigger of cataractogenesis in Cx46 knockout mice (Baruch et al. 2001). The nature of the connection between Cx46, gap junctional coupling, calcium and lens cataractogenesis has recently been identified by integrating data from the numerous studies described above with a novel method of calcium measurement in the intact lens (Gao et al. 2004).

Because Cx46 is the only connexin providing gap junctional coupling in the mature fiber cells (Baldo et al. 2001; Martinez-Wittinghan et al. 2004), it forms a critical component of the lens internal circulation system (Mathias et al. 1997; Donaldson et al. 2001). A brief review of the role played by junctional communication in the circulating current is necessary to fully

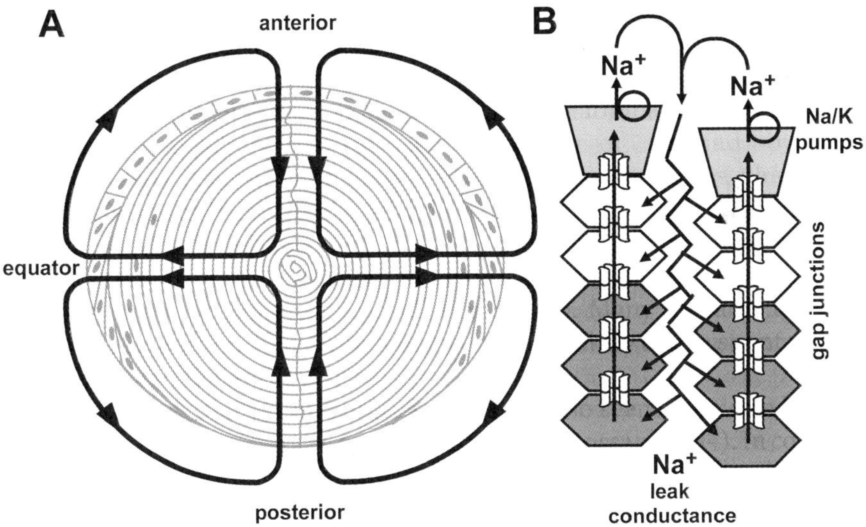

Fig. 8.4A, B. A model of how the lens internal circulation is generated. **A** In an intact lens, current flows in the pattern indicated by the *arrows*. **B** The major ion carrier of the circulating current appears to be sodium, which enters the lens along the extracellular spaces between cells, then moves down its electrochemical gradient into fiber cells, where it returns to the surface via gap junctions. The pattern of gap junction coupling in the differentiating fibers directs the intracellular current flow to the lens equator, where the epithelial cells transport it out of the lens using Na/K-ATPase activity

explain the Cx46 phenotype (Fig. 8.4). Current is primarily carried by Na^+ ions that enter the lens extracellularly at the surface and flow inward along the narrow extracellular spaces. As the Na^+ moves inward, it continuously crosses into fiber cells, driven by its transmembrane electrochemical gradient. Once in the intracellular compartment, Na^+ reverses direction and flows from cell to cell via Cx46 gap junctions back toward the lens surface. Gap junctional coupling is not uniformly distributed in the differentiating fibers, but is concentrated at the lens equator (Baldo and Mathias 1992). As the intracellular current nears the surface, it is directed toward the equatorial epithelial cells and away from the anterior and posterior poles. Na/K-ATPase activity is also concentrated in the equatorial epithelium (Gao et al. 2000; Candia and Zamudio 2002; Tamiya et al. 2003), where it transports the intracellular flux of Na^+ out of the lens and back into the extracellular space, completing the current loop.

The circulating current represents a circulation of solute that generates water flow in the same pattern (Donaldson et al. 2001). Both the fluid flow and the voltage gradients associated with the current flow, would also be expected to bring calcium into the extracellular spaces within the lens. The only path out of the lens for Ca^{2+} would be to move down its electrochemical gradient into fiber cells, then move from cell to cell by electrodiffusion through Cx46 gap junctions back to surface cells, following the same path as Na^+. Lens epithelial cells have Ca-ATPase activity and Na/Ca exchangers (Paterson and Delamere 2004) that would transport Ca^{2+} back into the aqueous or vitreous humors completing the pathway. This model predicts a circulation of Ca^{2+} in the lens, where the Cx46 containing gap junctions coupling the interior fiber cells to the surface cells would be an essential component of Ca-homeostasis.

This hypothesis was tested by studying calcium homeostasis in wild-type, Cx46 knockout, and knockin lenses, which exhibit different degrees of gap junction coupling in their mature fibers. Intracellular Ca^{2+} was measured by injecting FURA2 into fiber cells and mapping the gradient in calcium concentration from center to surface in each type of lens (Gao et al. 2004). In wild-type lenses the Ca^{2+} concentration varied smoothly from 700 nM in the center to 300 nM at the surface. In the knockin lenses, calcium varied from about 500 nM at the center to 300 nM at the surface. As described above, knockin lenses have about twice the coupling conductance in their mature fibers compared to wild type, due to the additional copies of Cx46 being expressed from the Cx50 gene locus. When the coupling conductance doubled, the Ca^{2+} concentration gradient halved, suggesting that Cx46 gap junctions represented the limiting factor in the Ca^{2+} efflux pathway. Data from Cx46 knockout lenses further supported this idea. Their mature fiber coupling conductance was zero, hence the efflux path was blocked and calcium accumulated to a concentration of $\sim 2\,\mu M$ in the mature fibers. As reviewed above, this accumulated Ca^{2+} in Cx46 knockout lenses correlated with a dense central cataract triggered by activation of the calcium-dependent protease, Lp82 (Gong et al. 1997; Baruch et al. 2001; Gao et al. 2004). Taken together, these data provide a mechanism describing each step leading to cataract formation: knockout of Cx46 causes loss of coupling of mature fiber cells; the efflux path for calcium is therefore blocked; calcium accumulates in the central cells; Lp82 is activated; Lp82 cleaves γ-crystallin; the cleaved γ-crystallin stimulates crystallin aggregation and precipitation.

One caveat to this mechanistic model was the observation that, while the knockin lenses had increased coupling in the central fibers, they also showed significantly reduced coupling in the outer fibers. This suggested that the magnitude of coupling in the differentiating fibers was not critical to the circulation, as long as coupling remained concentrated toward the lens equator. This was directly tested in knockin lenses by plotting coupling conductance as a function of the angular distance (θ) from the equator (Fig. 8.5A). In wild-type lenses, gap junctional conductance was known to be maximal at the lens equator ($\theta = 0°$) and minimal at the

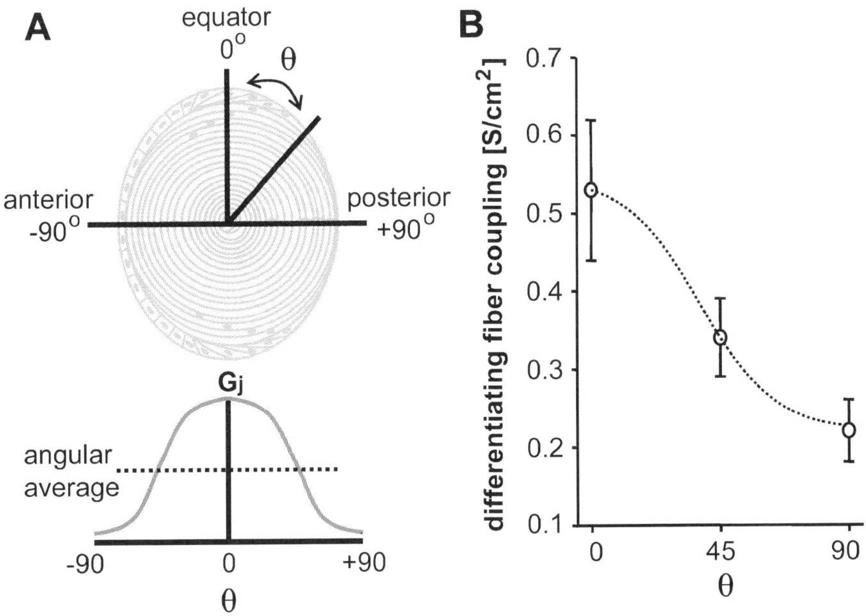

Fig. 8.5A, B. Angular variation in gap junctional coupling. **A** In wild-type animals, differentiating fiber coupling varies as a function of angular location, θ, decreasing from the equator to the poles, and reaching its angular average at ~ 45° (see Baldo and Mathias, 1992). **B** In knockin lenses, differentiating fiber coupling also varied as a function of angular location, despite a lower magnitude of coupling ($n = 5$, mean ± SEM)

anterior and posterior poles ($\theta = \pm 90°$;Baldo and Mathias 1992). In the knockin lenses, the radial variation in differentiating fiber conductance was preserved, with the equatorial differentiating fibers having a greater G_j than the posterior pole (Fig. 8.5B; F. Martinez-Wittinghan, T.W. White and R.T. Mathias, unpubl. data). Thus, the reduced magnitude of coupling in the differentiating fibers of knockin lenses did not limit the circulating current as long as the strong bias of coupling toward the equator was preserved.

8.5 How Does Cx50 Influence Lens Growth?

The most striking difference between the Cx46 and the Cx50 knockout animals was the defect in lens and eye growth produced by the loss of Cx50, but not Cx46 (Gong et al. 1997; White et al. 1998; Rong et al. 2002). In addition, unlike the cataract phenotypes, the genetic background of the animals did not influence the growth defect of Cx50 knockout mice (Gerido et al. 2003). Furthermore, replacement of Cx50 with Cx46 by genetic knockin or knockover failed to restore normal lens growth (White 2002; Martinez-Wittinghan et al. 2003). Taken together, these data suggest a growth defect that is universally present in all Cx50-deficient lenses. A link between Cx50 and epithelial cell mitosis has recently been identified as a major cause of this growth defect (Sellitto et al. 2004).

Reductions in lens growth rate in Cx50-deficient mice were highest between birth and postnatal day 7 (P7), after which all lenses grew at the same relative rates. However, the growth defect was less severe in knockin mice (34% reduction in lens mass) compared to knockout mice (46% reduction), but in both cases growth failed in the early postnatal period and the lenses remained smaller throughout life (Fig. 8.6A). Since the lens is a solid cellular cyst, reduced organ size could only result from either a decrease in the total number of cells, or a reduction in the unit cell size. Morphological and histological studies of the Cx50 knockout and knockin mice have established that cell dimensions in these lenses are not significantly different from wild type (White et al. 1998, 2002). Thus, reduced growth must result from fewer cells being present in the Cx50-deficient lenses.

Fewer cells, in turn, would arise from either an increased incidence of apoptosis, or a decrease in mitosis. Examination of apoptosis during the first postantal week by TUNEL labeling and immunostaining for cleaved caspase-3 revealed no significant differences in the number of apoptotic epithelial cells between wild-type, knockin and Cx50 knockout lenses (T.W. White and C. Sellitto, unpubl. data). In contrast, mitosis was dramatically altered in Cx50-deficient mice when the mitotic indices of wild-type and knockout lenses were compared during the first postnatal week by in

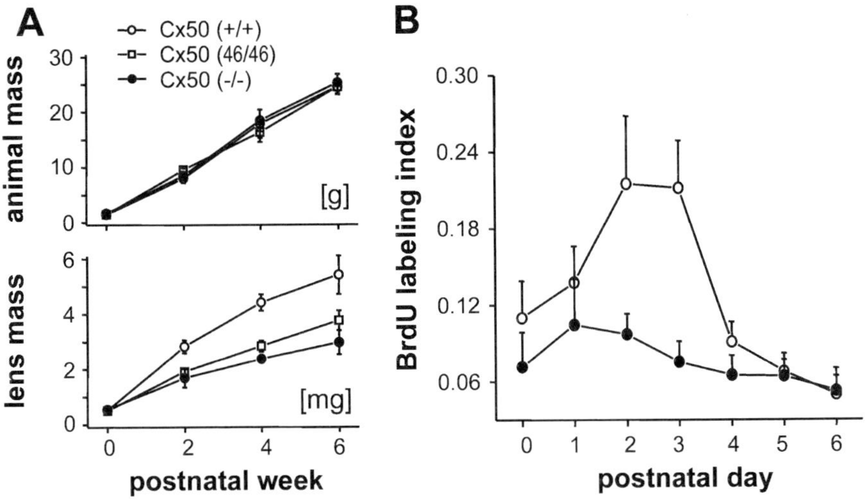

Fig. 8.6A, B. Normal lens growth requires Cx50. **A** Cx50 knockout or replacement with Cx46 by knockin produces mice with normal overall body growth, but with significantly reduced lens growth. Lens growth failure was maximal in the early postnatal period. **B** The growth defect is caused by a reduction in mitotic nuclei in early postnatal lenses. Mean labeling indices for wild-type and knockout lenses after a 1-h exposure to BrdU. In wild-type mice, cell division greatly increased on postnatal day 2 (P2) and P3. In knockout mice, the increase in P2 and P3 was absent, and the mitotic index was essentially constant (mean ± SD)

vivo bromodeoxyuridine (BrdU) labeling. On postnatal days (P) 2–3 wild-type lenses displayed a significantly increased mitotic index not evident in knockout lenses (Fig. 8.6B). In addition, genetic knockin of Cx46 only partially rescued the growth deficit. Reductions in the number of mitotic cells did not reflect a decrease in the rate of cell division, and temporally correlated with reduction in lens mass (Sellitto et al. 2004). Therefore, Cx50-deficient lenses were smaller because fewer cells were recruited into the cell cycle during the period between P2–P3, when wild-type lenses underwent a large transient increase in mitotic index.

Lens cell proliferation is modulated by growth factors (Brewitt and Clark 1988; McAvoy and Chamberlain 1989; Hyatt and Beebe 1993), and following receptor activation, many of these mitogenic signal transduction processes involve activation of the extracellular signal regulated kinase (ERK) pathway (Lovicu and McAvoy 2001; Menko 2002). However, the reduced mitotic indices in Cx50-deficient lenses did not result from a reduced transduction of mitogenic signals. Evaluation of the phosphorylation status of ERK1/2 in postnatal wild-type, knockout and knockin lenses on P2, a day where the differences in mitotic indices were maximal, and P6, when mitotic indices were identical showed that levels of total ERK and phospo-ERK were identical in wild-type and Cx50-deficient lens epithelia (Sellitto et al. 2004). These data suggest a positive mitogenic role for gap junctional communication that is connexin-specific and distinct from one of the major pathways of growth factor signaling.

8.6 Human Connexin Mutations Cause Cataract

Gap junctional communication plays important roles in multiple aspects of tissue homeostasis, a view validated by the association of connexin mutations with a wide variety of human genetic diseases (White and Paul 1999; Evans and Martin 2002; Willecke et al. 2002; Gerido and White 2004). This association of connexins with multiple human pathologies provides an unequivocal demonstration that gap junctional communication is crucial for a variety of physiological processes, as illustrated by the diverse subjects covered in this book. In addition, the specific connexin mutations underlying these human disorders are providing new insights into the roles of connexin diversity in vivo, particularly in the case of the lens where all three of the connexins expressed have been implicated in genetic disorders. Mutations in Cx43 cause oculodentodigital dysplasia (Paznekas et al. 2003), whereas mutations in either Cx46, or Cx50 result in autosomal dominant cataracts (Shiels et al. 1998; Mackay et al. 1999).

To date, human cataracts have not been directly linked to mutations in Cx43, but oculodentodigital dysplasia (ODDD, OMIM database #164200), a complex disorder that affects the face, eyes, teeth, and limbs, is caused by mutations in the Cx43 gene (Paznekas et al. 2003). In addition to a spectrum of facial abnormalities and malformation of the toes and fingers, ODDD can also include microphthalmia and other ocular defects (Frasson et al. 2004). It is currently not possible to determine if these defects are caused by a loss of Cx43 function in the lens, as Cx43 is also abundant in other tissues of the eye, including the cornea, retina and cilliary body (Bruzzone et al. 1996).

In contrast to Cx43, mutations in both Cx46 and Cx50 have been directly linked to cataracts in humans. Early evidence described a congenital zonular pulverulent human cataract (CZP1, OMIM #116200) that was later mapped to the long arm of human chromosome 1 which contains the Cx50 gene (Renwick and Lawler 1963; Church et al. 1995). Further analysis of this family revealed that mutations within the Cx50 gene were responsible for the observed cataracts. One such mutation is a C to T transition that resulted in the substitution of a serine residue at position 262 for a proline residue (Shiels et al. 1998). This mutation was only detected in individuals with cataracts in a large pedigree spanning several generations. Subsequently, additional mutations within the Cx50 gene were found to cause congenital cataracts. Like the P88S mutation, these variants R23T, D47A, I247 M, and E48 K all encoded single amino acid substitutions (Berry et al. 1999; Polyakov et al. 2001; Willoughby et al. 2003).

Mutations in the Cx46 gene have also been linked to zonular pulverulent cataracts in humans (CZP3 OMIM #601885). To date, six separate mutations have been identified in the Cx46 gene that results in congenital cataracts. Five of these mutations are single amino acid substitutions resulting in the F32L, P59L, N63S, N188T, and P187L Cx46 mutants (Mackay et al. 1999; Rees et al. 2000; Jiang et al. 2003; Bennett et al. 2004; Li et al. 2004). The sixth mutation was identified as an insertion of a cytosine at nucleotide position 1137 within the Cx46 gene resulting in a frameshift at amino acid 380 (Mackay et al. 1999).

While there is a strong correlation between data linking genetic cataracts to connexins, it should be noted that there are distinct differences in the inheritance patterns seen in humans and the knockout mice. Congenital cataracts in humans typically follow an autosomal dominant pattern of inheritance, whereas the phenotypes studied in knockout mice are recessive. To date, recessive cataracts caused by connexin mutation have not been described in humans, but dominant cataracts have been identified in mice treated with chemical mutagens (Steele, Jr. et al. 1998; Graw et al. 2001; Chang et al. 2002) and in heterozygous knockin animals (Martinez-Wittinghan et al. 2003). The finding of dominantly inherited cataracts in these mice may provide researchers with additional animal models that more closely resemble the cataract phenotypes seen in humans.

References

Baldo GJ, Mathias RT (1992) Spatial variations in membrane properties in the intact rat lens. Biophys J 63:518–529

Baldo GJ, Gong X, Martinez-Wittinghan FJ, Kumar NM, Gilula NB, Mathias RT (2001) Gap junctional coupling in lenses from alpha(8) connexin knockout mice. J Gen Physiol 118:447–456

Baruch A, Greenbaum D, Levy ET, Nielsen PA, Gilula NB, Kumar NM, Bogyo M (2001) Defining a link between gap junction communication, proteolysis, and cataract formation. J Biol Chem 276:28999–29006

Bassnett S (2002) Lens organelle degradation. Exp Eye Res 74:1–6

Bennett TM, Mackay DS, Knopf HL, Shiels A (2004) A novel missense mutation in the gene for gap-junction protein alpha3 (GJA3) associated with autosomal dominant "nuclear punctate" cataracts linked to chromosome 13q. Mol Vis 10:376–382

Berry V, Mackay D, Khaliq S, Francis PJ, Hameed A, Anwar K, Mehdi SQ, Newbold RJ, Ionides A, Shiels A, Moore T, Bhattacharya SS (1999) Connexin 50 mutation in a family with congenital "zonular nuclear" pulverulent cataract of Pakistani origin. Hum Genet 105:168–170

Bevans CG, Kordel M, Rhee SK, Harris AL (1998) Isoform composition of connexin channels determines selectivity among second messengers and uncharged molecules. J Biol Chem 273:2808–2816

Beyer EC, Kistler J, Paul DL, Goodenough DA (1989) Antisera directed against connexin43 peptides react with a 43-kD protein localized to gap junctions in myocardium and other tissues. J Cell Biol 108:595–605

Brewitt B, Clark JI (1988) Growth and transparency in the lens, an epithelial tissue, stimulated by pulses of PDGF. Science 242:777–779

Bruzzone R, White TW, Paul DL (1996) Connections with connexins: the molecular basis of direct intercellular signaling. Eur J Biochem 238:1–27

Candia OA, Zamudio AC (2002) Regional distribution of the Na(+) and K(+) currents around the crystalline lens of rabbit. Am J Physiol Cell Physiol 282:C252–C262

Chang B, Wang X, Hawes NL, Ojakian R, Davisson MT, Lo WK, Gong X (2002) A Gja8 (Cx50) point mutation causes an alteration of alpha 3 connexin (Cx46) in semi-dominant cataracts of Lop10 mice. Hum Mol Genet 11:507–513

Church RL, Wang JH, Steele E (1995) The human lens intrinsic membrane protein MP70 (Cx50) gene: clonal analysis and chromosome mapping. Curr Eye Res 14:215–221

Dahm R, van Marle J, Prescott AR, Quinlan RA (1999) Gap junctions containing alpha8-connexin (MP70) in the adult mammalian lens epithelium suggests a re-evaluation of its role in the lens. Exp Eye Res 69:45–56

Donaldson PJ, Roos M, Evans C, Beyer E, Kistler J (1994) Electrical properties of mammalian lens epithelial gap junction channels. Invest Ophthalmol Vis Sci 35:3422–3428

Donaldson PJ, Dong Y, Roos M, Green C, Goodenough DA, Kistler J (1995) Changes in lens connexin expression lead to increased gap junctional voltage dependence and conductance. Am J Physiol 269:C590–C600

Donaldson P, Kistler J, Mathias RT (2001) Molecular solutions to mammalian lens transparency. News Physiol Sci 16:118–123

Ebihara L, Steiner E (1993) Properties of a nonjunctional current expressed from a rat connexin46 cDNA in *Xenopus* oocytes. J Gen Physiol 102:59–74

Ebihara L, Xu X, Oberti C, Beyer EC, Berthoud VM (1999) Co-expression of lens fiber connexins modifies hemi-gap-junctional channel behavior. Biophys J 76:198–206

Evans WH, Martin PE (2002) Gap junctions: structure and function (Review). Mol Membr Biol 19:121–136

Fishman GI, Moreno AP, Spray DC, Leinwand LA (1991) Functional analysis of human cardiac gap junction channel mutants. Proc Natl Acad Sci USA 88:3525–3529

Frasson M, Calixto N, Cronemberger S, de Aguiar RA, Leao LL, de Aguiar MJ (2004) Oculodentodigital dysplasia: study of ophthalmological and clinical manifestations in three boys with probably autosomal recessive inheritance. Ophthalmic Genet 25:227–236

Gao J, Sun X, Yatsula V, Wymore RS, Mathias RT (2000) Isoform-specific function and distribution of Na/K pumps in the frog lens epithelium. J Membr Biol 178:89–101

Gao J, Sun X, Martinez-Wittinghan FJ, Gong X, White TW, Mathias RT (2004) Connections between connexins, calcium, and cataracts in the lens. J Gen Physiol 124:289–300

Gao Y, Spray DC (1998) Structural changes in lenses of mice lacking the gap junction protein connexin43. Invest Ophthalmol Vis Sci 39:1198–1209

Gerido DA, White TW (2004) Connexin disorders of the ear, skin, and lens. Biochim Biophys Acta 1662:159–170

Gerido DA, Sellitto C, Li L, White TW (2003) Genetic background influences cataractogenesis, but not lens growth deficiency, in Cx50-knockout mice. Invest Ophthalmol Vis Sci 44:2669–2674

Goldberg GS, Lampe PD, Nicholson BJ (1999) Selective transfer of endogenous metabolites through gap junctions composed of different connexins. Nat Cell Biol 1:457–459

Gong X, Li E, Klier G, Huang Q, Wu Y, Lei H, Kumar NM, Horwitz J, Gilula NB (1997) Disruption of alpha3 connexin gene leads to proteolysis and cataractogenesis in mice. Cell 91:833–843
Gong X, Baldo GJ, Kumar NM, Gilula NB, Mathias RT (1998) Gap junctional coupling in lenses lacking alpha3 connexin. Proc Natl Acad Sci USA 95:15303–15308
Gong X, Agopian K, Kumar NM, Gilula NB (1999) Genetic factors influence cataract formation in alpha 3 connexin knockout mice. Dev Genet 24:27–32
Goodenough DA (1992) The crystalline lens. A system networked by gap junctional intercellular communication. Semin Cell Biol 3:49–58
Graw J, Loster J, Soewarto D, Fuchs H, Meyer B, Reis A, Wolf E, Balling R, Hrabe dA (2001) Characterization of a mutation in the lens-specific MP70 encoding gene of the mouse leading to a dominant cataract. Exp Eye Res 73:867–876
Gruijters WT, Kistler J, Bullivant S, Goodenough DA (1987) Immunolocalization of MP70 in lens fiber 16–17-nm intercellular junctions. J Cell Biol 104:565–572
Hopperstad MG, Srinivas M, Spray DC (2000) Properties of gap junction channels formed by Cx46 alone and in combination with Cx50. Biophys J 79:1954–1966
Hyatt GA, Beebe DC (1993) Regulation of lens cell growth and polarity by an embryo-specific growth factor and by inhibitors of lens cell proliferation and differentiation. Development 117:701–709
Jiang H, Jin Y, Bu L, Zhang W, Liu J, Cui B, Kong X, Hu L (2003) A novel mutation in GJA3 (connexin46) for autosomal dominant congenital nuclear pulverulent cataract. Mol Vis 9:579–583
Jiang JX, Goodenough DA (1996) Heteromeric connexons in lens gap junction channels. Proc Natl Acad Sci USA 93:1287–1291
Kistler J, Kirkland B, Bullivant S (1985) Identification of a 70,000-D protein in lens membrane junctional domains. J Cell Biol 101:28–35
Kistler J, Schaller J, Sigrist H (1990) MP38 contains the membrane-embedded domain of the lens fiber gap junction protein MP70. J Biol Chem 265:13357–13361
Konig N, Zampighi GA (1995) Purification of bovine lens cell-to-cell channels composed of connexin44 and connexin50. J Cell Sci 108 (Pt 9):3091–3098
Le AC, Musil LS (1998) Normal differentiation of cultured lens cells after inhibition of gap junction-mediated intercellular communication. Dev Biol 204:80–96
Li Y, Wang J, Dong B, Man H (2004) A novel connexin46 (GJA3) mutation in autosomal dominant congenital nuclear pulverulent cataract. Mol Vis 10:668–671
Lin JS, Fitzgerald S, Dong Y, Knight C, Donaldson P, Kistler J (1997) Processing of the gap junction protein connexin50 in the ocular lens is accomplished by calpain. Eur J Cell Biol 73:141–149
Lovicu FJ, McAvoy JW (2001) FGF-induced lens cell proliferation and differentiation is dependent on MAPK (ERK1/2) signalling. Development 128:5075–5084
Mackay D, Ionides A, Kibar Z, Rouleau G, Berry V, Moore A, Shiels A, Bhattacharya S (1999) Connexin46 mutations in autosomal dominant congenital cataract. Am J Hum Genet 64:1357–1364
Martinez-Wittinghan FJ, Sellitto C, Li L, Gong X, Brink PR, Mathias RT, White TW (2003) Dominant cataracts result from incongruous mixing of wild-type lens connexins. J Cell Biol 161:969–978
Martinez-Wittinghan FJ, Sellitto C, White TW, Mathias RT, Paul D, Goodenough DA (2004) Lens gap junctional coupling is modulated by connexin identity and the locus of gene expression. Invest Ophthalmol Vis Sci 45:3629–3637
Mathias RT, Riquelme G, Rae JL (1991) Cell to cell communication and pH in the frog lens. J Gen Physiol 98:1085–1103

Mathias RT, Rae JL, Baldo GJ (1997) Physiological properties of the normal lens. Physiol Rev 77:21–50
McAvoy JW, Chamberlain CG (1989) Fibroblast growth factor (FGF) induces different responses in lens epithelial cells depending on its concentration. Development 107:221–228
Menko AS (2002) Lens epithelial cell differentiation. Experimental Eye Research 75:485–490
Miller TM, Goodenough DA (1986) Evidence for two physiologically distinct gap junctions expressed by the chick lens epithelial cell. J Cell Biol 102:194–199
Moreno AP, Saez JC, Fishman GI, Spray DC (1994) Human connexin43 gap junction channels. Regulation of unitary conductances by phosphorylation. Circ Res 74:1050–1057
Paterson CA, Delamere NA (2004) ATPases and lens ion balance. Exp Eye Res 78:699–703
Paul DL, Ebihara L, Takemoto LJ, Swenson KI, Goodenough DA (1991) Connexin46, a novel lens gap junction protein, induces voltage-gated currents in nonjunctional plasma membrane of *Xenopus* oocytes. J Cell Biol 115:1077–1089
Paznekas WA, Boyadjiev SA, Shapiro RE, Daniels O, Wollnik B, Keegan CE, Innis JW, Dinulos MB, Christian C, Hannibal MC, Jabs EW (2003) Connexin 43 (GJA1) mutations cause the pleiotropic phenotype of oculodentodigital dysplasia. Am J Hum Genet 72:408–418
Piatigorsky J (1981) Lens differentiation in vertebrates. A review of cellular and molecular features. Differentiation 19:134–153
Polyakov AV, Shagina IA, Khlebnikova OV, Evgrafov OV (2001) Mutation in the connexin 50 gene (GJA8) in a Russian family with zonular pulverulent cataract. Clin Genet 60:476–478
Rae JL, Bartling C, Rae J, Mathias RT (1996) Dye transfer between cells of the lens. J Membr Biol 150:89–103
Reaume AG, de Sousa PA, Kulkarni S, Langille BL, Zhu D, Davies TC, Juneja SC, Kidder GM, Rossant J (1995) Cardiac malformation in neonatal mice lacking connexin43. Science 267:1831–1834
Rees MI, Watts P, Fenton I, Clarke A, Snell RG, Owen MJ, Gray J (2000) Further evidence of autosomal dominant congenital zonular pulverulent cataracts linked to 13q11 (CZP3) and a novel mutation in connexin 46 (GJA3). Hum Genet 106:206–209
Renwick JH, Lawler SD (1963) Probable linkage between a congenital cataract locus and the Duffy blood group locus. Ann Hum Genet 27:67–84
Rong P, Wang X, Niesman I, Wu Y, Benedetti LE, Dunia I, Levy E, Gong X (2002) Disruption of Gja8 (alpha8 connexin) in mice leads to microphthalmia associated with retardation of lens growth and lens fiber maturation. Development 129:167–174
Sellitto C, Li L, White TW (2004) Connexin50 is essential for normal postnatal lens cell proliferation. Invest Ophthalmol Vis Sci 45:3196–3202
Shiels A, Mackay D, Ionides A, Berry V, Moore A, Bhattacharya S (1998) A missense mutation in the human connexin50 gene (GJA8) underlies autosomal dominant "zonular pulverulent" cataract, on chromosome 1q. Am J Hum Genet 62:526–532
Sosinsky GE (1996) Molecular organization of gap junction membrane channels. J Bioenerg Biomembr 28:297–309
Srinivas M, Costa M, Gao Y, Fort A, Fishman GI, Spray DC (1999) Voltage dependence of macroscopic and unitary currents of gap junction channels formed by mouse connexin50 expressed in rat neuroblastoma cells. J Physiol 517 (Pt 3):673–689
Steele EC, Jr., Lyon MF, Favor J, Guillot PV, Boyd Y, Church RL (1998) A mutation in the connexin 50 (Cx50) gene is a candidate for the No2 mouse cataract. Curr Eye Res 17:883–889
Tamiya S, Dean WL, Paterson CA, Delamere NA (2003) Regional distribution of Na, K-ATPase activity in porcine lens epithelium. Invest Ophthalmol Vis Sci 44:4395–4399
Unger VM, Kumar NM, Gilula NB, Yeager M (1999) Three-dimensional structure of a recombinant gap junction membrane channel. Science 283:1176–1180

Valiunas V, Beyer EC, Brink PR (2002) Cardiac gap junction channels show quantitative differences in selectivity. Circ Res 91:104–111

White TW (2002) Unique and redundant connexin contributions to lens development. Science 295:319–320

White TW, Bruzzone R (1996) Multiple connexin proteins in single intercellular channels: connexin compatibility and functional consequences. J Bioenerg Biomembr 28:339–350

White TW, Paul DL (1999) Genetic diseases and gene knockouts reveal diverse connexin functions. Annu Rev Physiol 61:283–310

White TW, Bruzzone R (2000) Intercellular communication in the eye: clarifying the need for connexin diversity. Brain Res Rev 32:130–137

White TW, Bruzzone R, Goodenough DA, Paul DL (1992) Mouse Cx50, a functional member of the connexin family of gap junction proteins, is the lens fiber protein MP70. Mol Biol Cell 3:711–720

White TW, Bruzzone R, Wolfram S, Paul DL, Goodenough DA (1994) Selective interactions among the multiple connexin proteins expressed in the vertebrate lens: the second extracellular domain is a determinant of compatibility between connexins. J Cell Biol 125:879–892

White TW, Goodenough DA, Paul DL (1998) Targeted ablation of connexin50 in mice results in microphthalmia and zonular pulverulent cataracts. J Cell Biol 143:815–825

White TW, Sellitto C, Paul DL, Goodenough DA (2001) Prenatal lens development in connexin43 and connexin50 double knockout mice. Invest Ophthalmol Vis Sci 42:2916–2923

Willecke K, Eiberger J, Degen J, Eckardt D, Romualdi A, Guldenagel M, Deutsch U, Sohl G (2002) Structural and functional diversity of connexin genes in the mouse and human genome. Biol Chem 383:725–737

Willoughby CE, Arab S, Gandhi R, Zeinali S, Arab S, Luk D, Billingsley G, Munier FL, Heon E (2003) A novel GJA8 mutation in an Iranian family with progressive autosomal dominant congenital nuclear cataract. J Med Genet 40:e124

Zampighi GA, Simon SA, Hall JE (1992) The specialized junctions of the lens. Int Rev Cytol 136:185–225

9 Connexin Modulators of Endocrine Function

Philippe Klee, Nathalie Boucard, Dorothée Caille, José Cancela, Anne Charollais, Eric Charpantier, Laetitia Michon, Céline Populaire, Manon Peyrou, Rachel Nlend Nlend, Laurence Zulianello, Jacques-Antoine Haefliger, Paolo Meda[1]

Abstract The emergence of multicellular organisms has necessitated the specialization of short- and long-range chemical signaling systems, including that provided by the endocrine system. Conversely, the existence of an endocrine system conceptually demands a multicellular organism, to which proper signaling usually also imposes a multicellular gland. Accordingly, the secretory cells of all endocrine glands have developed mechanisms for interacting with adjacent and distant cells. With evolution, such mechanisms have diversified and have been progressively integrated in a complex regulatory network, whereby individual endocrine cells sense the state of activity of their neighbors and regulate accordingly their own level of functioning. A consistent feature of this network is the expression of connexin-made channels between the hormone-producing cells of all glands so far investigated in vertebrates. In a few instances, these channels have also been documented between the endocrine cells and nearby target cells. Here, we have reviewed the distribution of connexins in the mammalian endocrine system, and have discussed the recent evidence pointing to the participation of these proteins in the functioning of endocrine cells, and on the action of hormones on specific target cells.

9.1 Introduction

The appearance of multicellular systems, some 800 million years ago, has been dependent on the ability of cells to develop nervous and endocrine systems that ensure chemical signaling between adjacent and distant cells. In turn, proper signaling by these systems to the multicellular organism has imposed the development of signaling units sizably larger than the unicellular exocrine systems, which are considered their primordial precursors (LeRoith 1990; Stoka 1999). Thus, most of the vital endocrine functions are now dependent on the proper functioning of multicellular glands, which implies the existence of a system coordinating the function of the hormone-producing cells. With evolution, such a coordination system has developed in a complex network in which signals diffusing in the intercellular spaces interplay with the signaling cascades dependent on the surface proteins that concentrate at cell contacts (Meda and Bosco 2001; Serre-Beinier et

[1]Department of Cell Physiology and Metabolism, University of Geneva, School of medicine C.M.U., 1 rue Michel Servet, CH-1211 Geneve 4, Switzerland, E-mail: Paolo.Meda@medecine.unige.ch

Gap Junctions in Development and Disease
(ed. by E. Winterhager)

al. 2002), so that individual hormone-producing cells sense the state of activity of their neighbors and regulate accordingly their own level of functioning. In vertebrates, a consistent feature of this network are the cell-to-cell channels made of connexin proteins (Meda et al. 1984; Meda 1996, 1997; Serre-Beinier et al. 2002). As compared to other forms of cell-to-cell communication, connexin-dependent cell-to-cell coupling achieves a rapid intercellular equilibration of electrochemical gradients of cytoplasmic ions and molecules which may be crucial for hormone secretion (Meda and Spray 2000). In turn, this equilibration ensures the recruitment and coordination of a population of cells that, taken individually, would otherwise function asynchronously (Bosco et al. 1989, 1991, 1998; Meda and Spray 2000). Although secretory cells of a given type are generally considered to form homogeneous populations, ample evidence indicates that they actually differ in several respects, including the ability to biosynthesize, store, and release hormones (Salomon and Meda 1986; Bosco et al. 1988, 1989, 1991).

A number of studies have addressed the function of gap junctions and coupling in endocrine glands and questioned the requirement, if not the advantage of the connexin-dependent signalling. Thus, a large body of circumstantial evidence has accumulated, indicating that gap junctions are required for the fine regulation of the biosynthesis, storage, and release of various hormones, as well as for their proper action on target cells (Petersen 1980; Meda et al. 1984; Meda 1996, 1997; Munari-Silem and Rousset 1996; Meda and Spray 2000; Serre-Beinier et al. 2002, Murray et al. 2003). The more recent unraveling of the connexin family, and the availability of novel cell and molecular biology tools to test the distribution and function of these proteins, now permit us to experimentally test the implications generated by the first studies. For example, different hormone-producing cells express distinct connexins, in a pattern that is highly conserved across species, while systematically excluding other connexin forms, and recent studies are beginning to unravel the molecular mechanisms that lead to such specific choices. Novel studies, using cell-specific alterations of connexins in transgenic mice, are also beginning to provide clues that the proper expression of connexins, and/or the ionic and metabolic cell coupling these proteins permit is implicated in the fine in vivo tuning of hormone production and secretion. More intriguingly, recent work implies that connexin signaling may also be involved between selected endocrine cells and their cognate, closely associated target cells, a hitherto unsuspected mechanism for the two-way interaction between signaling and effector cells. Here, we have reviewed this body of evidence, with particular regard to the studies which have investigated the connexin-dependent signaling of vertebrate endocrine cells since the publication of previous reviews on this topic. The reader is referred to these previous publications for a more complete coverage of the early studies on this field (Petersen 1980; Meda et al. 1984; Meda 1996, 1997; Munari-Silem and Rousset 1996; Meda and Spray 2000; Serre-Beinier et al. 2002; Murray et al. 2003).

9.2 Endocrine Cells Are Connected by Selected Connexins

The secretory cells of most endocrine glands, and of their putative precursor pheromone glands, usually express only one connexin isoform of either the α or the γ group (Table 9.1). Cx43 is the most widespread isoform. A few endocrine cells and most endocrine neurons have selected Cx36, whereas Cx40 is so far restricted to the renin-producing cells of the kidney (Table 9.1). With the notable exception of the thyroid, which embryologically develops as an exocrine gland and maintains throughout life an architectural organization reminiscent of this primordial exocrine function (Meda et al. 1993; Skoda 1999), no endocrine cell can be positively shown to express Cx32, even though a few of these cells have selected Cx26, another connexin of the β group (Table 9.1). Conversely, the parenchymal cells of exocrine glands usually express various combinations of β connexins, usually Cx32 (Meda et al. 1993). Recent studies provide clues about the mechanism underlying this different choice. Thus, the characterization of the 5' regulatory region of the human Cx36 gene has led to

Table 9.1. Distribution of connexins in endocrine glands

Type of hormone	Organ	Cell	Hormone	Connexin
Peptide	Kidney	Myoepithelial cells of afferent arteriole	Renin	Cx40
	Heart	Auricular myoendocrine cells	Atrial natriuretic hormone (ANH)	Cx43 Cx45
	Pancreas	β-Cells	Insulin	Cx36
		α-Cells	Glucagon	Cx36?, Cx43?
	Pineal	Pinealocytes	Melatonin	Cx26
	Thyroid	C cells	Calcitonin	Cx43
	Parathyroid	Principal cells	Parathormone	Cx43

Table 9.1. (*continued*)

Type of hormone	Organ	Cell	Hormone	Connexin
	Pituitary (adenohypophysis)	Acidophil cells	Growth hormonel (GH), prolactin (Prl)	Cx43, Cx26
		Basophil cells	Adrenocorticotropic hormone (ACTH)	Cx43
	Hypothalamus	Terminals of neuroendocrine cells in neurohypophysis	Oxytocin, Vasopressin	Cx36
		Neuroendocrine cells	Gonadotropin releasing hormone (GnRH), corticotropin releasing hormone (CRH), thyrotropin releasing hormone (TRH), growth hormone releasing hormone (GHRH), somatostatin, prolactin release inhibiting hormone (PIH), prolactin releasing factor (PRF),...	Cx36
Glycoprotein	Thyroid	Follicular cells	Triiodothyronine (T_3), thyroxine (T_4)	Cx43, Cx32, Cx26
	Pituitary (adenohypophysis)	Basophil cells	Follicle stimulating hormone (FSH); luteinizing hormone (LH); thyroid stimulating hormone (TSH)	Cx43
	Placenta	Throphoblast cells	Human chorionic gonadotropin (hCG), human chorionic somatomammotropin	Cx43
Steroid	Adrenal (cortex)	Cortical cells	Corticoids, mineralocorticoids, androgens	Cx43
	Testis	Leydig cells	Testosterone	Cx43
	Ovary	Thecal cells	Estradiol	Cx43
		Luteal cells	Progesterone	Cx43
Catecholamines	Adrenal (medulla)	Chromaffin cells	Epinephrine, norepinephrine	Cx36
Pheromone	Skin	Sebaceous cells	Sebum components	Cx43
	Preputial glands	Secretory cells	Farnesenes	Cx43

the identification of a promoter region sufficient to restrict the expression of the cognate connexin to insulin-secreting cells and neurons. Indeed, within this regulatory region, a conserved neuron-restrictive silencer element binds the NRSF/REST factor which functions as a potent repressor of neuronal genes. NRSF/REST is widely expressed in most cell types, but insulin-secreting cells and neurons, and their ectopic expression reduces that of Cx36 in insulin-producing cells (Martin D et al. 2003). In a parallel study, mice lacking the basic helix-loop-helix transcription factor Mist1, which is normally expressed in most exocrine cell types, were shown to feature altered architecture and function of pancreatic acini, due, at least in part, to a transcriptional downregulation of the Cx32 gene (Rukstalis et al. 2003). Hence, differential transcriptional controls of distinct connexin genes account for the different expression of gap-junction proteins by endocrine and exocrine cells (Meda et al. 1993).

9.3 Different Endocrine Cells Express Different Connexin Patterns

The reason why different connexin species have been selected by different endocrine cells is less clear. Thus, this selection cannot be directly related to either the embryological origin, chronology of development, or architecture of the adult glands. It cannot be related either to the biochemical nature of the hormones produced by each gland, the rate of their secretion, or the control of their release and biosynthesis (Table 9.1). However, since different connexins form channels with distinct conductance, permeability, and regulatory characteristics (Veenstra et al. 1998; Goldberg et al. 1999; Niessen et al. 2000; Harris et al. 2001), it is plausible that, during evolution, endocrine cells have selected the sets of channels that, by favouring the intercellular exchange of specific signals while preventing the diffusion of others, better fit the requirements of their own secretory machinery.

9.3.1 Cells Producing Peptide Hormones

The insulin-producing β-cells of pancreatic islets (Fig. 9.1) are coupled by Cx36 channels (Serre-Beinier et al. 2000; Degen et al. 2004; Theis et al. 2004), which are similar, though not identical to the homomeric Cx36 channels connecting neurons (Moreno et al. 2004), possibly as a result of the close interaction between Cx36 and ZO-1 within β-cell gap junctions (Li et al. 2005; Figs. 9.2, 9.3). Several lines of evidence indicate that β-cell

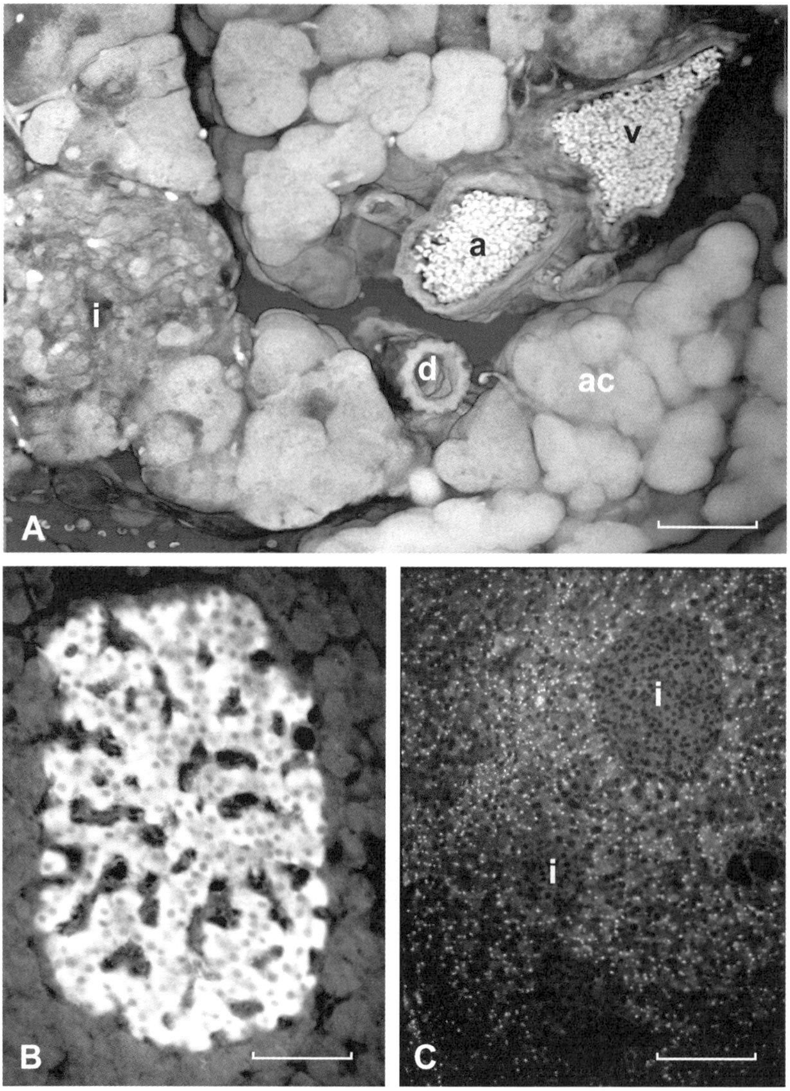

Fig. 9.1A–C. The endocrine cells of the pancreas are organized within microislets. **A** A three-dimensional confocal microscopy reconstruction of a slice of pancreas reveals that the gland closely associates exocrine acini (*ac*) secreting digestive enzymes, to about 1 million (in humans) islets of Langerhans (*i*), which collectively form the endocrine pancreas. An excretory duct (*d*), an artery (*a*) and a vein (*v*) are seen in the connective spaces joining the two secretory units. *Bar* 90 µm. **B** After immunostaining for insulin, a pancreatic islet reveals the close, cord-like association of the insulin-producing β-cells that form the bulk of the microorgan. The dark spaces between the β-cell cords are the capillaries into which islet hormones are secreted. *Bar* 130 µm. **C** After immunostaining for Cx32, abundant levels of this β-type connexin are observed in the exocrine acini. In contrast, the protein is not detected within the endocrine islets (*i*). *Bar* 270 µm

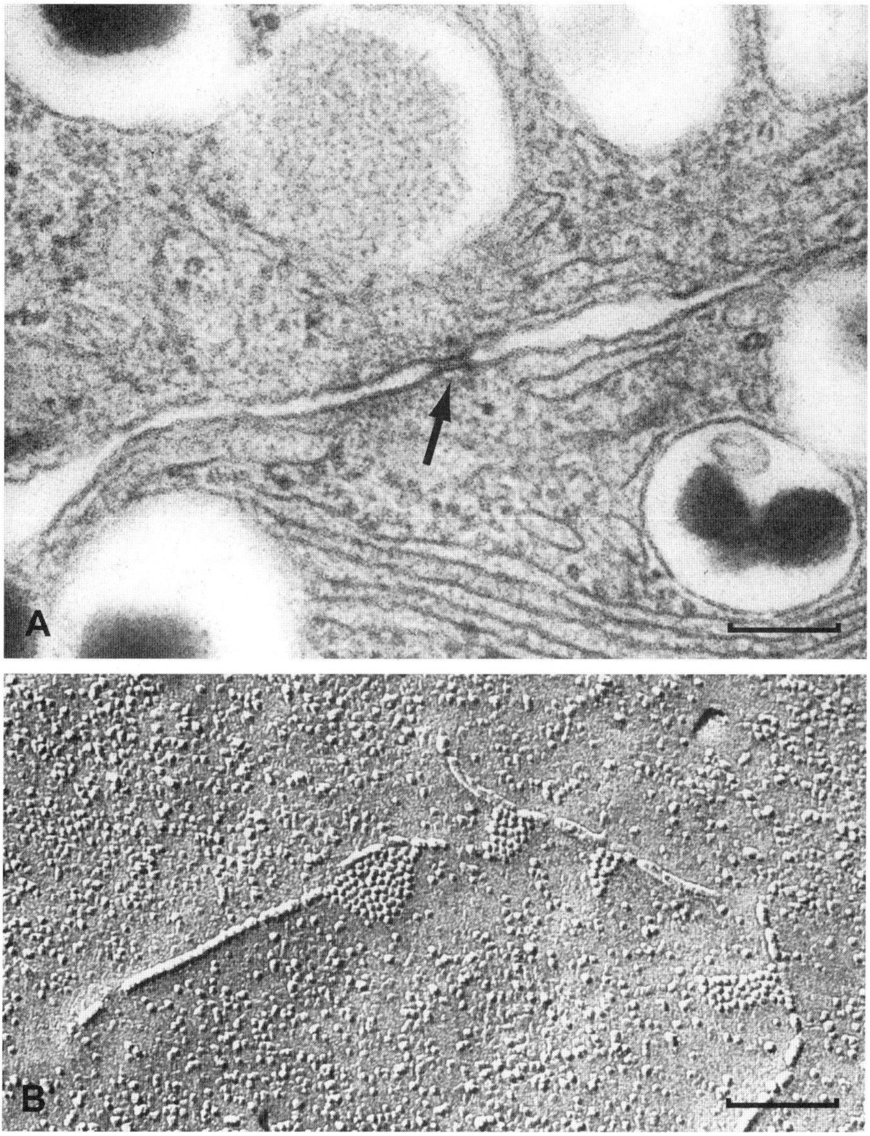

Fig. 9.2A, B. Pancreatic β-cells are connected by gap and tight junctions. **A** Electron microscopy shows that the membranes of adjacent pancreatic β-cells, identified by the typical ultrastructure of proinsulin- and insulin-containing secretory granules, come into contact at sites of minute gap junctions (*arrow*). *Bar* 200 nm. **B** Freeze-fracture electron microscopy reveals that the gap-junction plaques connecting β-cells typically contain a few dozen connexons, are clustered in restricted domains of the cell membrane, and are usually connected to tight junction fibrils. *Bar* 50 nm

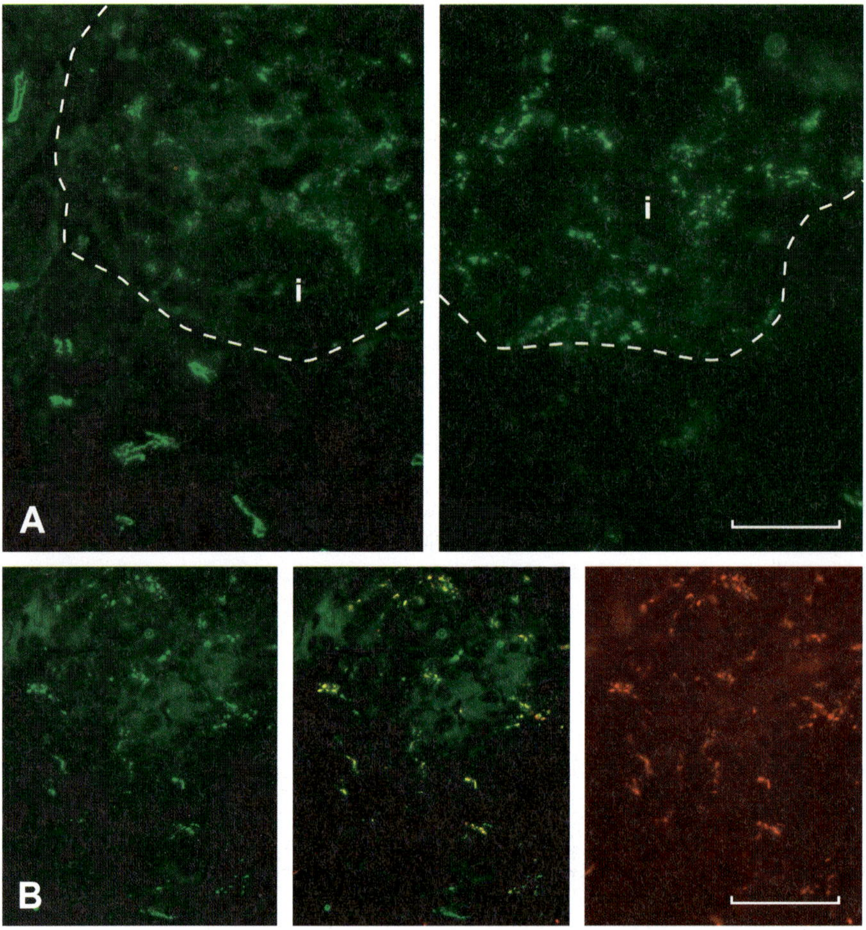

Fig. 9.3. Pancreatic *β*-cells coexpress Cx36 and ZO-1. **A** (*left*) Immunostaining reveals the punctate distribution of ZO-1 in the *β*-cells of a pancreatic islet (*i*; the border of the microorgan is indicated by the *dotted line*) and the continuous, linear arrangement of the tight junction-associated protein along the lumen of exocrine acini. **A** (*right*) Immunostaining reveals the punctate distribution of Cx36 in the *β*-cells of a pancreatic islet (*i*), but fails to detect the protein within the surrounding exocrine acini. *Bar* 40 μm. **B** The dual immunostaining of a pancreatic islet for ZO-1 (*right*) and Cx36 (*left*), reveals the close association of the two proteins within the *β*-cell membrane (*center*). *Bar* 55 μm

coupling is involved in the control of insulin secretion. Thus, cell lines featuring defective insulin secretion do not express connexins (Vozzi et al. 1995; Calabrese et al. 2003), and single *β*-cells show alterations in the transcription and secretion of insulin which are rapidly corrected after restoration of *β*-cell contacts (Salomon and Meda 1986; Bosco and Meda

1991; Philippe et al. 1992) and which are mimicked by the pharmacological blockade of connexin channels (Meda et al. 1990). Investigating Cx36-KO transgenic mice (Guldenagel et al. 2001), we observed that loss of Cx36 did not alter the normal development of pancreatic islets which, however, comprised uncoupled β-cells lacking gap junctions (Ravier et al. 2005). These cells also failed to show the normal intercellular synchronization of $[Ca^{2+}]_i$ transients, which is seen during glucose stimulation of β-cells expressing Cx36. As a result, islets lacking Cx36 did not release insulin in the normal pulsatile fashion (Ravier et al. 2005). Furthermore, islets lacking Cx36 showed an increased basal release of insulin, explaining why no significant increase in hormonal output was observed when the pancreas was challenged by concentrations of glucose expected in the postprandial period (Fig. 9.4). These alterations, which are reminiscent of those observed in type II diabetics, show that Cx36-dependent signaling is essential for proper regulation of insulin release (Ravier et al. 2005), thus extending to the native islet connexin and to the in vivo situation, the observations previously made after the transgenic expression of the islet-ectopic Cx32 (Charollais et al. 2000), or in several in vitro models (Calabrese et al. 2003, 2004; Caton et al. 2003; LeGurun et al. 2003). The latter experiments were instrumental to document the specificity of the secretion control achieved by the Cx36 signaling, which cannot be achieved either by other connexins (Caton et al. 2003), or by E-cadherin (Calabrese et al. 2004). They also revealed a tight relationship between adequate levels of Cx36 and proper insulin secretion, by showing that, in terms of secretory effects, the excess of Cx36 is as deleterious as the lack of the connexin (Caton et al. 2003; Calabrese et al. 2004).

These events have been less investigated in other peptide-producing endocrine glands. Thus, pinealocytes have been reported to be coupled by Cx26 channels that are upregulated in vitro by norepinephrine, possibly to improve melatonin secretion (Saez et al. 1991).

Hypothalamic neuroendocrine cells are also coupled (Taylor and Dudek 1982; Matesic et al. 1993, 1997), presumably by Cx36, the only connexin which has so far been convincingly shown to be expressed by neurons (Nagy et al. 2004). Electrophysiological studies testing gap-junction blockers have indicated that this coupling is implicated in the control of the pulsatile release of several neuropeptides, including GHRH (Shinohara et al. 2001), GnRH (Vazquez-Martinez et al. 2001) and LHRH (Terasawa 2001).

Eventually, the endocrine cell types that form the anterior pituitary are coupled (Morand et al. 1996; Fauquier et al. 2001; Levavi-Sivan et al. 2005) via channels made predominantly of Cx43 and Cx26 (Meda et al. 1993; Yamamoto et al. 1993). This coupling increases the intercellular synchronization of Ca^{2+} transients (Guerineau et al. 1998). Ca^{2+} waves are also propagated throughout the entire gland via the network of folliculo-

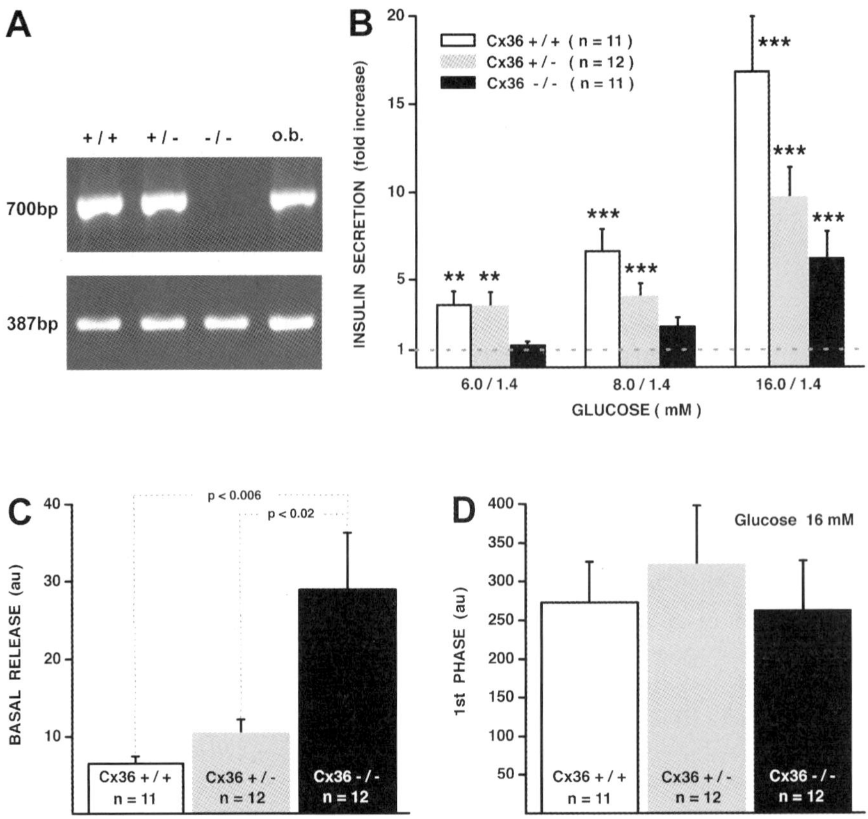

Fig. 9.4A–D. Pancreatic β-cells lacking Cx36 show selective defects of insulin secretion. **A** RT-PCR analysis of RNA extracted from pancreatic islets distinguishes mice knocked out for the Cx36 gene (−/−), from Cx36-expressing wild type (+/+) and heterozygous transgenic littermates (+/−). The Cx36 and cyclophillin transcripts are detected as a 700-bp and a 387-amplicon; respectively. Olfactory bulbs (*o.b.*) served as positive controls. **B** KO-Cx36 −/− mice, but not +/+ and +/− littermates ($^{**}\ p < 0.01$, $^{***}\ p < 0.001$), failed to significantly increase insulin release when stimulated with physiologically relevant (≤ 8.0 mM) glucose concentrations. **C** This defect resulted from a significantly higher basal output of the hormone in mice lacking Cx36. **B, D** However, these animals responded like +/+ and +/− littermates when stimulated by a supraphysiological (16.0 mM) glucose concentration

stellate cells that are extensively coupled by Cx43 channels (Fauquier et al. 2001, Shirasawa et al. 2004). Seasonal changes in the expression of pituitary Cx43 have been associated with changes in prolactin secretion (Vitale et al. 2001).

9.3.2
Cells Producing Glycoprotein Hormones

Contrasting with most, if not all other types of endocrine cells, the follicular cells of thyroid are coupled by Cx32-, Cx26-, and Cx43-made channels (Meda et al. 1993; Guerrier et al. 1995). Cx32 and coupling are lost with passages in culture, together with the ability of the cells to form follicular structures (Munari-Silem et al. 1994; Green et al. 2001). These alterations can be reverted by exposing cell monolayers to TSH (Munari-Silem et al. 1994) or by transfection of communication-incompetent thyrocytes with Cx32, but not Cx43 (Tonoli et al. 2000). While these data implicate Cx32 in the morphogenesis of thyroid follicles, no obvious thyroid defect has been reported in mice knockout for the Cx32 gene (Nelles et al. 1996; Houghton et al. 1999). The involvement of connexins in thyroid secretion is supported by the observation that TSH stimulation increases the coupling of thyrocytes in a time- and concentration-dependent manner (Munari-Silem et al. 1991), whereas loss of coupling, due to a Cx32 mutation, reduces the release of thyroxin (Green et al. 2001). Furthermore, transfection of cell lines for Cx32 also resulted in increased expression of the thyroglobulin gene (Statuto et al. 1997).

Cx43 connects the cells of the placental cytotrophoblast, and these cells to those of the syncytio-trophoblast (Risek and Gilula 1991; Winterhager et al. 1991; Cronier et al. 1994). Recent studies have reported that the pharmacological blockade of gap junctions, as well as the antisense interference with the Cx43 transcript uncouple cytotrophoblast cells, impairing their fusion and, thus, the formation of the syncytio-trophoblast (Cronier et al. 2003; Frendo et al. 2003). These alterations were associated with a decrease in both the expression of trophoblast-specific genes, including those coding for β-hCG and human chorionic somatomammotropin (Cronier et al. 2003; Frendo et al. 2003), and the secretion of the former placental hormone (Frendo et al. 2003).

9.3.3
Cells Producing Steroid Hormones

The endocrine cells of the adrenal cortex are coupled by Cx43 channels (Meda et al. 1993; Murray et al. 1995), particularly in the glucocorticoid- and the androgen-producing regions (Davis et al. 2002). In vitro experiments testing drugs thought to block Cx43 channels have documented an impaired secretion of cortisol from clusters of adrenal cells, but not single cells, stimulated by ACTH, but not cAMP (Munari-Silem et al. 1995). In agreement with these findings, transfection of adrenal cells with a Cx43 antisense construct also resulted in the inhibition of ACTH-induced steroid secretion

(Oyoyo et al. 1997). These data are consistent with the observation that ACTH increases the expression of Cx43 both in vivo and in vitro (Murray et al. 2003). However, and unexpectedly, another gap-junction blocker has been recently reported to stimulate steroid production, via the activation of both an extracellular signal-related kinase and a calcium/calmodulin-dependent kinase, i.e., by two pathways distinct from the protein kinase A-dependent pathway which normally controls steroidogenesis of adrenal cells (Huang et al. 2003).

Cells derived from the testosterone-producing Leydig cells of testis express Cx43 in culture and are coupled up to confluence, under which conditions Cx43 expression and coupling decrease, presumably as a result of activation of pathways dependent on protein kinase A and C (Goldenberg et al. 2003). Interestingly, a comparable decrease in coupling was observed during LH-induced stimulation of testosterone secretion, suggesting a tonic inhibitory influence of connexin signaling on the output of the steroid hormone (Goldenberg et al. 2003).

Cells of the ovarian corporea lutea are also connected by Cx43 (Mayerhofer and Garfield 1995; Khan-Dawood et al. 1996). Pharmacological treatments enhancing this coupling increase progesterone secretion, whereas treatments uncoupling luteal cells decrease steroid release (Grazul-Bilska et al. 2001). Consistent with a relationship between connexin signaling and steroid production, the experimental inhibition of Cx43 expression decreased the LH-induced steroid secretion of luteal cells (Khan-Dawood et al. 1998). Strikingly, mice lacking Cx37, the connexin that forms gap junctions between oocytes and granulosa cells, develop numerous abnormal corpora lutea (Simon et al. 1997).

The thecal cells that derive from the endocrine differentiation of the granulosa cells making the ovarian follicles are also connected by Cx43 (Risek et al. 1990; Simon et al. 1997). The expression of this connexin varies with the stage of follicle development, consistent with an involvement of connexins in the endocrine function of thecal cells (Johnson et al. 2002; Kidder and Mhawi 2002; Klinger and De Felici 2002).

9.3.4 Cells Producing Catecholamines

Chromaffin cells of the adrenal medulla are joined by Cx36-made gap junctions (A.O. Martin et al. 2001; Degen et al. 2004), which account for the cell-to-cell spreading of the $[Ca^{2+}]_i$ transients that are driven by action potentials (A.O. Martin et al. 2001). This spreading is enhanced after exposure of a few cells to nicotine, which also triggers the release of catecholamines, indicating that connexin signaling can amplify the secretion of epinephrine and norepinephrine induced by the synaptic activation of individual cells (A.O. Martin et al. 2001). Interestingly, both the pharmacological blockade of synaptic transmission, as well as the surgical denervation of the adrenal glands resulted in increased coupling of chromaffin cells, an effect which was also observed in newborn rats at a time when the synaptic transmission of the adrenal gland had not yet fully matured (A.O. Martin et al. 2003).These results indicate that the Cx36-dependent signaling of chromaffin cells is tonically inhibited by cholinergic synaptic inputs (A.O. Martin et al. 2003).

9.3.5 Cells Producing Pheromones

Pheromone-producing glands are considered to be the phylogenetic precursors of vertebrate multicellular endocrine systems, combining the ancestral exocrine mode of secretion with the evolved chemical signalling provided by hormones (LeRoith 1990; Stoka 1999). Whether Cx43, the only connexin so far described in the pheromone-producing cells of rodents (Meda et al. 1993) and of their human sebaceous gland counterparts (Salomon et al. 1994), plays any role in the secretion of these volatile chemical messages has not yet been investigated. However, when viewed in the background of the short half-life of connexins, the presence of unusually large levels of Cx43 between sebocytes argues for some essential physiological role throughout their life time. The easy accessibility of sebaceous glands should facilitate the direct experimental testing of this hypothesis.

9.4 Some Endocrine Cells Share Connexins with Their Targets

Most targets of hormones are at a sizable distance from the gland that produces the endocrine products, usually preventing the establishment of connexin-dependent signaling between the secretory cells and the cognate effector cells. However, connexins may be involved in the cross-talk between these two cell types in a few instances. The first instance is that of paracrine signaling, whereby an endocrine cell releases a hormone that acts on nearby cells. An example is provided by pancreatic islets, in which the insulin released by β-cells acts on receptors of close by, sometimes adjacent α-, δ- and PP-cells, and, vice versa, glucagon and somatostatin act on receptors of β-cells (Patel et al. 1982; Kawai et al. 1982). These different islet cell types are known to be joined by gap junctions (Orci et al. 1975) and to be electrically coupled (Meda et al. 1986; Quesada et al. 2003), even though this coupling appears low, and possibly inconsistent (Kanno et al. 2002). No study has as yet tested whether this heterologous coupling modulates the endocrine function of pancreatic islets.

A second challenging situation is that of the numerous cell types which form the juxtaglomerulus apparatus, the endocrine regions of the kidney nephrons that collectively play a central role in the control of blood pressure (Persson 2003). The renin-producing cells of the afferent arteriole, which are central players of this apparatus, share Cx40 junctions with both adjacent endocrine cell and endothelial cells (Haefliger et al. 2001). Rats made hypertensive after the clipping of one renal artery, showed a selective increase in kidney renin and Cx40, implying that the connexin is implicated in the control of renin secretion and/or in the vasomotor control of kidney vessels (Haefliger et al. 2001). Thus, connexin signaling may integrate the chemical and mechanical signals that concur to control blood pressure in this renin-dependent model. Since this model also results in changes of aortic Cx43 (Haefliger et al. 2004), we revisited it using transgenic mice in which the coding region of this connexin was replaced by that of Cx32 (Plum et al. 2000). As expected, wild-type and heterozygous mice, which showed a normal distribution of Cx43, became hypertensive as a result of increased plasma renin levels. In contrast, homozygous littermates, in which Cx32 had fully replaced Cx43, retained a normal blood pressure and control levels of circulating renin (Haeliger et al. 2005), providing additional support for a hitherto unsuspected mechanism, whereby altered connexin signaling between endothelial cells modifies the functioning of the renin-secreting cells.

The case of the juxtaglomerular apparatus may not be unique. Thus, in the pituitary, the non endocrine folliculo-stellate cells cross-talk with distant endocrine cells by establishing gap junctions that allow for a rapid and extensive propagation of waves of cytosolic calcium (Fauquier et al. 2001). This arrangement provides an efficient mechanism to orchestrate the function of endocrine cell types that are scattered throughout the gland (Fauquier et al. 2001).

9.5 Nonsecretory Functions of Connexins in Endocrine Glands

Analysis of transgenic mice lacking one or two of the connexins that are normally expressed by endocrine cells have failed to reveal obvious alterations in the pre-natal development, morphogenesis and differentiation of endocrine glands (Houghton et al. 1999; Guldenagel et al. 2001; Gros et al. 2004; Theis et al. 2004). However, altered growth of the endocrine pancreas was detected by morphometric analysis and measurements of hormone content in transgenic mice whose insulin-producing β-cells were selectively forced to ectopically express Cx32 (Charollais et al. 2000). The data suggest a role of connexin-dependent signaling in the postnatal growth of pancreatic islet cells, a hypothesis verified in still another transgenic mice model in which β-cells were selectively induced to over-express the native Cx36 (Klee et al. 2004).

Transfection of Cx32 in thyroid-derived, communication-incompetent cell lines markedly reduces their proliferation rate (Tonoli et al, 2000). A new report now indicates that this effect is somewhat specific, inasmuch as restoration of intercellular communication by stable expression of Cx43 did not modify the growth of the same cell types (Flachon et al. 2002). This study also indicates that the slow-down of cell proliferation induced by Cx32 was accounted for by a prolonged G1 phase, possibly mediated by proteins of the Cip/Kip cyclin-dependent kinase inhibitor family (Flachon et al. 2002).

Adrenocortical cells normally respond to ACTH stimulation by increasing steroid production and Cx43 expression, and by decreasing their growth (Murray et al. 2003). Pharmacological inhibition of their gap junctions produces the opposite effect, i.e., increases cell growth and decreases ACTH-stimulated steroidogenesis, without altering the levels of Cx43 (Shah et al. 2001). Furthermore, exposure of adrenal cells to a Cx43 antisense cDNA, which reduced these levels, also resulted in a decreased steroidogenic response, and in an increase in the cell growth rate (Shah et al. 2001). Thus, in different endocrine cells, connexins contribute to control hormone biosynthesis and release, as well as the size of the endocrine cell population. As yet, however, the relevance of the later control for primary endocrine cell in vivo has only been shown in the case of pancreatic islets (Charollais et al. 2000).

Innovative experiments have recently revealed that early-passage cultures of rat thyroid cells, which express functional Cx32 channels, are more resistant to γ irradiation than later-passage cultures which express a truncated, nonfunctional form of Cx32. Strikingly, this beneficial effect was not observed after proton irradiation (Green et al. 2002). These observations may be related to those made on insulin-producing β-cells after exposure to drugs that experimentally reproduce the massive cell death observed in pancreatic islets at the onset of type I diabetes (Klee et al. 2004). The data indicate a direct protective role of Cx36, Cx32 and Cx43 in vitro and, more importantly, provide the very first evidence that this protection operates also in vivo. Thus, whereas transgenic mice overexpressing either the native islet Cx36 or the islet ectopic Cx32, were fully protected against the cytotoxic effects of streptozotocin and alloxan, had a normal insulin content and remained normoglycemic, transgenic mice lacking Cx36 were highly sensitized to the effects of the drugs and became rapidly hyperglycemic due to the loss of most β-cells and pancreatic insulin (Klee et al., 2004). Strikingly, this sensitization appear to reflect a gene-dosage dependence, inasmuch as heterozygous Cx36-KO mice showed alterations intermediate between those of wild-type littermates expressing normal levels of the connexin and homozygous littermates lacking the protein (Klee et al. 2004). Together, these data strongly indicate that connexin channels significantly contribute to the resistance of endocrine cells to cytotoxic conditions, either by enhancing their resistance to damaging conditions and/or by favoring efficient cell repair after the insult.

9.6 Hormones and Connexins

Many hormones have been reported in a variety of models to affect the transcription, mRNA stability, translation and cytoplasmic trafficking of connexins, as well as the gating and regulation of the cell-to-cell channels formed by these proteins. The effects observed in different studies are variable, and sometimes opposite, presumably reflecting the essential influence of cell types, physiological states at the time of the experiment and different connexin patterns. Since the previous reviews, which provided a comprehensive coverage of this topic (Petersen 1980; Meda et al. 1984; Stagg and Fletcher 1990, Munari-Silem and Rousset 1996; Meda 1996, 1997; JC Saez et al. 1998; Murray et al. 2003), several observations have been made.

Implantation of mammalian embryos in the uterus occurs in the so-called receptive phase, during which steroid hormones and local embryonic signals suppress the expression of endometrial Cx26 and Cx43. Later, connexin expression is locally reinduced by the implanting blastocyst, presumably as a result of factors secreted by the growing throphoblast (Grummer et al. 1998). Ovariectomized rats treated with different ratios of 17beta-estradiol and progesterone revealed a dose- and time-dependent regulation of the expression of Cx26 (the main connexin of the endometrial epithelium) and Cx43 (the prominent connexin of uterine stroma and myometrium), but not Cx32, indicating that hormonal ratios mimicking pregnancy conditions differentially regulate α- and β-type connexins (Grummer et al. 1999). A recent study now indicates that, during preimplantation, this induction is due to estrogens, via the activation of a pathway initiated at estrogen receptors alpha. However, during the subsequent embryo implantation and decidualization, endometrial connexins are upregulated by an estrogen receptor-independent pathway, possibly implicating catechol estrogen, prostaglandin F(2α) and interleukin-1β (Grummer et al. 2004).

LH and FSH control the coupling of ovarian thecal cells, by regulating Cx43 expression at the transcriptional, translational and post-translational levels (Granot and Dekel 2002). Strikingly, the effects of the two gonadotrophins vary as a function of the developmental stage of the ovarian follicles (Johnson et al. 2002).

Throughout the estrous cycle, luteal cells are coupled by Cx43-made channels that are modulated by agonists and antagonists of cytosolic calcium, cAMP and protein kinase C (Grazul-Bilska et al. 2001). Thus, post-translational phosphorylation of Cx43 is likely to be relevant for the function of luteal cell channels, as it is the case in many other cell types (Saez et al. 1998; Warn-Cramer and Lau 2004).

The levels of Cx43 increase with gestational age in the adrenal cortex of sheep fetuses, and cortex size is stimulated by ACTH, but not dexamethasone infusion (McDonald et al. 2003). Under these conditions, the cellular distribution of Cx43 is also altered, possibly implicating connexin signaling in the long-term effects of ACTH, which are critical for both the maturation of adrenal glands and initiation of parturition (McDonald et al. 2003).

Administration of steroids to castrated rats resulted in different changes in the number of gap junctions connecting folliculo-stellate cells of the anterior pituitary, depending on the type of steroid tested and the sex of the animals (Sakuma et al. 2003). In a parallel study, hydrocortisone was further reported to delay gap-junction formation, a change that was concomitant with an altered phenotype of folliculo-stellate cells (Sakuma et al. 2001).

9.7 The Future

With evolution, the release of hormones by endocrine glands has come under the control of a complex network that integrates multiple regulatory mechanisms. The studies summarized above provide compelling evidence that, in this network, the cell-to-cell coupling mediated by connexins plays a physiologically relevant role in the proper regulation of hormone gene expression, biosynthesis and release, as well as in the maintenance of adequate numbers of hormone-producing cells. However, the reason why connexin-dependent coupling is required, if not obligatory for proper endocrine function remains to be elucidated, as is the mechanism underlying this requirement. Recent work on insulin- (Calabrese et al. 2003; Ravier et al. 2005) and catecholamine-producing cells (A.O. Martin et al. 2001, 2003) implicates the intercellular synchronization of secretagogue-induced transients of free cytosolic Ca^{2+}, as a key event in this mechanism.

The challenge for the years to come is manifold. First, we have to identify the signals that endocrine cells exchange through connexin channels and that affect their function. Many of the endogenous signal molecules (cAMP,

Ca^{2+}, K^+, IP_3,...) and metabolites (glycolytic intermediates, nucleotides,...) that permeate connexin channels are also important players for many types of secretion (Petersen 1980; Stagg and Fletcher 1990; Meda 1996, 1997; Meda and Spray 2000; Niessen et al. 2000; Meda and Bosco 2001), complicating the identification of the signal(s) that specifically couple the stimulus to an endocrine cell to the activation of the connexin-dependent pathway, and this pathway, in turn, to the secretion of a hormone. Furthermore, recent work indicates that connexins may control cell functions independently of their mediation of cell-to-cell communication (Stout et al. 2004). The underlying mechanism (hemi-connexin channels for the passage of selected second messengers? interaction of connexins with other proteins and/or specific transcriptions factors? other?) remains to be elucidated. It is certainly conceivable that connexins may not need to assemble into functional gap junction channels to exert a biological effect. However, much additional data are still required for the validation of this putative mechanism. Another challenge is to determine the hierarchical relationship of the connexin-dependent signalling, relative to the many other mechanisms that interplay to control endocrine function. The finding that chronic alterations in the connexins of pancreatic β-cells can sufficiently perturb pancreas function to reproduce the defects in insulin secretion which are observed in type II diabetes (Charollais et al. 2000; Ravier et al. 2005), indicate a prominent role of connexin signaling, and raise the intriguing possibility that connexin defects may contribute to the disease. Therefore, still another challenge is to determine whether connexins are implicated in the pathogenesis of endocrine disorders. In addition to an increasing body of indirect, circumstantial association studies, a single case report has so far documented a link between a connexin alteration and an endocrine syndrome (Houang et al. 2002). The elucidation of the structure of the human connexin genes together with the mapping of hitherto unidentified sites of polymorphism that are linked to human diseases (Mas et al. 2004), provide novel tools that should facilitate the screening of this link in patients. If alterations in connexins and/or in the direct communications these proteins permit, are linked to endocrine disorders, one would be left with the problem of identifying a means to restore a normal situation in vivo. The task is formidable since connexins are widespread, are expressed in patterns that overlap in different cell systems, and are controlled by a large variety of factors. Thus, methods will have to be developed to specifically interact with individual proteins, on individual cell types. While awaiting the availability of genetic approaches that could efficiently correct qualitative and quantitative defects in connexin expression, a conceivable goal in the short term is the development of a pharmacology targeted to selected connexin channels. For example, antidiabetogenic sulfonylureas promote the assembly of Cx36 channels (Meda et al. 1980), whereas antimalarial drugs block them with a certain degree of selectivity (Cruikshank et al. 2004). Eventually, an innovative approach to the treatment of endocrine disorders that, like type I diabetes, are due to a major loss of endocrine cells, is the transplantation of surrogate hormone-producing and secretory competent cells. In view of the many functions of endocrine cells in which connexins participate, it is most likely that proper use of these gap-junction proteins will be found useful, if not instrumental for many steps of the production of such cells, including their secretory differentiation from stem cell precursors, their expansion in vitro and their survival in the host environment after transplantation.

Acknowledgements. We thank Drs J.-C. Henquin (University of Louvain, Belgium), P. Herrera (University of Geneva, Switzerland), T. Pedrazzini (University of Lausanne, Switzerland), L.S. Satin (Virginia Commonwealth University, USA), E. Scemes (Albert Einstein College of Medicine, USA), K. Willecke (University of Bonn, Germany) and C.B. Wollheim (University of Geneva, Switzerland) and all the members of their groups for most fruitful collaboration, Dr. S. Arnaudaud (Geneva Bioimaging Core Facility) and Mrs P. Severi-De Marco for technical assistance. Our research is supported by grants from the SNSF (31–67788), the JDRF (1–2001–622, 5–2004–255 and 1–2005–46), the EU (QLRT-2001–01777), and the NIH (1RO1 DK-63443–01).

References

Bosco D, Meda P (1991) Actively synthesizing B-cells secrete preferentially during glucose stimulation. Endocrinology 129:3157–3166

Bosco D, Meda P (1998) Reconstructing islet function in vitro. Adv Exp Med Biol 426:285–298

Bosco D, Orci L, Meda P (1989) Homologous but not heterologous contact increases the insulin secretion of individual secretory cells. Exp Cell Res 184:72–80

Calabrese A, Serre-Beinier V, Zhang M, Caton D, Mas C, Satin LS, Meda P (2003) Connexin 36 controls synchronization of Ca^{2+} oscillations and insulin secretion of MIN6 cells Diabetes 52:417–424

Calabrese A, Caton D, Meda P. (2004) Differentiating the effects of Cx36 and E-cadherin for proper insulin secretion of MIN6 cells. Exp Cell Res 294:379–391

Caton D, Calabrese A, Mas C, Serre-Beinier V, Charollais A, Caille D, Zufferey R, Trono D, Meda P (2003) Lentivirus-mediated transduction of connexin cDNAs shows level- and isoform-specific alterations in insulin secretion of primary pancreatic beta-cells. J Cell Sci 116:2285–2294

Charollais A, Gjinovci A, Huarte J, Bauquis J, Nadal A, Martin F, Andreu E, Sanchez-Andres JV, Calabrese A, Bosco D, Soria B, Wollheim CB, Herrera PL, Meda P (2000) Junctional communication of pancreatic beta cells contributes to the control of insulin secretion and glucose tolerance. J Clin Invest 106:235–243

Cronier L, Bastide B, Herve JC, Deleze J, Malassine A (1994) Gap junctional communication during human trophoblast differentiation: influence of human chorionic gonadotropin. Endocrinology 135:402–408

Cronier L, Frendo JL, Defamie N, Pidoux G, Bertin G, Guibourdenche J, Pointis G, Malassine A (2003) Requirement of gap junctional intercellular communication for human villous trophoblast differentiation. Biol Reprod 69:1472–1480

Cruikshank SJ, Hopperstad M, Younger M, Connors BW, Spray DC, Srinivas M (2004) Potent block of Cx36 and Cx50 gap junction channels by mefloquine. Proc Natl Acad Sci USA 101 :12364–12369

Davis KT, Prentice N, Gay VL, Murray SA (2002) Gap junction proteins and cell-cell communication in the three functional zones of the adrenal gland. J Endocrinol 173:13–21

Degen J, Meier C, Van Der Giessen RS, Sohl G, Petrasch-Parwez E, Urschel S, Dermietzel R, Schilling K, De Zeeuw CI, Willecke K (2004) Expression pattern of lacZ reporter gene representing connexin36 in transgenic mice. J Comp Neurol 473:511–525

Fauquier T, Guerineau NC, McKinney RA, Bauer K, Mollard P (2001) Folliculostellate cell network: a route for long-distance communication in the anterior pituitary. Proc Natl Acad Sci USA 98:8891–8896

Flachon V, Tonoli H, Selmi-Ruby S, Durand C, Rabilloud R, Rousset B, Munari-Silem Y (2002) Thyroid cell proliferation in response to forced expression of gap junction proteins. Eur J Cell Biol 81:243–252

Frendo JL, Cronier L, Bertin G, Guibourdenche J, Vidaud M, Evain-Brion D, MalassineA (2003). Involvement of connexin 43 in human trophoblast cell fusion and differentiation. J Cell Sci 116:3413–3421

Goldberg GS, Lampe PD, Nicholson BJ (1999) Selective transfer of endogenous metabolites through gap junctions composed of different connexins. Nat Cell Biol 1:457–459

Goldenberg RC, Fortes FS, Cristancho JM, Morales MM, Franci CR, Varanda WA, Campos de Carvalho AC (2003) Modulation of gap junction mediated intercellular communication in TM3 Leydig cells. J Endocrinol 177:327–335

Granot I, Dekel N (2002) The ovarian gap junction protein connexin43: regulation by gonadotropins. Trends Endocrinol Metab 13:310–313

Grazul-Bilska AT, Reynolds LP, Bilski JJ, Redmer DA (2001) Effects of second messengers on gap junctional intercellular communication of ovine luteal cells throughout the estrous cycle. Biol Reprod 65:777–783

Green LM, Murray DK, Tran DT, Nelson GA, Shah MM, Luben RA (2001) A spontaneously arising mutation in connexin32 with repeated passage of FRTL-5 cells coincides with increased growth rate and reduced thyroxine release. J Mol Endocrinol 27:145–163

Green LM, Tran DT, Murray DK, Rightnar SS, Todd S, Nelson GA (2002) Response of thyroid follicular cells to gamma irradiation compared to proton irradiation: II. The role of connexin 32. Radiat Res 158:475–485

Gros D, Dupays L, Alcolea S, Meysen S, Miquerol L, Theveniau-Ruissy M (2004) Genetically modified mice: tools to decode the functions of connexins in the heart-new models for cardiovascular research. Cardiovasc Res 62:299–308

Grummer R, Winterhager E (1998) Regulation of gap junction connexins in the endometrium during early pregnancy. Cell Tissue Res 293:189–194

Grummer R, Traub O, Winterhager E (1999) Gap junction connexin genes cx26 and cx43 are differentially regulated by ovarian steroid hormones in rat endometrium. Endocrinology 140:2509–2516

Grummer R, Hewitt SW, Traub O, Korach KS, Winterhager E. (2004) Different regulatory pathways of endometrial connexin expression: preimplantation hormonal-mediated pathway versus embryo implantation-initiated pathway. Biol Reprod 71:273–281

Guerineau NC, Bonnefont X, Stoeckel L, Mollard P (1998) Synchronized spontaneous Ca^{2+} transients in acute anterior pituitary slices. J Biol Chem 273:10389–10395

Guerrier A, Fonlupt P, Morand I, Rabilloud R, Audebet C, Krutovskikh V, Gros D, Rousset B, Munari-Silem Y (1995) Gap junctions and cell polarity: connexin32 and connexin43 expressed in polarized thyroid epithelial cells assemble into separate gap junctions, which are located in distinct regions of the lateral plasma membrane domain. J Cell Sci 108:2609–2617

Guldenagel M, Ammermuller J, Feigenspan A, Teubner B, Degen J, Sohl G, Willecke K, Weiler R (2001) Visual transmission deficits in mice with targeted disruption of the gap junction gene connexin36. J Neurosci 21:6036–6044

Haefliger JA, Demotz S, Braissant O, Suter E, Waeber B, Nicod P, Meda P (2001) Connexins 40 and 43 are differentially regulated within the kidneys of rats with renovascular hypertension. Kidney Int 60:190–201

Haefliger JA, Nicod P, Meda P (2004) Contribution of connexins to the function of the vascular wall. Cardiovasc Res 62:345–356

Haefliger JA, Krattinger N, Martin D, Pedrazzini T, Capponi A, Döring B, Plum A, Charollais A, Willecke K, Meda P (2005) Connexin43 controls renin-dependent hypertension in mice. (in press)

Harris AL (2001) Emerging issues of connexin channels: biophysics fills the gap. Q Rev Biophys 34:325–472

Houang M, Gourmelen M, Moatti L, Le BY, Garabedian EN, Denoyelle F (2002) Hypogonadotrophic hypogonadism associated with prelingual deafness due to a connexin 26 gene mutation. J Pediatr Endocrinol Metab 15:219–223

Houghton FD, Thonnissen E, Kidder GM, Naus CC, Willecke K, Winterhager E (1999) Doubly mutant mice, deficient in connexin32 and -43, show normal prenatal development of organs where the two gap junction proteins are expressed in the same cells. Dev Genet 24:5–12

Huang SH, Wu JC, Hwang RD, Yeo HL, Wang SM (2003) Effects of 18beta-glycyrrhetinic acid on the junctional complex and steroidogenesis in rat adrenocortical cells. J Cell Biochem 90:33–41

Johnson ML, Redmer DA, Reynolds LP, Bilski JJ, Grazul-Bilska AT (2002) Gap junctional intercellular communication of bovine granulosa and thecal cells from antral follicles: effects of luteinizing hormone and follicle-stimulating hormone. Endocrine 18:261–270

Kanno T, Gopel SO, Rorsman P, Wakui M (2002) Cellular function in multicellular system for hormone-secretion: electrophysiological aspect of studies on alpha-, beta- and delta-cells of the pancreatic islet. Neurosci Res 42:79–90

Kawai K, Ipp E, Orci L, Perrelet A, Unger RH (1982) Circulating somatostatin acts on the islets of Langerhans by way of a somatostatin-poor compartment. Science 218:477–478

Khan-Dawood FS, Yang J, Dawood MY (1996) Expression of gap junction protein connexin-43 in the human and baboon (*Papio anubis*) corpus luteum. J Clin Endocrinol Metab 81:835–842

Khan-Dawood FS, Yang J, Dawood MY (1998) Hormonal regulation of connexin-43 in baboon corpora lutea. J Endocrinol 157:405–414

Kidder GM, Mhawi AA (2002) Gap junctions and ovarian folliculogenesis. Reproduction 123:613–620

Klee P, Charollais A, Caille D, Meda P (2004) Role of connexin proteins in the protection of pancreatic β-cells against cytotoxic attacks. Proceedings Prien post-EASD meeting: http://www.mh-hannover.de/institute/clinbiochemistry/Post_EASD2004/poster.html

Klinger FG, De Felici M (2002) In vitro development of growing oocytes from fetal mouse oocytes: stage-specific regulation by stem cell factor and granulosa cells. Dev Biol 244:85–95

Le Gurun S, Martin D, Formenton A, Maechler P, Caille D, Waeber G, Meda P, Haefliger JA. (2003) Connexin-36 contributes to control function of insulin-producing cells. J Biol Chem 278:37690–37697

LeRoith D (1990) Are all cells "endocrine"? In: Becker KL et al (eds) Principles and practice of endocrinology and metabolism, JB Lippincott, Philadelphia, pp 10–13

Levavi-Sivan B, Bloch CL, Gutnick MJ, Fleidervish IA (2005) Electrotonic coupling in the anterior pituitary of a teleost fish. Endocrinology (in press)

Li X, Olson C, Lu S, Nagy JI (2005) Association of connexin36 with zonula occludens-1 in HeLa cells, betaTC-3 cells, pancreas, and adrenal gland. Histochem Cell Biol 122:485–498

Martin AO, Mathieu MN, Chevillard C, Guerineau NC (2001) Gap junctions mediate electrical signaling and ensuing cytosolic Ca^{2+} increases between chromaffin cells in adrenal slices: A role in catecholamine release. J Neurosci 21:5397–5405

Martin D, Tawadros T, Meylan L, Abderrahmani A, Condorelli DF, Waeber G, Haefliger JA. (2003) Critical role of the transcriptional repressor neuron-restrictive silencer factor in the specific control of connexin36 in insulin-producing cell lines. J Biol Chem 278:53082–53089

Martin AO, Mathieu MN, Guerineau NC (2003) Evidence for long-lasting cholinergic control of gap junctional communication between adrenal chromaffin cells. J Neurosci 23:3669–3678

Mas C, Taske N, Deutsch S, Guipponi M, Thomas P, Covanis A, Friis M, Kjeldsen MJ, Pizzolato GP, Villemure JG, Buresi C, Rees M, Malafosse A, Gardiner M, Antonarakis SE, Meda P (2004) Association of the connexin36 gene with juvenile myoclonic epilepsy. J Med Genet 41:e93–e98

Matesic DF, Germak JA, Dupont E, Madhukar BV (1993) Immortalized hypothalamic luteinizing hormone-releasing hormone neurons express a connexin26-like protein and display functional gap junction coupling assayed by fluorescence recovery after photobleaching. Neuroendocrinology 58:485–492

Matesic D, Attardi B, Dellovade T, Pfaff D, Germak J (1997) Differential LHRH secretion, dye coupling, and protein expression in two morphologically distinct cell types identified in GT1–7 cultures. J Neuroendocrinol 9:467–478

Mayerhofer A, Garfield RE (1995) Immunocytochemical analysis of the expression of gap junction protein connexin 43 in the rat ovary. Mol Reprod Dev 41:331–338

McDonald TJ, Li C, Massmann GA, Figueroa JP (2003) Connexin 43 ontogeny in fetal sheep adrenal glands. Steroids 68:613–620

Meda P (1996) The role of gap junction membrane channels in secretion and hormonal action. J Bioenerg Biomembr 28:369–377

Meda P (1997) Intercellular communication and insulin secretion. In: Zahnd GR, Wollheim CB (eds) Contributions of physiology to the understanding of diabetes. Springer, Berlin Heidelberg New York, pp 24–42

Meda P, Spray DC (2000) Gap junction function. Adv Mol Cell Biol 30:263–322

Meda P, Bosco D (2001) Communication of islet cells: molecules, mechanisms, functions. In: Habener JF, Hussein M (eds) Molecular basis of endocrine pancreas development and functions, Kluwer, Norwell, pp 138–159

Meda P, Halban P, Perrelet A, Renold AE, Orci L (1980) Gap junction development is correlated with insulin content in the pancreatic B cell. Science 209:1026–1028

Meda P, Perrelet A, Orci, L (1984) Gap junctions and cell-to-cell coupling in endocrine glands. In: Satir BH (ed) Modern cell biology. Alan Liss, New York, vol 3, pp 131–196

Meda P, Santos RM, Atwater I (1986) Direct identification of electrophysiologically monitored cells within intact mouse islets of Langerhans. Diabetes 35:232–236

Meda P, Bosco D, Chanson M, Giordano E, Vallar L, Wollheim C, Orci L (1990) Rapid and reversible secretion changes during uncoupling of rat insulin-producing cells. J Clin Invest 86:759–768

Meda P, Pepper MS, Traub O, Willecke K, Gros D, Beyer E, Nicholson B, Paul D, Orci L (1993) Differential expression of gap junction connexins in endocrine and exocrine glands. Endocrinology 133:2371–2378

Morand I, Fonlupt P, Guerrier A, Trouillas J, Calle A, Remy C, Rousset B, Munari-Silem Y (1996) Cell-to-cell communication in the anterior pituitary: evidence for gap junction-mediated exchanges between endocrine cells and folliculostellate cells. Endocrinology 137:3356–3367

Moreno AP, Berthoud VM, Perez-Palacios G, Perez-Armendariz EM (2005) Biophysical evidence that connexin 36 forms functional gap junction channels between pancreatic mouse beta-cells. Am J Physiol Endocrinol Metab (in press)

Munari-Silem Y, Audebet C, Rousset B (1991) Hormonal control of cell to cell communication: regulation by thyrotropin of the gap junction-mediated dye transfer between thyroid cells. Endocrinology 128:3299–3309

Munari-Silem Y, Guerrier A, Fromaget C, Rabilloud R, Gros D, Rousset B (1994) Differential control of connexin-32 and connexin-43 expression in thyroid epithelial cells: evidence for a direct relationship between connexin-32 expression and histiotypic morphogenesis. Endocrinology 135:724–734

Munari-Silem Y, Lebrethon MC, Morand I, Rousset B, Saez JM (1995) Gap junction-mediated cell-to-cell communication in bovine and human adrenal cells. A process whereby cells increase their responsiveness to physiological corticotropin concentrations. J Clin Invest 95:1429–1439

Munari-Silem Y, Rousset B (1996) Gap junction-mediated cell-to-cell communication in endocrine glands – molecular and functional aspects: a review. Eur J Endocrinol 135:251–264

Murray SA, Oyoyo UA, Pharrams SY, Kumar NM, Gilula NB (1995) Characterization of gap junction expression in the adrenal gland. Endocr Res 21:221–229

Murray SA, Davis K, Gay V (2003) ACTH and adrenocortical gap junctions. Microsc Res Tech 61:240–246

Nagy JI, Dudek FE, Rash JE (2004) Update on connexins and gap junctions in neurons and glia in the mammalian nervous system. Brain Res Rev 47:191–215

Nelles E, Butzler C, Jung D, Temme A, Gabriel HD, Dahl U, Traub O, Stumpel F, Jungermann K, Zielasek J, Toyka KV, Dermietzel R, Willecke K (1996). Defective propagation of signals generated by sympathetic nerve stimulation in the liver of connexin32-deficient mice. Proc Natl Acad Sci USA 93:9565–9570

Niessen H, Harz H, Bedner P, Kramer K, Willecke K (2000) Selective permeability of different connexin channels to the second messenger inositol 1,4,5-trisphosphate. J Cell Sci 113:1365–1372

Orci L, Malaisse-Lagae F, Ravazzola M, Rouiller D, Renold AE, Perrelet A, Unger R (1975) A morphological basis for intercellular communication between alpha- and beta-cells in the endocrine pancreas. J Clin Invest 56:1066–1070

Oyoyo UA, Shah US, Murray SA (1997) The role of alpha1 (connexin-43) gap junction expression in adrenal cortical cell function. Endocrinology 138:5385–5399

Patel YC, Amherdt M, Orci L (1982) Quantitative electron microscopic autoradiography of insulin, glucagon, and somatostatin binding sites on islets. Science 217:1155–1156

Persson PB (2003) Renin: origin, secretion and synthesis. J Physiol 552:667–671

Petersen OH (1980) The electrophysiology of gland cells. Academic Press, London

Philippe J, Giordano E, Gjinovci A, Meda P (1992) Cyclic adenosine monophosphate prevents the glucocorticoid-mediated inhibition of insulin gene expression in rodent islet cells. J Clin Invest 90:2228–2233

Plum A, Hallas G, Magin T, Dombrowski F, Hagendorff A, Schumacher B, Wolpert C, Kim J, Lamers WH, Evert M, Meda P, Traub O, Willecke K (2000) Unique and shared functions of different connexins in mice. Curr Biol 10:1083–1091

Quesada I, Fuentes E, Andreu E, Meda P, Nadal A, Soria B (2003) On-line analysis of gap junctions reveals more efficient electrical than dye coupling between islet cells. Am J Physiol Endocrinol Metab 284:E980–E987

Ravier MA, Güldenagel M, Charollais A, Gjinovci A, Caille D, Söhl G, Wollheim CB, Willecke K, Henquin JC, Meda P (2005) Loss of Connexin36 channels alters β-cell coupling, synchronization of glucose-induced Ca^{2+} and insulin oscillations, and basal insulin release. Diabetes (in press)

Risek B, Gilula NB (1991) Spatiotemporal expression of three gap junction gene products involved in fetomaternal communication during rat pregnancy. Development 113:165–181

Risek B, Guthrie S, Kumar N, Gilula NB (1990) Modulation of gap junction transcript and protein expression during pregnancy in the rat. J Cell Biol 110:269–282

Rukstalis JM, Kowalik A, Zhu L, Lidington D, Pin CL, Konieczny SF (2003) Exocrine specific expression of Connexin32 is dependent on the basic helix-loop-helix transcription factor Mist1. J Cell Sci 116:3315–3325

Saez JC, Berthoud VM, Kadle R, Traub O, Nicholson BJ, Bennett MV, Dermietzel R (1991) Pinealocytes in rats: connexin identification and increase in coupling caused by norepinephrine. Brain Res 568:265–275

Saez JC, Martinez AD, Branes MC, Gonzalez HE (1998). Regulation of gap junctions by protein phosphorylation. Braz J Med Biol Res 31:593–600

Sakuma E, Soji T, Herbert DC (2001) Effects of hydrocortisone on the formation of gap junctions and the abnormal growth of cilia within the rat anterior pituitary gland: possible role of gap junctions on the regulation of cell development. Anat Rec 262:169–175

Sakuma E, Herbert DC, Soji T (2003) The effects of sex steroids on the formation of gap junctions between folliculo-stellate cells; a study in castrated male rats and ovariectomized female rats. Arch Histol Cytol 66:229–238

Salomon D, Meda P (1986) Heterogeneity and contact-dependent regulation of hormone secretion by individual B cells. Exp Cell Res 162:507–520

Salomon D, Masgrau E, Vischer S, Ullrich S, Dupont E, Sappino P, Saurat JH, Meda P (1994) Topography of mammalian connexins in human skin. J Invest Dermatol 103:240–247

Serre-Beinier V, Le Gurun S, Belluardo N, Trovato-Salinaro A, Charollais A, Haefliger JA, Condorelli DF, MedaP (2000) Cx36 preferentially connects beta-cells within pancreatic islets. Diabetes 49:727–734

Serre-Beinier V, Mas C, Calabrese A, Caton D, Bauquis J, Caille D, Charollais A, Cirulli V, Meda P (2002) Connexins and secretion. Biol Cell 94:477–492

Shah US, Murray SA (2001) Bimodal inhibition of connexin 43 gap junctions decreases ACTH-induced steroidogenesis and increases bovine adrenal cell population growth. J Endocrinol 171:199–208

Shinohara K, Funabashi T, Mitushima D, Kimura F (2000) Effects of gap junction blocker on vasopressin and vasoactive intestinal polypeptide rhythms in the rat suprachiasmatic nucleus in vitro. Neurosci Res 38:43–47

Shirasawa N, Mabuchi Y, Sakuma E, Horiuchi O, Yashiro T, Kikuchi M, Hashimoto Y, Tsuruo Y, Herbert DC, Soji T (2004) Intercellular communication within the rat anterior pituitary gland: X. Immunohistocytochemistry of S-100 and connexin 43 of folliculo-stellate cells in the rat anterior pituitary gland. Anat Rec 278:462–473

Simon AM, Goodenough DA, Li E, Paul DL (1997) Female infertility in mice lacking connexin 37. Nature 385:525–529

Stagg RB, Fletcher WH (1990) The hormone-induced regulation of contact-dependent cell–cell communication by phosphorylation. Endocr Rev 11:302–325

Statuto M, Audebet C, Tonoli H, Selmi-Ruby S, Rousset B, Munari-Silem Y (1997) Restoration of cell-to-cell communication in thyroid cell lines by transfection with and stable expression of the connexin-32 gene. Impact on cell proliferation and tissue-specific gene expression. J Biol Chem 272:24710–24716

Stoka AM (1999) Phylogeny and evolution of chemical communication: an endocrine approach. J Mol Endocrinol 22:207–225
Stout C, Goodenough DA, Paul DL (2004) Connexins: functions without junctions. Curr Opin Cell Biol 16:507–512
Taylor CP, Dudek FE (1982) A physiological test for electrotonic coupling between CA1 pyramidal cells in rat hippocampal slices. Brain Res 235:351–357
Terasawa E (2001) Luteinizing hormone-releasing hormone (LHRH) neurons: mechanism of pulsatile LHRH release. Vitam Horm 63:91–129
Theis M, Mas C, Doring B, Degen J, Brink C, Caille D, Charollais A, Kruger O, Plum A, Nepote V, Herrera P, Meda P, Willecke K 2004) Replacement by a lacZ reporter gene assigns mouse connexin36, 45 and 43 to distinct cell types in pancreatic islets. Exp Cell Res 294:18–29
Tonoli H, Flachon V, Audebet C, Calle A, Jarry-Guichard T, Statuto M, Rousset B, Munari-Silem Y (2000). Formation of three-dimensional thyroid follicle-like structures by polarized FRT cells made communication competent by transfection and stable expression of the connexin-32 gene. Endocrinology 141:1403–1413
Vazquez-Martinez R, Shorte SL, Boockfor FR, Frawley LS (2001) Synchronized exocytotic bursts from gonadotropin-releasing hormone-expressing cells: dual control by intrinsic cellular pulsatility and gap junctional communication. Endocrinology 142:2095–2101
Veenstra RD, Beblo DA, Wang HZ (1998) In: Werner R (ed) Gap junctions, IOS Press, Amsterdam, pp 40–44
Vitale ML, Cardin J, Gilula NB, Carbajal ME, Pelletier RM (2001) Dynamics of connexin 43 levels and distribution in the mink (*Mustela vison*) anterior pituitary are associated with seasonal changes in anterior pituitary prolactin content. Biol Reprod 64:625–633
Vozzi C, Ullrich S, Charollais A, Philippe J, Orci L, Meda P (1995) Adequate connexin-mediated coupling is required for proper insulin production. J Cell Biol 131:1561–1572
Warn-Cramer BJ, Lau AF (2004) Regulation of gap junctions by tyrosine protein kinases. Biochim Biophys Acta 1662:81–95
Winterhager E, Stutenkemper R, Traub O, Beyer E, Willecke K (1991) Expression of different connexin genes in rat uterus during decidualization and at term. Eur J Cell Biol 55:133–142
Yamamoto T, Hossain MZ, Hertzberg EL, Uemura H, Murphy LJ, Nagy JI (1993) Connexin43 in rat pituitary: localization at pituicyte and stellate cell gap junctions and within gonadotrophs. Histochemistry 100:53–64

10 Roles of Gap Junctions in Ovarian Folliculogenesis: Implications for Female Infertility

Gerald M. Kidder[1]

10.1 Introduction

In developing follicles, gap junctions couple the growing oocyte and its surrounding follicle (granulosa) cells into a functional syncytium. In this chapter, evidence is summarized pertaining to the expression of various connexins, the constituent proteins of gap-junction channels, in developing follicles, and the likely roles that some of them play based on experiments with connexin mutant mice. Gap junctions between granulosa cells of preantral follicles of the mouse are composed of connexin43 (Cx43). Although this connexin has also been detected in a small minority of gap junctions at the oocyte surface using immunoelectron microscopy, its absence does not significantly restrict coupling between the oocyte and granulosa cells. The importance of Cx43 for granulosa cell function is demonstrated by the fact that follicles from knockout mice lacking this connexin are growth-restricted and produce oocytes that fail to achieve meiotic competence. Connexin37 (Cx37) appears to be the only connexin contributed by oocytes to the gap junctions coupling them with granulosa cells, and loss of this connexin ablates oocyte-granulosa cell coupling resulting in oocyte loss and premature luteinization of the follicles. Granulosa cells also express Cx37, but do not assemble it into gap junctions unless they are in contact with an oocyte. Additional connexins have been identified in antral follicles. The expression of multiple connexins in developing follicles is proposed to reflect multiple functions served by gap junctional communication in folliculogenesis. The expression of Cx43 in follicles of humans and mice, and the reproductive defects of female mice with connexin mutations, make it likely that some forms of human female infertility involve defects in gap junctional communication within developing follicles.

[1] Department of Physiology and Pharmacology, The University of Western Ontario, London, Ontario N6A 5C1, Canada, E-mail: gerald.kidder@fmd.uwo.ca

Gap Junctions in Development and Disease
(ed. by E. Winterhager)

10.2 Ovarian Follicle Development

The ovarian follicle is the main functional unit of the mammalian ovary (Fig. 10.1). The follicle consists of a single oocyte, layers of surrounding somatic cells (granulosa cells), and an outer rim of theca cells. The follicle grows primarily through proliferation of granulosa cells, adding additional layers until a fluid-filled cavity, the antrum, develops. The antrum separates the granulosa cells into two subpopulations, the cumulus granulosa surrounding the oocyte and the mural granulosa adjacent to the theca layers. Within the follicle, the oocyte itself grows and develops competence to undergo meiosis and be fertilized. In growing follicles, receptors for follicle-stimulating hormone (FSH) are expressed on granulosa cells, whereas receptors for luteinizing hormone (LH) are expressed on theca cells. In response to LH stimulation, theca cells produce testosterone; granulosa cells then aromatize the testosterone to estradiol. During each

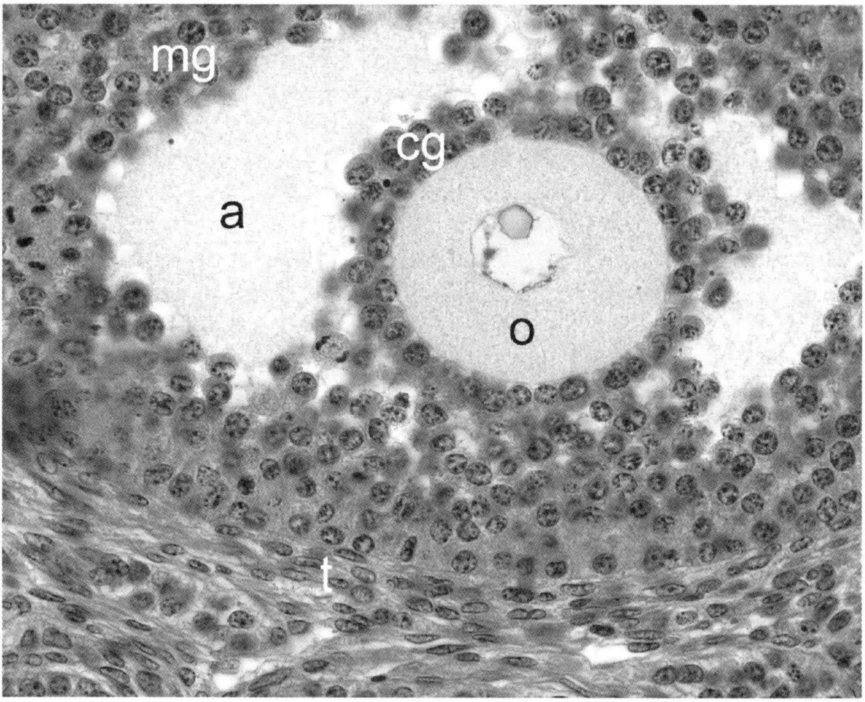

Fig. 10.1. Anatomy of the growing follicle. The oocyte (*o*; with germinal vesicle) develops in continuous gap junctional communication with the surrounding cumulus granulosa cells (*cg*), which are themselves in continuous gap junctional communication with the mural granulosa cells (*mg*). The antral cavity (*a*) and theca cell layers (*t*) are also indicated

cycle, depending on the species, one or more growing follicles will become mature with all others undergoing atresia (death by cellular apoptosis). During their growth, follicles pass through primary (unilaminar) and secondary (multilaminar) stages until becoming tertiary (antral) follicles. A fully grown antral follicle ready for ovulation is called a preovulatory or Graafian follicle.

Despite its structural simplicity, ovarian folliculogenesis requires complex regulatory mechanisms involving both extrinsic (endocrine) and intrinsic (intra-ovarian) signalling pathways. In the latter case, numerous peptides, members of several growth factor families, have been implicated as influencing follicle growth and maturation through autocrine/paracrine signaling (reviewed by Udoff and Adashi 1999). Gap junctions provide another intra-ovarian signaling pathway. Recent findings from gene expression studies in several species and gene targeting in mice have implicated gap junctional intercellular communication in follicular development and have suggested its involvement in female infertility, making this an important research area.

Gap-junction channels allow inorganic ions, second messengers, and small metabolites to pass from cell to cell and this permeability is assumed to underlie the physiological roles played by gap junctions. The developing ovarian follicle is a particularly good example of the involvement of gap junctions in intercellular transfer of nutrients and regulatory molecules. During the growth of mammalian oocytes, there is continuous coupling with the surrounding granulosa cells via gap junctions (reviewed by Eppig 1991 and Eppig et al. 1996). According to morphological evidence, this coupling begins as primordial follicles form and expands as folliculogenesis proceeds through primary, secondary, and tertiary (antral) follicle stages (Mitchell and Burghardt 1986). Amino acids, glucose metabolites, and nucleotides are among the molecules known to be transferred to the growing oocyte through gap junctions (Eppig 1991). In addition, signals that regulate meiotic maturation of fully grown oocytes pass through the oocyte-granulosa cell gap junctions (Fagbohun and Downs 1991; Coskun and Lin 1994; Granot and Dekel 1994; Downs 1995).

10.3
Connexins in Developing Follicles

Multiple connexins are expressed within ovarian follicles, in some cases within the same cell type. For example, both Cx32 and Cx43 have been detected in gap junctions joining the cumulus granulosa cells of fully grown mouse follicles (Valdimarsson et al. 1993). Cx43 has been detected in fetal mouse ovaries as early as embryonic day 12.5, when ovarian differentiation has just begun (Pérez-Armendariz et al. 2003), and Cx43 (but not Cx32) gap junctions were seen in what appeared to be pre-granulosa cells on the day of birth, when the first primordial follicles were forming (Juneja et al. 1999). Hence in the mouse, this connexin is involved in ovarian development throughout its time course. Cx45 is present as a minor component in granulosa cell gap junctions of mouse and rat, colocalizing with Cx43 in some gap junctions (Okuma et al. 1996; Alcoléa et al. 1999; Wright et al. 2001). Immunogold colabeling for Cx43 and Cx45 demonstrated that Cx45 is restricted to small regions of some Cx43-containing gap junction plaques between cumulus granulosa cells (Kidder and Mhawi 2002). A fourth ovarian connexin, Cx37, is present in gap junctions from the primary follicle stage onward, restricted to the interface between oocytes and granulosa cells [Simon et al. 1997; it should be noted, however, that Wright et al. (2001) also detected some Cx37 immunoreactivity between granulosa cells in antral follicles]. Thus in the mouse, connexins 32, 37, 43, and 45 are implicated in the process of folliculogenesis by virtue of their expression within developing and/or mature follicles. Additional connexins have been detected in the ovarian follicles of other species, including Cx30.3 in pig and rat granulosa and theca cells (Itahana et al. 1996, 1998) and Cx26 in sheep and cow oocytes and pig theca cells (Itahana et al. 1996; Grazul-Bilska et al. 1998; Johnson et al. 1999). According to one report, Cx26 and Cx32 are also expressed in mouse oocytes and theca cells (Wright et al. 2001), but attempts to detect Cx26 mRNA in mouse oocytes and granulosa cells by RT-PCR (G.M. Kidder and J.J. Eppig, unpubl. data), or in rat ovaries by Northern blotting (Risek et al. 1990) were unsuccessful. This discrepancy remains unresolved. It is unlikely that Cx32 plays an important role in mouse folliculogenesis because knockout mice lacking this connexin are fertile (Nelles et al. 1996). Likewise, despite a report of Cx57 mRNA in mouse ovaries (Manthey et al. 1999), Cx57 knockout mice do not display any ovarian defects (Hombach et al. 2004). In the pig, on the other hand, its ortholog Cx60 is expressed in cumulus and theca cells (Itahana et al. 1998).

In light of the important role played by oocyte-granulosa cell gap junctions in oogenesis, it was of interest to determine which connexins contribute to those heterocellular junctions. This has been most carefully examined in the mouse. As noted above, Cx37 can clearly be seen by immunofluorescence microscopy in gap junction-like structures on the surface of mouse oocytes, beneath the zona pellucida (Simon et al. 1997; Veitch et al. 2004). In contrast, several studies using light microscope immunostaining failed to convincingly reveal the presence of Cx32 or Cx43 in that location (Valdimarsson et al. 1993; Simon et al. 1997; Wright et al. 2001; Veitch et al. 2004) although one other study reported numerous Cx43-immunoreactive foci along the oocyte plasma membrane (Wong et al. 2000). This discrepancy also remains unresolved. At the electron microscope level, both Cx43 and Cx45 can be detected in gap junctions close to the oocyte surface and, in the case of Cx43, on the cumulus cell side of gap junctions between cumulus granulosa cell projections and the oocyte plasma membrane (A.A. Mhawi, R.R. Shivers, and G.M. Kidder, unpubl. data). This raises the possibility that at least a few of the gap junctions connecting the oocyte with granulosa cells may be heterotypic junctions formed from Cx37, contributed by the oocyte, docking with Cx43 contributed by the cumulus cells. Cx37 is presumed to be the only connexin contributed by the oocyte because ablation of this connexin by gene targeting removed all gap junctions from the oocyte surface and disrupted oocyte-granulosa cell coupling (Simon et al. 1997). Ablation of Cx43, however, did not disrupt oocyte-granulosa cell coupling, suggesting that Cx43, though present at or very near the oocyte surface, does not contribute to heterocellular coupling in an important way (Veitch et al. 2004).

The identity of the dominant granulosa cell connexin docking with Cx37 from the oocyte was finally revealed in mice through studies involving coculture of granulosa cells and denuded oocytes (Veitch et al. 2004). Oocytes of different genotypes (wild type, Cx37 null, or Cx43 null) were seeded onto freshly isolated monolayers of granulosa cells of the same range of genotypes and allowed time to reestablish heterocellular coupling. Preloading the oocytes with calcein dye allowed the restoration of intercellular coupling between the two cell types to be monitored. It was found that the passage of calcein from preloaded oocytes to recipient granulosa cells occurred only when both cell types were capable of synthesizing Cx37; their ability to synthesize Cx43 was irrelevant. This finding implied that granulosa cells must contain Cx37 connexons, despite their absence from gap junctions with other granulosa cells, where Cx43 predominates. The presence of Cx37 mRNA in granulosa cells was confirmed by RT-PCR and selective recruitment (or stabilization) of granulosa cell Cx37 into gap junction-like plaques in response to oocyte contact in vitro was confirmed by immunostaining. Thus, granulosa cells express both Cx37 and Cx43 and target them differentially to regions of cell contact, with Cx37 localizing in contacts with the oocyte and Cx43 localizing in contacts with other granulosa cells. Localization of granulosa cell Cx37 in membrane plaques did not occur in response to Cx37 null mutant oocytes (Veitch et al. 2004), suggesting that homotypic connexon docking may be the critical factor in selectively stabilizing Cx37 at the oocyte–granulosa cell interface. An obvious corollary is that the majority of gap junctions between the mouse oocyte and adjacent granulosa cells are homomeric, homotypic Cx37 junctions, as illustrated in Fig. 10.2, but this must be confirmed directly through analysis of their biophysical properties.

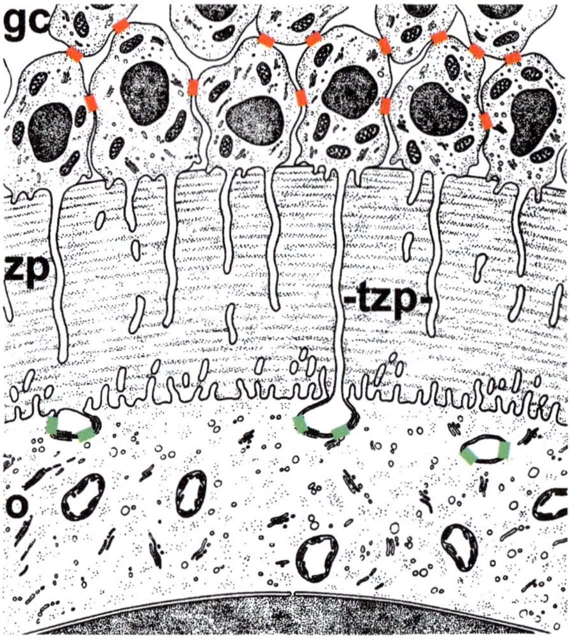

Fig. 10.2. Connexin contributions to gap junctions in mouse preantral follicles. Gap junctions at the surface of the oocyte (*o*) are located at the ends of transzonal projections (*tzp*) from the granulosa cells (*gc*). On the basis of recent evidence (Veitch et al, 2004), these are thought to be homomeric, homotypic gap junctions with Cx37 (*green*) being contributed by both cell types. Gap junctions between granulosa cells, at least in preantral follicles, are formed exclusively of Cx43 (*red*; Gittens et al. 2003), indicating that granulosa cells express both connexins, but localize them to different cell contact regions. *Zp* Zona pellucida (adapted from Austin and Short 1982, Chap. 2)

10.4 Roles of Individual Connexins in Folliculogenesis

Connexin knockout mice have provided new insights into the roles of gap junctions in ovarian development and folliculogenesis (Fig. 10.3). Mice lacking Cx37 are viable and ovarian folliculogenesis proceeds in apparently normal fashion until a late preantral stage (Simon et al. 1997; Carabatsos et al. 2000). Mature preovulatory follicles were never seen in the mutant ovaries, however, and ovulation could not be induced by gonadotropin stimulation. The mutation abolishes intercellular coupling between granulosa cells and oocytes and, consistent with the role of granulosa cell–oocyte coupling in transferring nutrients to the oocyte, causes oocyte growth to arrest at 74% of normal size (Carabatsos et al. 2000). In addition, oocytes developing without Cx37 fail to achieve full meiotic competence. Even-

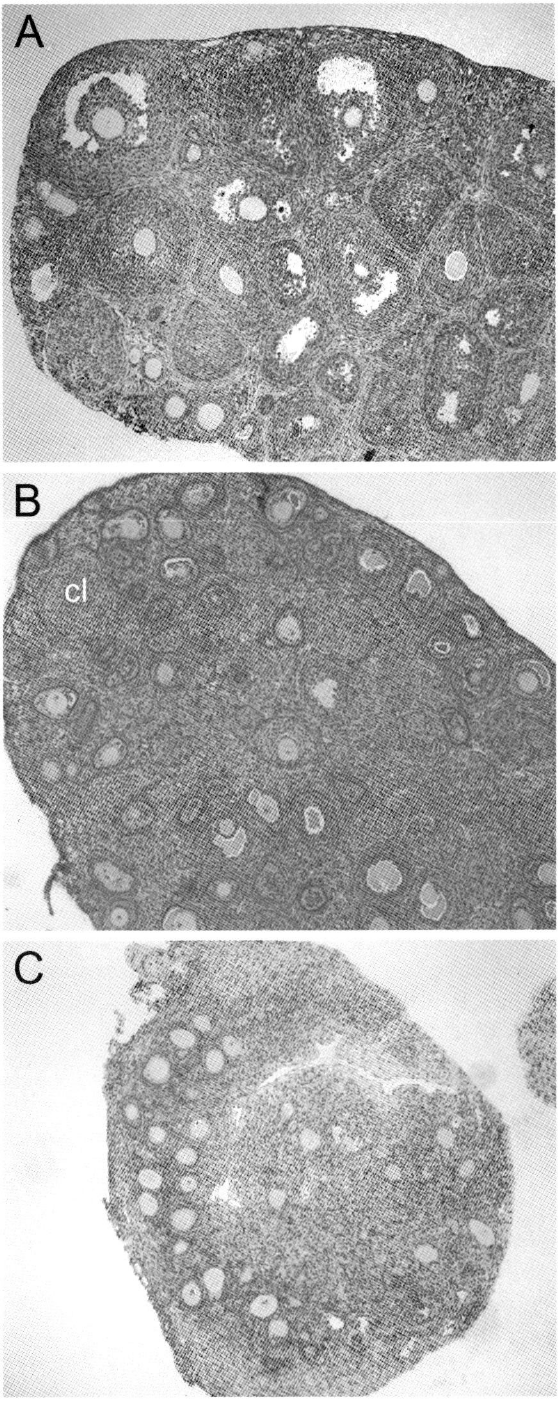

Fig. 10.3. Disruption of folliculogenesis in C57BL/6 knockout mice lacking Cx37 or Cx43. In contrast to wild type ovaries in which the full range of follicle stages is present (**A**), Cx37 null mutant ovaries (**B**) show no evidence of large antral or preovulatory follicles, but contain structures resembling corpora lutea (*cl*). Cx43 null mutant ovaries (**C**) have fewer follicles (due to a fetal germ cell deficiency) and very few of them develop beyond the unilaminar stage or acquire an antrum

tually, the oocytes degenerate and the mutant ovaries become filled with structures resembling corpora lutea, as though the granulosa cells differentiate prematurely to become luteal cells (Simon et al. 1997). Thus, it is likely that gap junctional coupling between oocytes and granulosa cells via Cx37 channels is required to maintain the differentiated state of the granulosa cells, preventing them from luteinizing before ovulation.

A very different result was obtained when the gene encoding Cx43 was ablated by gene targeting. Offspring (either male or female) homozygous for the mutation have very few germ line cells (Juneja et al. 1999). This is evident from as early as 11.5 dpc, when the first primordial germ cells (PGCs) have taken up residence in the genital ridges, suggesting that the deficiency arises during the period of PGC migration. However, the gonads of Cx43 null mutant fetuses do contain germ cells, about 10% of the normal number. Hence, it was of interest to determine whether the few oocytes in the ovaries of Cx43 null mutant neonates could participate in folliculogenesis. Because the absence of Cx43 causes neonatal lethality due to a severe heart defect (Reaume et al. 1995), and because ovarian folliculogenesis in mice is a postnatal process, it was necessary to graft late fetal ovaries into the kidney capsules of ovariectomized adult mice to allow postnatal development (Ackert et al. 2001). This procedure has been shown to permit the development of fully grown, meiotically competent, fertilizable oocytes (Eppig and Wigglesworth 2000). Whereas wild-type ovaries developed a full range of follicles from primordial through large, antral follicles within a 3-week grafting period, folliculogenesis in mutant ovaries of the C57BL/6 strain did not proceed in most cases beyond the primary, unilaminar stage. More extensive, but still incomplete, folliculogenesis was seen in grafted mutant ovaries of the CD1 strain, indicating the influence of strain-specific genetic modifiers in folliculogenesis. These results were interpreted as indicating a requirement for coupling among granulosa cells via Cx43 channels to sustain granulosa cell proliferation (Ackert et al. 2001). Failure of the Cx43-deficient follicles to develop multiple layers of granulosa cells was correlated with reduced growth of the oocytes, which were morphologically abnormal, meiotically incompetent, and could not be fertilized. The inability of granulosa cells to proliferate in the absence of Cx43 is consistent with the presence of this connexin from the onset of folliculogenesis and implies that Cx43 is either the only connexin expressed in granulosa cells of primary follicles or that it plays an indispensable role that cannot be fulfilled by other connexins that might be present (the question of connexin redundancy will be considered in the next section). It was later confirmed through electron microscopy, dye injection, and electrical coupling experiments that Cx43 is the only connexin that contributes to coupling between granulosa cells of mouse preantral follicles (Gittens et al. 2003).

What specific roles are played by Cx37 and Cx43 in developing mouse follicles? In general terms, it is clear that communication between the oocyte and its associated granulosa cells via Cx37 channels is required to maintain the health of the oocyte and the stage-appropriate function of the granulosa cells, and that communication among the granulosa cells via Cx43 channels is required to maintain the growth of the granulosa cell layers which will allow the oocyte to grow and mature. Recently, the role played by Cx43 in promoting granulosa cell growth has come under closer scrutiny. In addition to the roles played by pituitary gonadotropins in follicular development, numerous intraovarian paracrine factors have been identified, some of which promote granulosa cell proliferation and/or differentiation (Udoff and Adashi 1999). One such paracrine factor is GDF9 (growth/differentiation factor-9), a secreted TGFβ superfamily member that, within the mouse ovary, is expressed exclusively in growing oocytes (reviewed by Vitt and Hsueh 2002). In vitro studies have shown that GDF9 promotes proliferation and influences differentiation of granulosa cells. The importance of GDF9 in the mouse ovary became evident when the gene was knocked out: the mutant females were viable, but sterile because folliculogenesis was arrested in the primary follicle stage, demonstrating that GDF9 is an essential mitogen for granulosa cells (Dong et al. 1996). GDF9-deficient oocytes are morphologically abnormal, exhibit reduced meiotic competence, and eventually degenerate (Dong et al. 1996; Carabatsos et al. 1998). Thus, GDF9 is a key factor in ensuring that granulosa cell proliferation stays in step with oocyte growth.

The relationship between paracrine stimulation of granulosa cell proliferation via oocyte-derived GDF9 and the apparent role of Cx43 in granulosa cell proliferation was investigated through a combination of in vivo and in vitro experiments with ovaries of both mutant lines (Gittens et al. 2005). First, it was demonstrated that Cx43 expression and gap junctional communication among preantral granulosa cells are maintained in the absence of GDF9, and GDF9 expression is maintained in ovaries lacking Cx43. Granulosa cell proliferation is reduced, however, in Cx43 null mutant follicles as it is in GDF9 null mutant follicles. Furthermore, recombinant GDF9 added to follicle culture medium could restore the proliferation rate of granulosa cells from Cx43 null mutant follicles to equal that of wild-type granulosa cells. This indicated that, while not responding to endogenous GDF9 coming from the oocyte, the mutant granulosa cells could still respond to exogenous GDF9. In other words, the proliferation deficit in Cx43 null mutant follicles is due at least in part to reduced responsiveness to oocyte-derived GDF9. Gap junctional and paracrine signaling pathways thus interact in folliculogenesis. It was suggested that one role of Cx43 gap junctions in granulosa cells is to propagate signals arising downstream of GDF9 action to granulosa cells farther away to ensure responsiveness throughout the entire expanding population. This hypothesis remains to be tested experimentally.

10.5 Connexin Redundancy in Ovarian Follicles

The distinct locations of gap junctions composed of Cx37 and Cx43 in developing ovarian follicles, the differential targeting of granulosa cell Cx37 to contacts with the oocyte, and the distinct ovarian phenotypes of the two connexin knockouts could be interpreted to indicate that the two connexins play distinct roles in folliculogenesis. One possibility is that the different permeabilities of Cx37 and Cx43 homotypic channels (Elfgang et al. 1995; Weber et al. 2004) are utilized to differentiate coupling between granulosa cells from that between granulosa cells and the oocyte, a form of communication compartmentalization. Since we do not have the ability to screen specific cell types for the molecules passing through their gap junctions, it is not possible to test this possibility directly. However, it has been possible to test the hypothesis that Cx43 or Cx37 plays a unique role in folliculogenesis that cannot be fulfilled by another connexin. The first such test involved "knockin" mice in which the coding region of the *Gja1* gene encoding Cx43 was replaced by Cx32 or Cx40 cDNA (Plum et al. 2000). In both of these lines, the heart defect that causes neonatal lethality in the Cx43 germline knockout was rescued and the homozygous knockin mice were viable. Histological examination of their ovaries revealed the full range of stages of folliculogenesis and the females ovulated, but only two of seven Cx32 knockin females and no Cx40 knockin females produced offspring. The two Cx32 knockin females bore only single pups that were born dead. Thus, both Cx32 and Cx40 can substitute for Cx43 in granulosa cells to promote follicular growth, but the quality of the oocytes developing within the knockin follicles may have been adversely affected by the substitution. Interestingly, neither Cx32 nor Cx40 could substitute for Cx43 in the knockin testes, which closely matched the "Sertoli cell-only" phenotype of Cx43 knockout testes (Roscoe et al. 2001).

We have taken a different approach to explore connexin redundancy in ovarian follicles: the use of oocyte-specific connexin substitution via transgenesis (T.Y. Li, K.J. Barr, and G.M. Kidder, in prep.). Transgenic mice were produced in which expression of a Cx43 transgene was targeted to growing oocytes by the *Zp3* promoter (Rankin et al. 1998). These mice were crossed with the Cx37 knockout line to generate female offspring in which Cx37 was replaced by Cx43 specifically in oocytes. Females homozygous for the Cx37 null allele, but carrying at least one copy of the Cx43 transgene had their fertility restored. Thus, Cx43 is functionally equivalent to Cx37 in developing oocytes. This surprising result must mean that homotypic Cx43 gap junctions can serve the same purpose as homotypic Cx37 gap junctions in conveying small molecules between oocytes and adjacent granulosa cells during oocyte and follicular growth. It leaves unexplained the fact that Cx37 is uniquely involved in oocyte–granulosa cell coupling.

10.6 Implications for Understanding Human Female Infertility

Although there is little information about connexins in the human ovary, correlative evidence points to the involvement of aberrant gap junctional coupling in female infertility. For example, a common diagnosis for infertile women is premature ovarian failure (POF). POF is defined as cessation of ovarian function before age 40 and can be associated with both primary and secondary amenorrhea (for reviews see Anasti 1998 or Gosden and Faddy 1998). Primary amenorrhea (absence of menses since puberty) is indicative of a fundamental defect of ovarian function, whereas secondary amenorrhea (cessation or interruption of menses once established) is often ascribed to premature attrition of oocytes. This can result from an inadequate supply of oocytes from birth or loss of oocytes due to injury (such as radiation or chemical toxicity), or disease. POF is heterogeneous in its etiology and has been ascribed to a variety of nongenetic and genetic factors, the latter category including mutations/rearrangements affecting the X chromosome and certain autosomes (reviewed by Christin-Maitre et al. 1998). For example, POF has been conclusively linked with mutations in the genes encoding gonadotropins and their receptors (reviewed by Achermann et al. 2001). Many cases of POF, however, are of undiscovered cause (idiopathic), and a proportion of those (28.5% in one study; Vegetti et al. 2000) are familial. Because connexin loss-of-function mutations in mice cause female infertility involving both insufficient numbers of oocytes (the Cx43 null mutation) and defects in follicular development (both the Cx37 and Cx43 null mutations), it is possible that abnormalities in gap junctional coupling underlie certain cases of familial POF in women. The case is strongest for Cx43, which is known to be expressed in granulosa cells of human ovaries (Furger et al. 1996). There is evidence of a correlation between the Cx43 mRNA level in human granulosa cells and the developmental competence of the associated oocytes (Tsai et al. 2003). It has been suggested that an important contribution of gap junctional coupling during folliculogenesis is to supply the growing oocyte with glucose to support glycolysis in the hypoxic environment of the follicle (Gregory and Leese1996). If this is correct, then an impairment of oocyte quality would be an expected outcome if gap junctional coupling within the follicle were impaired.

Involvement of connexin mutations in female infertility is supported by analysis of mutant mice carrying a point mutation that causes a single amino acid substitution (G60S) in the first extracellular loop of Cx43 (Flenniken et al. submitted). The mutant allele (*Gja1*^{Jrt}) acts in dominant fashion to severely restrict, but not eliminate, gap junctional coupling among granulosa cells. The mutant females, though not infertile, have small litters with poor survival of pups, suggesting a possible effect on oocyte quality. Interestingly, the broad phenotype of the mutant mice resembles that of humans with an inherited disorder, oculodentodigital dysplasia (ODDD), caused by other point mutations in *GJA1* (Paznekas et al. 2003; Richardson et al. 2004). It remains to be determined whether ODDD females have unrecognized reproductive problems related to the mutations.

10.7 Conclusions

Ovarian folliculogenesis and the production of fertilizable oocytes depend on gap junctional intercellular communication within both the developing and the mature follicle. Multiple connexins, including but probably not limited to Cx26, Cx30.3, Cx32, Cx37, Cx40, Cx43, and Cx45 are expressed within the oocyte–granulosa cell complex depending on the species, but it is not yet clear whether each has an essential role to play. In the mouse, the different phenotypes caused by the absence of Cx43 and Cx37 do indicate that these connexins play distinct and essential roles consistent with their differential localizations, with Cx43 in gap junctions coupling granulosa cells and Cx37 in gap junctions coupling the oocyte with surrounding granulosa cells. The possibility must be entertained that molecules passing through the heterologous gap junctions at the oocyte surface are different from those passing through gap junctions between granulosa cells, but the ability of Cx43 to substitute for Cx37 in oocytes implies that this would not necessarily be due to differential permeability of the two types of channels. Identification of the permeant molecules and elucidation of their functions in oogenesis continue to be research priorities.

Acknowledgements. Work from the author's laboratory cited in this review was funded by the Canadian Institutes of Health Research.

References

Achermann JC, Weiss J, Lee E-J, Jameson JL (2001) Inherited disorders of the gonadotropin hormones. Mol Cell Endocrinol 179:89–96

Ackert CL, Gittens JEI, O'Brien MJ, Eppig JJ, Kidder GM (2001) Intercellular communication via connexin43 gap junctions is required for ovarian folliculogenesis in the mouse. Dev Biol 233:258–270

Alcoléa S, Théveniau-Ruissy M, Jarry-Guichard T, Marics I, Tzouanacou E, Chauvin J-P, Briand J-P, Moorman AFM, Lamers WH, Gros D (1999) Downregulation of connexin45 gene products during mouse heart development. Circ Res 84:1365–1379

Anasti JN (1998) Premature ovarian failure: an update. Fertil Steril 70:1–15

Austin CR, Short RV (1982) Reproduction in Mammals: 1 Germ cells and fertilization. Cambridge University Press, Cambridge, UK

Carabatsos MJ, Elvin J, Matzuk MM, Albertini DF (1998) Characterization of oocyte and follicle development in growth differentiation factor-9-deficient mice. Dev Biol 204:373–384

Carabatsos MJ, Sellitto C, Goodenough DA, Albertini DF (2000) Oocyte-granulosa cell heterologous gap junctions are required for the coordination of nuclear and cytoplasmic meiotic competence. Dev Biol 226:167–179

Christin-Maitre S, Vasseur C, Portnoï M-F, Bouchard P (1998) Genes and premature ovarian failure. Mol Cell Endocrinol 145:75–80

Coskun S, Lin YC (1994) Effects of transforming growth factors and activin-A on in vitro porcine oocyte maturation. Mol Reprod Dev 38:53–159

Dong J, Albertini DF, Nishimori K, Kumar TR, Lu N, Matzuk MM (1996) Growth differentiation factor-9 is required during early ovarian folliculogenesis. Nature 383:531–535

Downs SM (1995) The influence of glucose, cumulus cells, and metabolic coupling on ATP levels and meiotic control in the isolated mouse oocyte. Dev Biol 167:502–512

Elfgang C, Eckert R, Lichtenberg-Fraté H, Butterweck A, Traub O, Klein RA, Hülser DF, Willecke K (1995) Specific permeability and selective formation of gap junction channels in connexin-transfected HeLa cells. J Cell Biol 129:805–817

Eppig JJ (199l) Intercommunication between mammalian oocytes and companion follicle cells. BioEssays 13:569–574

Eppig JJ, Wigglesworth K (2000) Development of mouse and rat oocytes in chimeric reaggregated ovaries after interspecific exchange of somatic and germ cell components. Biol Reprod 63:1014–1023

Eppig JJ, O'Brien M, Wigglesworth K (1996) Mammalian oocyte growth and development in vitro. Mol Reprod Dev 44:260–273

Fagbohun CF, Downs SM (1991) Metabolic coupling and ligand-stimulated meiotic maturation in the mouse cumulus-oocyte complex. Biol Reprod 45:851–859

Furger C, Cronier L, Poirot C, Pouchelet M (1996) Human granulosa cells in culture exhibit functional cyclic AMP-regulated gap junctions. Mol Human Reprod 2:541–548

Gittens JEI, Mhawi AA, Lidington D, Ouellette Y, Kidder GM (2003) Functional analysis of gap junctions in ovarian granulosa cells: a distinct role for connexin43 in early stages of folliculogenesis. Am J Physiol Cell Physiol 284:C880–C887

Gittens JEI, Barr KJ, Vanderhyden BC, Kidder GM (2005) Interplay between paracrine signaling and gap junctional coupling in ovarian follicles. J Cell Sci 118:113–122

Gosden RG, Faddy MJ (1998) Biological bases of premature ovarian failure. Reprod Fert Dev 10:73–78

Granot I, Dekel N (1994) Phosphorylation and expression of connexin-43 ovarian gap junction protein are regulated by luteinizing hormone. J Biol Chem 269:30502–30509

Grazul-Bilska AT, Redmer DA, Bilski JJ, Jablonka-Shariff A, Doraiswamy V, Reynolds LP (1998) Gap junctional proteins, connexin 26, 32, and 43 in sheep ovaries throughout the estrous cycle. Endocrine 8:269–279

Gregory L, Leese HJ (1996) Determinants of oocyte and preimplantation embryo quality: metabolic requirements and the potential role of cumulus cells. Hum Reprod 11 Suppl:96–102

Hombach S, Janssen-Bienhold U, Söhl G, Schubert T, Büssow H, Ott T, Weiler R, Willecke K (2004) Functional expression of connexin57 in horizontal cells of the mouse retina. Eur J Neurosci 19:633–640

Itahana K, Morikazu Y, Takeya T (1996) Differential expression of four connexin genes, Cx26, Cx30.3, Cx32, and Cx43, in the porcine ovarian follicle. Endocrinology 137:5036–5044

Itahana K, Tanaka T, Morikazu Y, Komatu S, Ishida N, Takeya T (1998) Isolation and characterization of a novel connexin gene, Cx-60, in porcine ovarian follicles. Endocrinology 139:320–329

Johnson ML, Redmer DA, Reynolds LP, Grazul-Bilska AT (1999) Expression of gap junctional proteins connexin 43, 32, and 26 throughout follicular development and atresia in cows. Endocrine 10:43–51

Juneja SC, Barr KJ, Enders GC, Kidder GM (1999) Defects in the germ line and gonads of mice lacking connexin43. Biol Reprod 60:1263–1270

Kidder GM, Mhawi AA (2002) Gap junctions and ovarian folliculogenesis. Reproduction 123:613–620

Manthey D, Bukauskas F, Lee CG, Kozak CA, Willecke K (1999) Molecular cloning and functional expression of the mouse gap junction gene connexin57 in human HeLa cells. J Biol Chem 274:14716–14723

Mitchell PA, Burghardt RC (1986) The ontogeny of nexuses (gap junctions) in the ovary of the fetal mouse. Anat Rec 214:283–288

Nelles E, Bützler C, Jung D, Temme A, Gabriel H-D, Dahl U, Traub O, Stümpel F, Jungermann K, Zielasek J, Toyka KV, Dermietzel R, Willecke K (1996) Defective propagation of signals generated by sympathetic nerve stimulation in the liver of connexin32-deficient mice. Proc Natl Acad Sci USA 93:9565–9570

Okuma A, Kuraoka A, Iida H, Inai T, Wasano K, Shibata Y (1996) Colocalization of connexin43 and connexin45 but absence of connexin40 in granulosa cell gap junctions of rat ovary. J Reprod Fertil 107:255–264

Paznekas WA, Boyadjiev SA, Shapiro RE, Daniels O, Wollnik B, Keegan CE, Innis JW, Dinulos MB, Christian C, Hannibal MC, Jabs EW (2003) Connexin43 (*GJA1*) mutations cause the pleiotropic phenotype of oculodentodigital dysplasia. Am J Hum Genet 72:408–418

Pérez-Armendariz EM, Sáez JC, Bravo-Moreno JF, López-Olmos V, Enders GC, Villalpando I (2003) Connexin43 is expressed in mouse fetal ovary. Anat Rec Part A 271A:360–367

Plum A, Hallas G, Magin T, Dombrowski F, Hagendorff A, Schumacher B, Wolpert C, Kim J-S, Lamers WH, Evert M, Meda P, Traub O, Willecke K (2000) Unique and shared functions of different connexins in mice. Curr Biol 10:1083–1091

Rankin TL, Tong ZB, Castle PE, Lee E, Gore-Langton R, Nelson LM, Dean J (1998) Human ZP3 restores fertility in Zp3 null mice without affecting order-specific sperm binding. Development 125:2415–2424

Reaume A, De Sousa PA, Kulkarni S, Langille BL, Zhu D, Davies TC, Juneja SC, Kidder GM, Rossant J (1995) Cardiac malformation in neonatal mice lacking connexin43. Science 267:1831–1834

Richardson RR, Donnai D, Meire F, Dixon MJ (2004) Expression of *Gja1* correlates with the phenotype observed in oculodentodigital syndrome/type III syndactyly. J Med Genet 41:60–67

Risek B, Guthrie S, Kumar NM, Gilula NB (1990) Modulation of gap junction transcript and protein expression during pregnancy in the rat. J Cell Biol 110:269–282
Roscoe WA, Barr KJ, Mhawi AA, Pomerantz DK, Kidder GM (2001) Failure of spermatogenesis in mice lacking connexin43. Biol Reprod 65:829–838
Simon AM, Goodenough DA, Li E, Paul DL (1997) Female infertility in mice lacking connexin 37. Nature 385:525–529
Tsai M-Y, Lan K-C, Huang K-E, Huang F-J, Kung F-T, Chang S-H (2003) Significance of mRNA levels of connexin37, connexin43, and connexin45 in luteinized granulosa cells of controlled hyperstimulated follicles. Fertil Steril 80:1437–1443
Udoff LC, Adashi EY (1999) Autocrine/paracrine regulation of the ovarian follicle. Endocrinologist 9:99–106
Valdimarsson G, De Sousa PA, Kidder GM (1993) Coexpression of gap junction proteins in the cumulus-oocyte complex. Mol Reprod Dev 36:7–15
Vegetti W, Marozzi A, Manfredini E, Testa G, Alagna F, Nicolosi A, Caliari I, Taborelli M, Tibiletti MG, Dalprà L, Crosignani PG (2000) Premature ovarian failure. Mol Cell Endocrinol 161:53–57
Veitch GI, Gittens JEI, Shao Q, Laird DW, Kidder GM (2004) Selective assembly of connexin37 into gap junctions at the oocyte/granulosa cell interface. J Cell Sci 117:2699–2707
Vitt UA, Hsueh AJW (2002) Stage-dependent role of growth differentiation factor-9 in ovarian follicle development. Mol Cell Endocrinol 186:211–217
Weber PA, Chang HC, Spaeth KE, Nitsche JM, Nicholson BJ (2004) The permeability of gap junction channels to probes of different size is dependent on connexin composition and permeant-pore affinities. Biophys J 87:958–973
Wong J, Luckers L, Okawara Y, Pelletier R-M, Taketo T (2000) Follicular development and atresia in the B6.Y^{TIR} sex-reversed mouse ovary. Biol Reprod 63:756–762
Wright CS, Becker DL, Lin JS, Warner AE, Hardy K (2001) Stage-specific and differential expression of gap junctions in the mouse ovary: connexin-specific roles in follicular regulation. Reproduction 121:77–88

11 Placental Connexins of Mice and Men

Caroline Dunk, Mark Kibschull, Alexandra Gellhaus,
Elke Winterhager, Stephen Lye[1]

11.1 Introduction

The placenta is the first differentiated embryonic organ to form a physical connection between the mother and the embryo. Impairment of placental development results in clinical abnormalities, such as intrauterine growth restriction, preeclampsia and HELLP syndrome or death of the conceptus. The embryonic cell populations responsible for establishing the contact with the mother are the trophoblast cells. Trophoblast cells originate from the outer epithelial layer of the blastocyst, the trophectoderm, which is responsible for the attachment of the blastocyst to the uterine epithelium and the initiation of the implantation process. Later, these cells differentiate into the placental organ along the trophoblast cell lineage. The generation of distinct trophoblast cell types is required to fulfill the complex functions of this transient organ. Trophoblast cells generate structures known as villi with a large surface area for the controlled exchange of nutrients and waste between fetus and mother. Furthermore, they interact closely with the maternal compartment by invading into the decidual tissue and the uterine arteries to establish the nutrition route. Trophoblast cells also are able to adapt to the immunological response of the mother. As embryonic and fetal mortality and women's health is affected by impaired trophoblast development, the molecular control of this differentiation process has received increasing interest in recent years (reviews: Cross 2000; Cross et al. 2003; Loregger et al. 2003).

The study of mutant mice gives insight into the molecular signaling pathways controlling this cell lineage differentiation and appropriate physiology of the different trophoblast cell populations. Though placenta from different species are morphologically rather different (Wooding and Flint 1994), the basic function of a rodent placenta and a human placenta needs to be the same. Both belong to the hemochorial type of placenta, meaning that the maternal blood comes into direct contact with the transport and barrier trophoblast (Cross et al. 2003). These are comprised in rodents

[1]Samuel Lunenfeld Research Institute, Mount Sinai Hospital, 600 University Avenue, Toronto, Ontario M5G 1X5, Canada, E-mail: lye@mshri.on.ca

Gap Junctions in Development and Disease
(ed. by E. Winterhager)

by the three-layered labyrinthine trophoblast and in humans by the villous trophoblast which contains one syncytial layer with an underlying cytotrophoblast layer. The invasive and endocrine trophoblast in rodents consists of the spongiotrophoblast, which generates invasive glycogen cells and giant cells (Cross et al. 2003). In humans, the extravillous trophoblast represents this function with floating islands or cell columns anchored to the decidua. These cell columns generate similar to the spongiotrophoblast and giant cells in rodents. In rodents and humans the polyploid giant cells are invasive and erode maternal vessels and replace endothelial cells (endovascular trophoblast) to establish the uteroplacental circulation, probably by using similar molecular mechanisms (for comparison of the two placental types, see Fig. 11.1).

Thus, different placental structures serve the same function and as a consequence the results obtained in mice may be useful for understanding reproductive problems and placental dysfunction in humans. The trophoblast

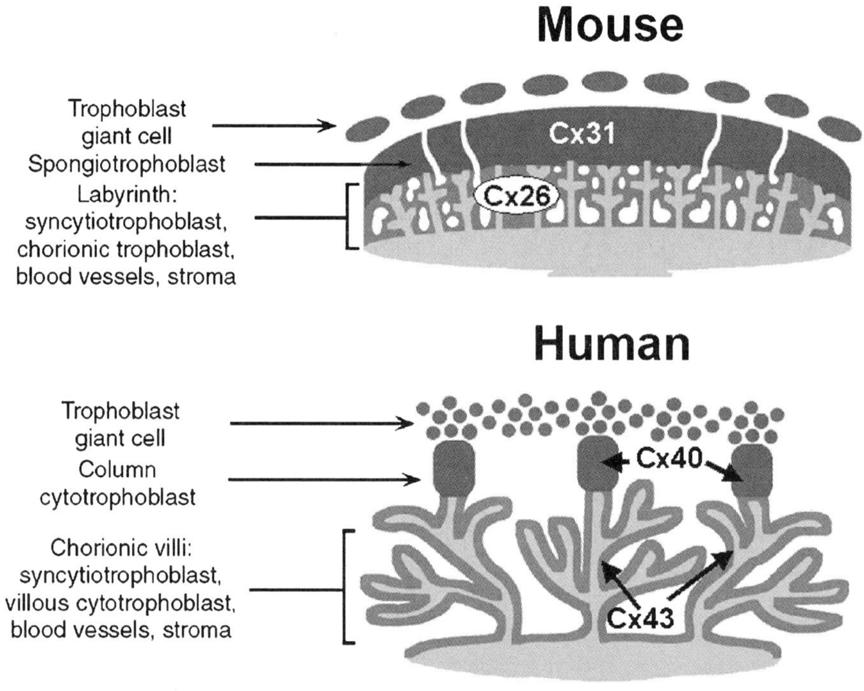

Fig. 11.1. Schematic representation of mouse and human placental architecture. In the mouse placenta, Cx31 is expressed by the spongiotrophoblast layer while the syncytiotrophoblast of the labyrinth expresses Cx26. In contrast, in the human placenta, the extravillous trophoblast cell columns express Cx40 while the syncytiotrophoblast expresses Cx43. (Cross 2003, with permission)

populations in human and mice share common gene expression patterns (Hemberger and Cross 2001). Among all these genes responsible for placental development and function, discovered by generating knockout mice, transcription factors such as *Mash2*, *Gcm1* and *Tef5* are also expressed in comparable functional compartments of the human placenta (reviewed by Hemberger and Cross 2001). Furthermore, some other proteins such as the matrix metalloproteinase 9 and the cell adhesion molecule $\alpha 1\beta 1$ integrin characterize both human and murine giant cell populations (Hemberger and Cross 2001). Within this broad field of molecules involved in placental development, we have focused on the cell–cell communication channels, which are known to be responsible for coordinating developmental processes.

11.1.1 Temporal and Spatial Pattern of Connexins in Placentae of Mice and Men

Screening of the different connexin isoforms have revealed that in the ectoplacental cone which gives rise to the placenta, only the Cx31 channel is expressed, whereas the mature mouse placenta reveals different connexins characterizing the different trophoblast subpopulations. Cx26 is located in the labyrinth, Cx31 characterizes the spongiotrophoblast population and Cx43 is found from midgestation onwards coexpressed with Cx31 (Grummer et al. 1996; Fig. 11.1).

In contrast to rodents, multiple connexins have been detected in the early human placenta, RT-PCR analysis has shown that transcripts for Cx32, Cx37, Cx40, Cx43, and Cx45 can be detected between 6 and 14 weeks of gestation (Nishimura et al. 2004). Of particular interest are Cx40 and Cx43, which localize to the different trophoblast subpopulations and appear to play a functional role in regulating their differentiation. Cx43 protein localizes to the differentiating cytotrophoblasts as they fuse to form syncytium (Cronier et al. 1994), and Cx40 to the extravillous trophoblast cell columns (Winterhager et al. 1999; Nishimura et al. 2004; Fig. 11.1). Thus, the comparable placental compartments in mice and men do not share the same connexin channels.

11.2 The Role of Gap Junction Connexins in the Murine Placenta

It is well established that intercellular communication via gap junctions is required to coordinate developmental processes in the mammalian placenta evidenced by the generation of different connexin-deficient mice.

The phenotypes of these mice have suggested that Cx26 and Cx31, but not Cx43, play an obligate role during placental development (Reaume et al., 1995; Gabriel et al. 1998; Plum et al. 2001). Each of these channels shows functional specializations within this tissue.

11.2.1 Connexin 26 Facilitates Transport Across the Placental Barrier

The Cx26-deficient mouse was the first knockout mouse to indicate that the connexin gene family was involved in placental development and function. All embryos missing the Cx26 channel died in utero around day 11 pc. The embryos were significantly smaller, did not show malformation, and the placenta revealed no dramatic morphological defects. The trophoblast cell lineage derivatives, the labyrinth, the spongiotrophoblast, and giant cells were present and arranged in the usual architecture (Gabriel et al. 1998). Several groups have documented in former investigations that in rat and mice gap junctions were abundantly expressed in the labyrinthine part. In contrast to the human villous trophoblast, the labyrinth trophoblast is composed of three cell layers. The outer trophoblast layer lining the maternal blood space looks similar to a fenestrated endothelium. The effective maternal–fetal barrier, however, is established by the syncytiotrophoblast layer I and syncytiotrophoblast layer II (Fig. 11.2).

Cx26 has been identified to connect both syncytial layers of the labyrinth (Risek and Gilua 1991; Grummer et al. 1996). Takata (1994) and Shin et al. (1996) had already suggested that Cx26 was involved in the transport of glucose across this placental barrier. They found that GLUT1 localized at the outer membrane of the maternal blood side of syncytial layer I and to the fetal blood side of syncytial layer II. According to the localization, the GLUT1 transporters are responsible for uptake of glucose from the maternal blood and release to the fetal compartment, however, for crossing the syncytial layers the Cx26 channels appeared to be needed (Fig. 11.2).

Indeed, mice lacking the Cx26 gene failed to pass glucose across the placental barrier in sufficient amounts, as demonstrated by a significantly reduced uptake of 3-*O*-[^{14}C] methylglucose into the embryonic compartment (Fig. 11.2). Embryonic lethality was not primarily caused by severe malformations or placental dysmorphogenesis, but by dysfunction of the placental transport, leading to starvation of the rapidly developing embryo (Gabriel et al. 1998). This hypothesis was supported by the fact that during this critical time of placental development the chorioallantoic placenta takes over the function of the early yolk sac placenta to mediate nutrition and waste disposal of the embryo (Poelmann and Mentink 1983).

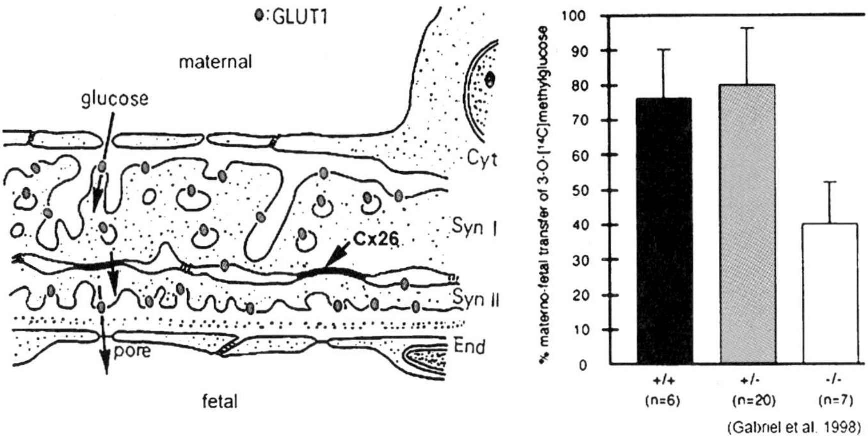

Fig. 11.2. A Schematic drawing of the trophoblast layers of the labyrinth in the murine placenta. The localization of gap junctional Cx26 channels and glucose transporter GLUT1 proteins as well as the likely route of glucose transfer from maternal to fetal blood vessels are illustrated. Glucose enters via GLUT1; the cytoplasm of the syncytiotrophoblast I diffuses mainly via gap-junction channels into syncytiotrophoblast layer II and is released via GLUT1 at the basal side of the trophoblast membrane into fetal blood circulation through fenestrated endothelial cells. *M* Maternal blood, *Cyt* cytotrophoblast, *Syn I* syncytiotrophoblastic layer I, *GJ* Cx26-containing gap junctions, *Syn II* syncytiotrophoblastic layer II, *BL* basal lamina, *E* endothelial cell, *F* fetal blood circulation. **B** Uptake of 3-*O*-[^{14}C]methylglucose into whole embryos at day 10 pc from maternal blood across the placenta. Significantly decreased accumulation ($p < 0.01$) of radioactivity was measured in homozygous Cx26-defective embryos compared with heterozygous and wild-type embryos (+/+). *n* designates the number of embryos investigated. (Adapted from Gabriel et al. 1998)

Taken together, these data suggest that Cx26 may serve as a diffusion channel for the exchange of small metabolites, predominantly glucose diffusion along the gradient from the mother to the fetus.

11.2.2
Connexin 31 Regulates Murine Trophoblast Cell Lineage Development

The inactivation of the Cx31 gene results in a transient placental dysmorphogenesis that leads to the intrauterine death of the embryos between day 10.5 and 13.5 post coitum (Plum et al. 2001). Morphological evaluation of the Cx31-deficient placenta phenotype revealed a severe failure in trophoblast differentiation on day 9.5 pc. The labyrinth and the spongiotrophoblast were barely developed, whereas a strong increase in the number of trophoblast giant cells could be detected. As a consequence of this dysmorphogenesis, the placenta failed to establish a nutrition route sufficient for embryonic growth and the embryos start to die from day 10.5 pc onwards, when the chorionic placenta takes over the transport of nutrients. Thus, the absence of the Cx31 channel leads to an imbalance in trophoblast differentiation in favor of giant cells during placental development. This enhanced differentiation of the trophoblast stem cells along the trophoblast cell lineage to giant cells is accompanied by a stop in cell proliferation. However, this impairment seems only to be a transient phenomenon since only 60% of the Cx31-deficient embryos died, whereas the placenta of the others showed developmental recovery leading to normal embryos without an obvious phenotype. The reason for this recovery is not clear yet. It has been assumed that the expression of Cx43, which starts at day 10.5 pc in the spongiotrophoblast, just when the Cx31 knockout embryos start dying, could compensate for the loss of the Cx31 channel (Plum et al. 2001). This hypothesis has been negated by the generation of the Cx31/Cx43 double-deficient mouse that showed viable embryos. The double knockout revealed both phenotypes, first the transient placental phenotype of the Cx31 knockout mouse and second the early postnatal lethality of the Cx43-deficient mouse (Reaume et al. 1995), but demonstrated no combined effect. The placental morphology of these embryos did not differ from the Cx31 knockout placenta. Thus, Cx43 cannot serve for the recovery of the Cx31 knockout placenta.

The role of the Cx43 in mouse placental development and physiology remains unclear since the placenta lacking the Cx43 channel does not demonstrate any morphological alterations.

Taken together, each connexin defines a specific subpopulation of the trophoblast: Cx26 the syncytiotrophoblast of the labyrinth; Cx31 the early trophoblast cell lineage and the placental spongiotrophoblast; Cx43 the late spongiotrophoblast and the trophoblast giant cells. Of special interest is Cx31 because it is associated with differentiation in the trophoblast cell lineage.

11.2.3 TS Cells as a Model to Investigate the Role of Connexins in Placental Development

The analysis of the molecular function of genes and proteins in the trophoblast cell lineage and the placenta has been advanced by the generation of trophoblast stem cells (Tanaka et al. 1998). Trophoblast stem (TS) cells can be generated from the trophectoderm of the blastocyst in the presence of FGF4 and embryonic mouse fibroblasts resulting in a continuous proliferating trophoblast specific cell line. These cells have the capacity to develop in all trophoblast subpopulations in vivo shown by aggregation experiments (Tanaka et al. 1998). Upon removal of FGF4, TS cells differentiate in vitro to all placental trophoblast subpopulations, as demonstrated by the expression of trophoblast marker genes (Tanaka et al. 1998; Hughes et al. 2004), and also by the expression of connexins (Kibschull et al. 2004). Undifferentiated TS cells only express Cx31 at protein level, on differentiation of TS cells the expression of both Cx26 and Cx43 is induced in different subpopulations of the culture.

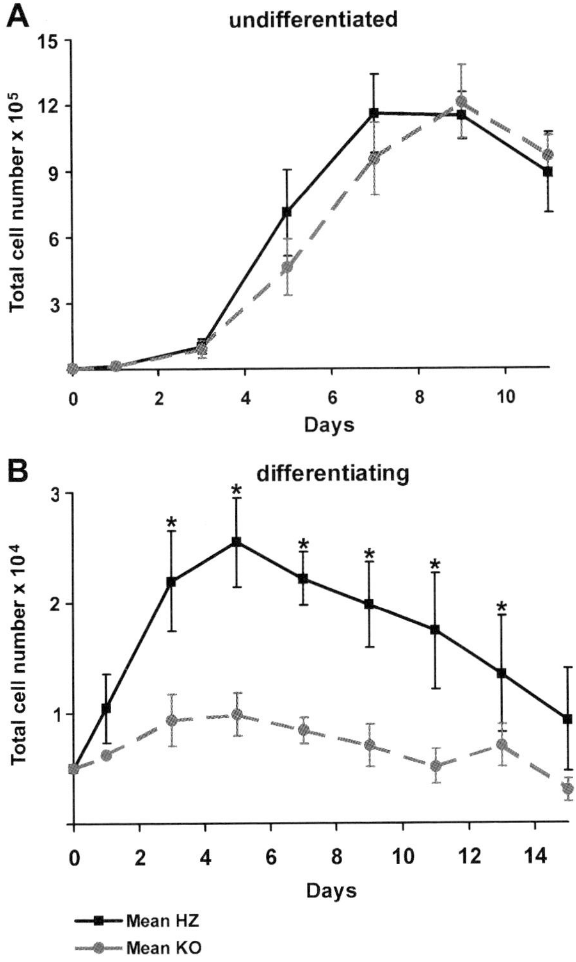

Fig. 11.3A,B. Reduced proliferation of differentiating Cx31−/− TS cells. Proliferation of Cx31+/− and Cx31−/− TS cell lines was investigated under undifferentiated (**A**) and differentiated (**B**) conditions by measuring total cell numbers every second day as indicated. Mean and standard deviation were calculated for each of the heterozygous and homozygous TS cell lines. Time points where values differed significantly (* *p* V 0.05) are indicated

Thus, TS cells are an appropriate in vitro model to analyze the function of genes in trophoblast differentiation. Unlike the placental tissue, TS cells are free from contamination with other maternal or fetal tissues. Moreover, TS cell lines generated from knockout mice provide a valuable system to study the effects of a gene knockout as has been shown by several groups (Hemberger et al. 2004; Hughes et al. 2004).

We have generated Cx31-deficient TS cell lines from blastocysts of the knockout strain to analyze their differentiation capacity in comparison to controls. The expression profiles of trophoblast marker genes and the connexins showed that the lack of the Cx31 channel results in an enhanced differentiation process indicated by the earlier induction of Tpbpa, Pl-1 and Cx26 compared to controls. An increased number of giant cells and most interestingly, a stop in proliferation (Fig. 11.3) accompany this enhanced differentiation process. Hence, the Cx31-deficient TS cells show the same phenotype as the Cx31 knockout placenta: first, an imbalance in trophoblast differentiation in favor of giant cell formation and a stop in proliferation (Kibschull et al. 2004). Thus, these cell lines display a favored model for further analysis of signal cascades of the connexins in trophoblast differentiation. Moreover, such a model provides the chance to discriminate channel capacity for differentiation from the interaction of the protein itself with intracellular signaling cascades by transfecting different connexin mutants.

11.3 The Role of Gap Junction Connexins in the Human Placenta

During development of the human placenta, two connexin channels characterize the two separate trophoblast subpopulations. The villous cytotrophoblasts express Cx43 in their apical membrane, while the extravillous trophoblast cell columns are characterized by Cx40.

11.3.1 Cx43 Supports Villous Cytotrophoblast Fusion to Form Syncytium

As discussed earlier, the human villous trophoblasts are comprised of a single layer of cytotrophoblasts overlaid by a single syncytiotrophoblast layer that is in direct contact with the maternal blood and provides the transport route between the maternal and fetal circulations. The syncytiotrophoblast layer is a fused terminally differentiated structure that is generated and maintained throughout pregnancy by the continuous fusion of

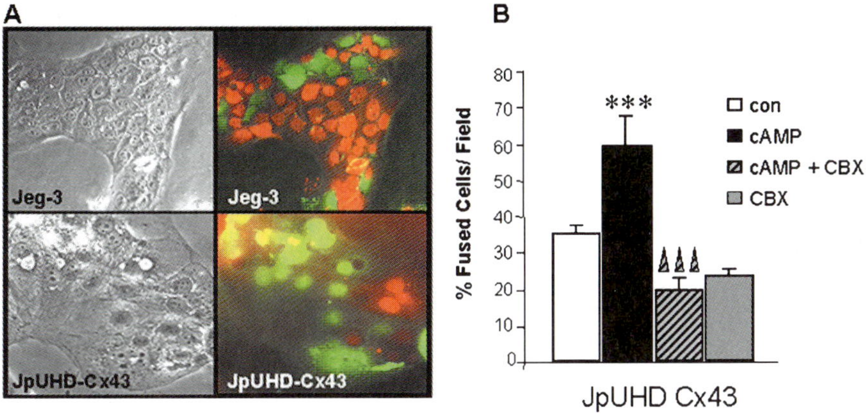

Fig. 11.4. A Treatment of Cx43 expressing cells with dibutyl cAMP induces cell fusion, shown by the merging of green and red dyes. Wild-type Jeg-3 cells do not express connexins and do not fuse. **B** Cx43-mediated cell fusion was completely inhibited by blockade of the Cx43 channel by carbenoxylone, demonstrating the requirement for GJIC

the underlying cytotrophoblasts. Several groups have demonstrated that gap junctional intercellular communication (GJIC) mediated by the Cx43 channel is integral to this process. Isolated cytotrophoblast cells are known to undergo spontaneous differentiation to syncytiotrophoblasts in vitro and upregulate Cx43 expression during this process (Cronier et al. 1994, 2003). Moreover, stable transfection of the transformed choriocarinoma cell line with Cx43 restores the ability of these proliferative and invasive cells to fuse and terminally differentiate (Fig. 11.4; Dunk et al. 2003). The Cx43-mediated cell fusion can be attributed to GJIC as treatment with Carbenoxelone or heptanol (GJIC blockers), inhibits cell fusion in both normal and malignant cells (Cronier et al. 2003, Dunk et al. 2003), suggesting it is the Cx43-mediated cell–cell communication that is important in this process (Fig. 11.4). However, before the cytotrophoblasts differentiate and fuse into the syncytium they must stop proliferating, and new data suggest an additional novel role for Cx43 in this process (Gellhaus et al. 2004).

11.3.2 A Novel Role for Cx43 in Intracellular Signaling

Recent studies have suggested a channel-independent role for connexins in intracellular signaling through interactions with other intracellular proteins. The connexin proteins are composed of four transmembrane groups that contribute to the channel and a C-terminal intracellular cytoplasmic tail. Cx43's tail contains numerous phosphorylation residues and has been shown to interact with a number of intracellular proteins such as ZO-1 (Giepmans and Moolenaar 1998), Src (Giepmans et al. 2001) and α and β tubulin (Giepmans et al. 2001). Moreover, it has been shown that the Cx43 protein, independent of channel formation, can affect growth control, cell adhesion and migration (Omori and Yamasaki 1998; Xu et al. 2001; Lin et al. 2002; Dang et al. 2003; Olbina and Eckhart 2003.) In a recent study we have shown that induction of Cx43 in the malignant trophoblast cell line Jeg-3 was able to reduce cell proliferation in vitro and tumor growth in vivo (Gellhaus et al. 2004). It was further demonstrated that the induction of Cx43 resulted in: the upregulation of the negative growth regulatory gene *nov*; a translocation of NOV from the nucleus to the cytoplasmic membrane, where it interacted with the Cx43 C-terminal tail (Gellhaus et al. 2004). These data suggest that Cx43 may mediate both the arrest of cytotrophoblast proliferation, through intracellular signaling, and the fusion of the cytotrophoblast into the syncytium, through cell–cell communication.

11.3.3 Connexin 40 Regulates Human Trophoblast Extravillous Cell Lineage Development

In the human placenta, the extravillous trophoblast (EVT) cell columns form at the tips of the anchoring villi, which attach the placenta to the uterine wall. At the base of the column, the EVT are proliferative; at the distal edge of the column, the EVT differentiate to become one of two populations of invasive EVT; either interstitial trophoblast or endovascular trophoblast; similar to the mouse spongiotrophoblast and giant cells. The interstitial trophoblasts invade the decidual stroma as far as the superficial layer of the myometrium: in contrast, the endovascular trophoblast arise as groups of trophoblast cells, which detach from the trophoblastic shell and invade the lumen of the decidual uterine spiral arteries. The EVT mediate substantial vascular remodeling that results in an increase in maternal blood flow to the intervillous space.

Cx40 is prominently expressed in anchoring cell column EVT cells throughout the first trimester in vivo (Winterhager et al. 1999; Fig. 11.5A). We have shown in an in vitro placental explant model of the anchoring cell column that Cx40 is similarly expressed in the proliferating EVT outgrowth, analogous to the proximal cell column (Nishimura et al. 2004; Fig. 11.5B). When placental explants were treated with GJIC inhibitors such as CBX or heptanol, or Cx40 expression was attenuated by antisense oligonucleotide treatment, the compact and organized outgrowth EVT cells became scattered, rounded and migrated from the villous tip, reminiscent of a more invasive phenotype (Nishimura et al. 2004). We further demonstrated that

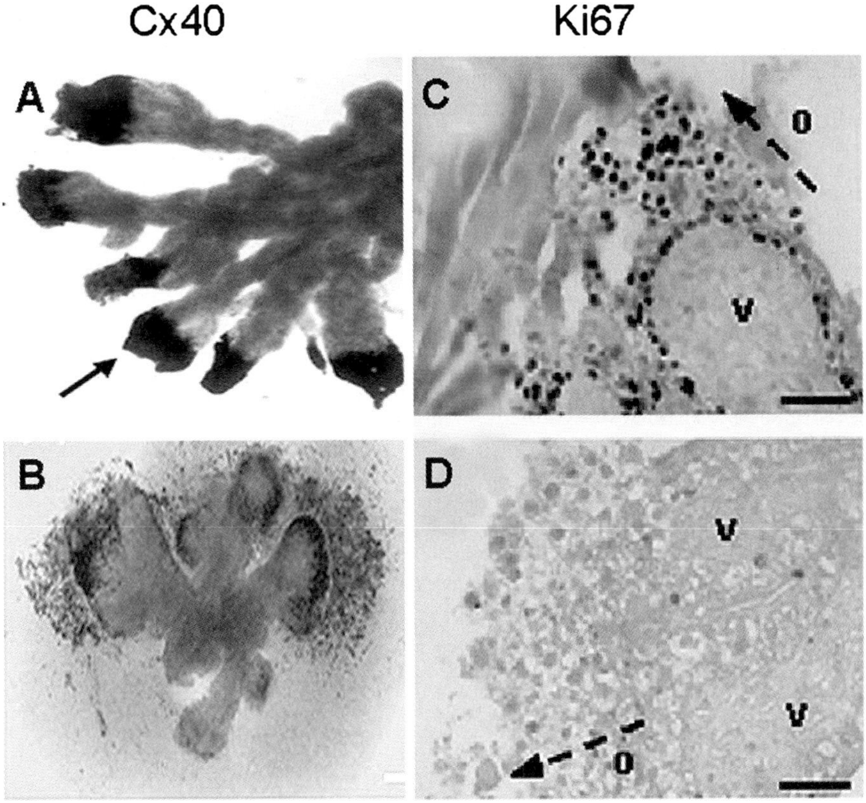

Fig. 11.5. **A** Cx40 mRNA is expressed by the extravillous trophoblast cell columns in vivo and **B** is maintained in vitro by the outgrowth extravillous trophoblast in first trimester villous explants. **C** Cx40 localizes to the proliferating extravillous trophoblast demonstrated by positive immunostaining for the proliferation marker Ki67. **D** Blockade of GJIC using carbenoxylone results in detachment and scattering of the trophoblast outgrowth and a loss of Ki67 expression (**D** compared to **C**), suggesting an exit of the trophoblast from the proliferative anchoring cell column phenotype

the blockade of GJIC in placental explants induces an exit from the proliferative EVT phenotype with a decrease in Ki67 (Fig. 11.5C,D) and MMP-2 expression and the induction of an invasive phenotype, as evidenced by the induction of $\alpha 1$ integrin expression in the EVT (Nishimura et al. 2004). Similarly, Cx40 protein is absent from EVT cells that detach from the distal cell column in vivo and invade into the decidua (Winterhager et al. 1999) and gap junction structures are also absent in these cells, as determined by electron microscopic cytological examination (Enders et al. 2001). Together, these data imply that Cx40-mediated GJIC is important in column integrity and organized migration similar to the role played by

Cx31 in mouse giant cell differentiation. Both of these connexins maintain a proliferative phenotype and prevent the cells from an enhanced differentiation along the invasive pathway. GJIC may also transduce changes in intracellular signaling molecules along a cell column from distal EVT cells responding to decidual-derived chemotactic ligands. We have new data that suggest that it may indeed be decidual-derived factors responsible as decidual-conditioned medium results in the downregulation of Cx40 in the distal EVT and the induction of the invasive EVT phenotype in the outgrowth EVT (Dunk et al. 2004).

As the EVT invade the uterine wall they approach the myometrium and must undergo a final differentiation state to become noninvasive and nonproliferative. It is suggested that the connexins are re-expressed during this process as the mononuclear interstitial EVT in cell aggregates within the decidua possess gap junction structures composed of Cx43 and Cx32, suggestive of a role in the fusion of interstitial EVT to form giant cells (Al Lamki et al. 1999). Furthermore, endothelial cells also express Cx40; it has been suggested that the endovascular EVT may re-express Cx40 to make homomeric gap junctions when they contact the maternal endothelial cells and acquire an endothelial phenotype through cell–cell communication during vessel remodeling, though this remains to be confirmed.

11.4 Analogous Functions of Murine and Human Connexins in Placental Development

As mentioned before, the human villous trophoblast is functionally analogous to the mouse labyrinth in that both structures predominantly provide the fetus with nutrients. However, the morphological architecture of the three-layered trophoblast comprising two syncytial layers in rodents and the cytotrophoblast and one syncytial layer in humans require different connexin channels responsible for different functions. The Cx26 channel is specific for the rodent placenta and supports the transplacental uptake of nutrients across the two syncytial layers, which are missing in humans, while the cytotrophoblast fusion to form the human syncytium appears to be mediated by the Cx43 channel. Interestingly, when mouse Cx26 is stably transfected into the human choriocarinoma cell line Jeg-3, it too, like Cx43, can promote and permit cAMP-stimulated cell fusion (unpubl. data). These data suggest that Cx26 and Cx43 are functional analogues for syncytial formation and may utilize similar signaling pathways. Similarly, an analogous functional role can be suggested for Cx31 and Cx40 in the differentiation of the extravillous trophoblast. Cx31 characterizes the diploid spongiotrophoblast in the mouse placenta as they differentiate to the invasive giant cells. In humans, Cx40 characterizes the extravillous trophoblast, regulating both proliferation and differentiation into invasive trophoblast cells. In the mouse, TS model and the human placental EVT model trophoblast proliferation and the switch to an invasive phenotype is directly regulated by Cx31 and Cx40, respectively, thereby providing the temporal and spatial control necessary for successful placental development.

In conclusion, these data imply a central role for the members of the gap junction family in regulating placental development, through control of the differentiation of the different trophoblast subpopulations in both mice and men.

References

al-Lamki RS, Skepper JN, Burton GJ (1999) Are human placental bed giant cells merely aggregates of small mononuclear trophoblast cells? An ultrastructural and immunocytochemical study. Hum Reprod 14:496–504

Cronier L, Bastide B, Herve JC, Deleze J, Malassine A (1994) Gap junctional communication during human trophoblast differentiation: influence of human chorionic gonadotropin. Endocrinology 135:402–408

Cronier L, Frendo JL, Defamie N, Pidoux G, Bertin G, Guibourdenche J, Pointis G, Malassine (2003) A. Requirement of gap junctional intercellular communication for human villous trophoblast differentiation. Biol Reprod 5:1472–1480

Cross JC (2000). Genetic insights into trophoblast differentiation and placental morphogenesis. Semin Cell Dev Biol 11:105–113

Cross JC (2003) the genetics of pre-eclampsia: a feto-maternal problem? Clin Genet 64:96–103

Cross JC, Baczyk D, Dobric N, Hemberger M, Hughes M, Simmons DG, Yamamoto H, Kingdom JC (2003) Genes, development and evolution of the placenta. Placenta 24:123–130

Dang X, Doble BW, Kardami E (2003) The carboxy-tail of connexin-43 localizes to the nucleus and inhibits cell growth. Mol Cell Biochem 242:35–38

Dunk C, Pavri S, Gellhaus, Winterhager E, Lye S (2003) Connexin 43 expression increases trophoblast cell fusion: a role in syncytilisation. J Soc Gynecol Investig 10:198A

Dunk C, Nishimura T, Winterhager E, Rossant J, Lye S (2004) Decidual conditioned medium downregulates Cx40 and induces trophoblast differentiation and invasion. J Soc Gynecol Invest 11:300A

Enders AC, Blankenship TN, Fazleabas AT, Jones CJ (2001) Structure of anchoring villi and the trophoblastic shell in the human, baboon and macaque placenta. Placenta 22:284–303

Gabriel HD, Jung D, Bützler C, Temme A, Traub O, Winterhager E Willecke K (1998) Transplacental uptake of glucose is decreased in embryonic lethal connexin26 deficient mice. J Cell Biol 140:1453–1461

Gellhaus A, Dong X, Propson S, Maass K, Klein-Hitpass L, Kibschull M, Traub O, Willecke K, Perbal B, Lye SJ, Winterhager E (2004) Connexin43 interacts with NOV: a possible mechanism for negative regulation of cell growth in choriocarcinoma cells. J Biol Chem 279:36931–36942

Grummer R, Reuss B, Winterhager E (1996) Expression pattern of different gap junction connexins is related to embryo implantation. Int J Dev Biol 40:361–367

Hemberger M, Cross JC (2001) Genes governing placental development. Trends Endocrinol Metab 12:162–168

Hemberger M, Hughes M, Cross JC (2004) Trophoblast stem cells differentiate in vitro into invasive trophoblast giant cells. Dev Biol 271:362–371

Hughes M, Dobric N, Scott IC, Su L, Starovic M, St-Pierre B, Egan SE, Kingdom JC, Cross JC (2004) The Hand1, Stra13 and Gcm1 transcription factors override FGF signaling to promote terminal differentiation of trophoblast stem cells. Dev Biol 271:26–37

Kibschull M, Nassiry M, Dunk C, Gellhaus A, Quinn JA, Rossant J, Lye SJ, Winterhager E (2004) Connexin31-deficient trophoblast stem cells: a model to analyze the role of gap junction communication in mouse placental development. Dev Biol 273:63–75

Lin JH, Takano T, Cotrina ML, Arcuino G, Kang J, Liu S, Gao Q, Jiang L, Li F, Lichtenberg-Frate H, Haubrich S, Willecke K, Goldman SA, Nedergaard M (2002) Connexin 43 enhances the adhesivity and mediates the invasion of malignant glioma cells. J Neurosci1 22:4302–4311

Loregger T, Pollheimer J, Knofler M (2003) Regulatory transcription factors controlling function and differentiation of human trophoblast-a review. Placenta 24 Suppl A:S104–110

Nishimura T, Dunk C, Lu Y, Feng X, Gellhaus A, Winterhager E, Rossant J, Lye SJ (2004) Gap junctions are required for trophoblast proliferation in early human placental development. Placenta 25:595–607

Olbina G, Eckhart W (2003) Mutations in the second extracellular region of connexin 43 prevent localization to the plasma membrane, but do not affect its ability to suppress cell growth. Mol Cancer Res 1:690–700

Omori Y, Yamasaki H (1998) Mutated connexin43 proteins inhibit rat glioma cell growth suppression mediated by wild-type connexin43 in a dominant-negative manner. Int J Cancer 78:446–53

Plum A, Winterhager E, Pesch J, Lautermann J, Hallas G, Rosentreter B, Traub O, Herberhold C Willecke K (2001) Connexin31-null mutation in mice causes transient placental dysmorphogenesis but does not impair hearing and skin differentiation. Dev Biol 231:334–347

Poelmann RE, Mentink MMT (1982) Parietal yolk sac in early gestation mouse embryos: structure and function. Bibl Anat 22:123–127

Reaume AG, de Sousa PA, Kulkarni S, Langille BL, Zhu D, Davies TC, Juneja SC, Kidder GM, Rossant J (1995) Cardiac malformation in neonatal mice lacking connexin43. Science 267:1831–1834

Risek B, Gilula NB (1991) Spatiotemporal expression of three gap junction gene products involved in fetomaternal communication during rat pregnancy. Development 113:165–81

Shin BC, Suzuki T, Matsuzaki T, Tanaka S, Kuraoka A, Shibata Y, Takata K (1996) Immunolocalization of GLUT1 and connexin 26 in the rat placenta. Cell Tissue Res 285:83–89

Takata K, Kasahara T, Kasahara M, Ezaki O, Hirano H (1994) Immunolocalization of glucose transporter GLUT1 in the rat placental barrier: possible role of GLUT1 and the gap junction in the transport of glucose across the placental barrier. Cell Tissue Res 276:411–418

Tanaka S, Kunath T, Hadjantonakis AK, Nagy A, Rossant J (1998) Promotion of trophoblast stem cell proliferation by FGF4. Science 282:2072–2075

Winterhager E, Von Ostau C, Gerke M, Gruemmer R, Traub O, Kaufmann P (1999) Connexin expression patterns in human trophoblast cells during placental development. Placenta 20:627–638

Xu X, Li WE, Huang GY, Meyer R, Chen T, Luo Y, Thomas MP, Radice GL, Lo CW (2001) Modulation of mouse neural crest cell motility by N-cadherin and connexin 43 gap junctions. J Cell Biol 154:217–230

12 Connexins in Growth Control and Cancer

Christian C. Naus[1], Gary S. Goldberg[2], Wun Chey Sin[1]

12.1 Introduction

Gap junctions, and their constituent connexin proteins, have long been considered to play a role in the control of cell growth and carcinogenesis. The current review will attempt to summarize the evidence supporting the long-held view that gap-junction channels themselves play a preeminent role in controlling cell proliferation and preventing carcinogenesis through mechanisms of direct intercellular communication. In addition, it is timely to address emerging roles of connexin proteins in their own right, not directly in their channel-forming function, i.e., nonjunctional functions.

In the years since the initial report by Lowenstein and Kanno (1966), which attributed a growth-suppressive role to gap junctions, understanding the mechanisms by which such growth is controlled has remained elusive. Substantial evidence linking decreased connexin expression and gap junctional coupling in association with an increased tumorigenic phenotype can readily be found. However, several well-documented exceptions exist. Connexin expression occurs in some tumor cells and is associated with gap junctional coupling. Following transfection into tumor cells, connexins display variability in their tumor-suppressing ability. In many cases, there is a discrepancy between in vivo and in vitro growth effects of enhanced connexin expression. In some cases, there has been a surprising absence of enhanced gap junctional coupling in spite of increased connexin expression and growth suppression. These discrepancies have motivated the search for other mechanisms by which connexins may alter growth. One avenue recently pursued concerns connexin-mediated alteration in expression of other genes which may be involved in growth control. This has been complemented by the identification of connexin interacting proteins and their possible role in growth suppression.

[1]Department of Cellular and Physiological Sciences, The University of British Columbia, Vancouver, BC V6T 1Z3, Canada, E-mail: cnaus@interchange.ubc.ca
[2]Department of Molecular Biology, University of Medicine and Dentistry of New Jersey, Stratford, NJ 08084

Gap Junctions in Development and Disease
(ed. by E. Winterhager)

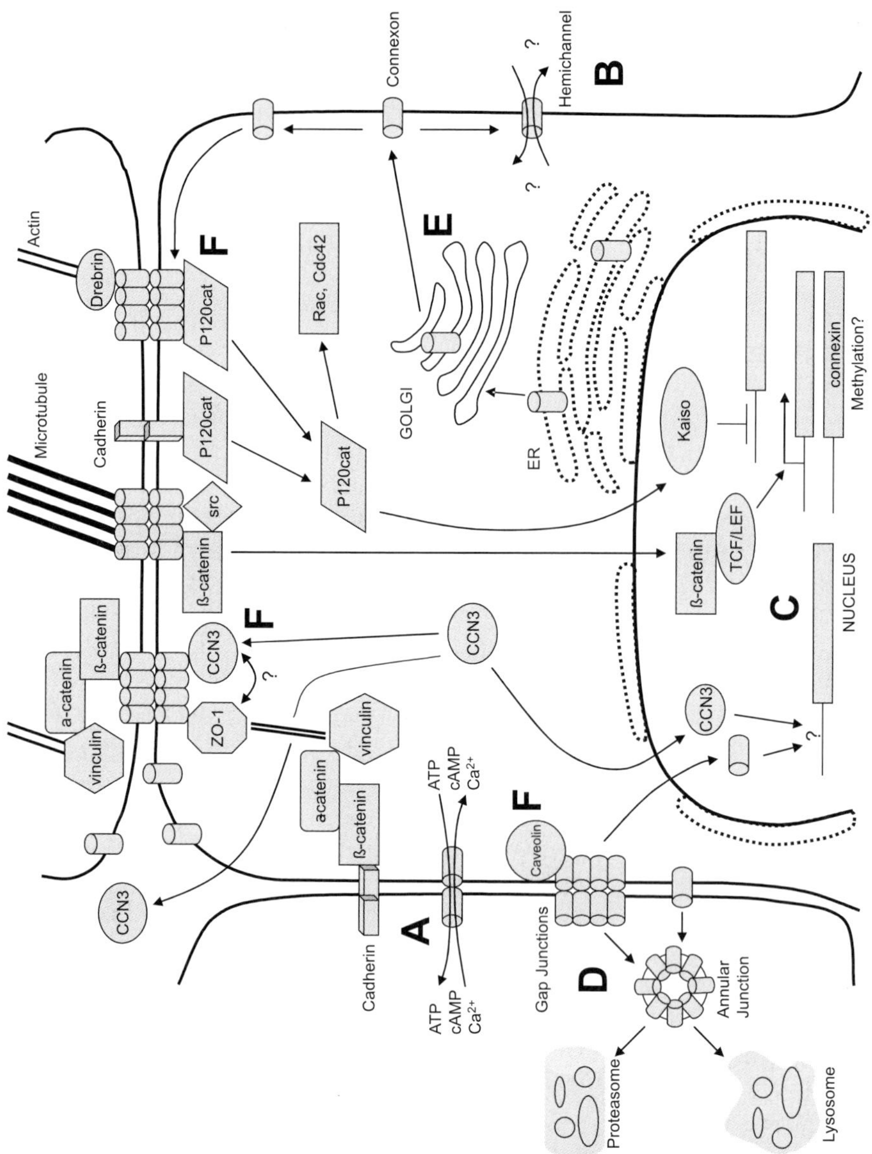

Fig. 12.1A–F. Summary of the various mechanisms which may be involved in connexin-mediated regulation of cell growth control. These include: **A** small signaling molecules passing through gap-junction channels; **B** hemichannels; **C** direct regulation of connexin genes; **D** regulation of connexin protein synthesis and trafficking; **E** connexin protein degradation; **F** mechanisms involving connexin-interacting proteins

In this review, we consider these various discrepancies and provide some insight into how some of these might be reconciled (summarized in Fig. 12.1). Nevertheless, it is clear that much is yet to be discovered regarding connexin functions in growth control.

12.2 Connexins and Gap Junctions

Gap junctions are a unique class of channels that directly connect the cytoplasm of adjacent cells, permitting the exchange of ions and molecules smaller than 1 kDa in size such as amino acids, second messengers and metabolites (Simon and Goodenough 1998). As such, they are positioned to coordinate a variety of cellular activities which can influence tissue and organ function. Gap junctions are formed by the end-to-end apposition of connexons, which are proteinaceous cylinders spanning the plasma membrane with a hydrophilic channel. Each connexon hemichannel is composed of a hexameric arrangement of proteins called connexins, members of a multigene family consisting of at least 20 members in vertebrates (Sohl and Willecke, 2004). The connexin protein spans the plasma membrane four times, and is oriented so that both its N- and C-termini face the cytoplasm, with extracellular domains involved in connexon–connexon interactions. Much of the regulation of the gap junction channels occurs via the C-terminus.

12.3 Gap Junctions and Tumor Suppression

12.3.1 Endogenous Expression of Connexins in Tumors

Gap junctions have long been implicated in cellular growth regulation and the control of tumor progression. Since the early report that gap junctional coupling was decreased in tumor cells as compared to their normal counterparts (Loewenstein and Kanno 1966), downregulation of gap junctions has been repeatedly observed in a variety of tumor cell lines and cancer tissues (see recent reviews, Mesnil 2002; Vine and Bertram 2002). This altered gap junctional coupling may take the form of aberrant homologous and/or heterologous communication (Yamasaki and Naus 1996).

The expression of connexins and the presence of gap junction channels can be readily altered by exogenous factors. In the context of carcinogenesis, many tumor-promoting agents have the ability to downregulate connexin expression and gap junctions. The best-studied example of this is the phorbol ester tumor promoter 12-*O*-tetradecanoylphorbol 13-acetate

(TPA; Enomoto et al. 1981; Swierenga and Yamasaki 1992; Lampe 1994). Several other tumor promoters have been extensively examined in this regard (reviewed by Trosko and Ruch 2002). Conversely, antineoplastic agents such as retinoic acid, carotenoids, vitamin D, and flavonoids are capable of reducing tumor cell growth in association with enhanced connexin expression and increased gap junctional coupling (Mehta et al. 1989; Hossain and Bertram 1994; Chaumontet et al. 1997;Trosko and Chang 2001). This responsiveness in the context of connexin expression is consistent with the proposed role of gap junctional intercellular communication (GJIC) in maintaining cellular homeostasis (De Maio et al. 2002).

12.3.2 Regulation of Endogenous Connexin Expression

Connexin Gene Expression

The most direct mechanism to increase connexin protein resides at the level of gene expression. Connexin gene expression has been shown to be increased by numerous endogenous factors, including growth factors (Reuss and Unsicker 1998) as illustrated in Fig. 12.1C. In this regard, analysis of the promoter regions of connexin genes has indicated the presence of a number of specific binding sites to enhance or suppress transcription (Yu et al. 1994; Echetebu et al. 1999; Teunissen et al. 2003). We have identified a ciliary neurotrophic factor (CNTF)-response element in the *Cx43* promoter, and shown that CNTF administered with its receptor, CNTFRα, enhanced Cx43 expression, GJIC, and suppressed growth of glioma cells in vitro (Ozog et al. 2002; Ozog et al. 2004). We have also shown that bone morphogenetic proteins (BMPs) increase Cx43 expression in P19 embryonal carcinoma cells (Bani-Yaghoub et al. 2000). Consistent with these findings, Smad binding elements responsive to BMP treatment have recently been described in osteosarcoma cells (Chatterjee et al. 2003). Cx43 expression is also regulated by Wnt1 at the transcriptional level, potentially via the TCF/LEF binding element (van der Heyden et al. 1998).

CpG island hypermethylation is known to be associated with gene silencing in cancer (Das and Singal 2004). Methylation of *Cx43* and *Cx32* has been examined in hepatoma cells, leading to the suggestion that hypermethylation suppresses connexin gene expression in tumor cells (Piechocki et al. 1999). In addition, inhibition of methylation was found to increase connexin expression in breast cancer cells (Tan et al. 2002), human renal carcinoma cells (Hirai et al. 2003) and HeLa cells (King et al. 2000). Hypermethylation of Cx32 has also been observed to occur only in cancerous regions of the human kidney and this was postulated to contribute to the early stage of renal carcinogenesis in hemodialysis patients (Yano et al. 2004). On the other hand, other investigators have found little evidence of involvement of hypermethylation in the suppressed expression of Cx26 in breast cancer cell lines (Singal et al. 2000), or human esophageal tumor cell lines (Loncarek et al. 2003). In an indirect fashion, 5' CpG island methylation of E-cadherin in endometrial cancer cells downregulated GJIC by altering the localization of Cx26 from cell borders to the cytoplasm (Nishimura et al. 2003).

Connexin Synthesis, Assembly, Degradation

The intracellular level of connexin proteins is tightly regulated by various signaling mechanisms (Musil et al. 2000). Cx43, for instance, has a rapid turnover rate (i.e., half-life of 1.5–5 h) and its degradation is believed to involve both the lysosomal and the ubiquitin-proteasomal pathway (Laing et al. 1997; Musil et al. 2000) as illustrated in Fig. 12.1D. Thus, the proteasome can be involved in regulating the number of functional gap junctions at the plasma membrane. By using proteasome inhibitors, gap junction "assembly-incompetent" cells were made to increase the number of gap junction plaques at the membrane concomitant with increased gap junctional coupling (Musil et al. 2000). Such regulation of connexin degradation has been suggested as a novel approach to increase intercellular communication.

Conversely, factors which stimulate connexin degradation would result in decreased gap junctional coupling. For example, the mitogenic growth factor EGF was shown to induce hyperphosphorylation of Cx43 and binding of ubiquitin, resulting in a rapid decrease in GJIC by degradation of Cx43 via the MAPK cascade (Leithe and Rivedal 2004). In addition, various tumor cell lines are characterized by defective connexin trafficking. Impaired connexin trafficking in endosomes has been suggested as a neoplastic marker for testicular tumors (Segretain et al. 2003). Intracellular accumulation of connexins due to impaired trafficking was observed in several androgen-independent prostate cancer cell lines (Govindarajan et al. 2002). Such trafficking defects have also been reported for a number of connexin mutations (Deschenes et al. 1997; Thomas et al. 2003). However, as noted below, there have been very limited reports of any connexin mutations identified in cancer.

12.3.3 Gap Junctions and Oncogenic Transformation

Inhibition of gap junctions by tumor promoters and oncogenes has been implicated in promoting carcinogenesis by removing the suppression of growth mediated by intercellular communication with neighboring cells (Esinduy et al. 1995). Overexpression of oncogenes such as Src, Neu, Ras, and Myc/Ras in cultured cells downregulates gap junctional communication and an inverse correlation exists in the levels of intercellular communication and the tumorigenicity of the transformed cells (Azarnia et al. 1989; Kalimi et al. 1992; Jou et al. 1995; Defeijter et al. 1996; Hayashi et al. 1998). Myc-transformed cells maintained high gap junctional coupling, but doubly transformed Myc/Ras or Myc/Raf cells had the lowest coupling and the highest tumorigenic potential. Conversely, forced expression of the tumor suppressor *BRMS1* (breast cancer metastasis suppressor 1) increases GJIC in tumor cells (Saunders et al. 2001; Shevde et al. 2002; Kapoor et al. 2004). Oncogenes are believed to mediate their effects on GJIC primarily by activating kinase cascades that result in the phosphorylation of connexin proteins. Many serine and tyrosine phosphorylation sites have been identified in the cytoplasmic tail of connexins (Lampe and Lau 2004). Src, PKC, MAPK, casein kinase I and phosphatases have all been implicated in the gatings of gap junction channels, although the EGFR-Ras-MAPK signaling pathway appears to be one of the predominant mechanisms modulating gap junctional coupling (Herve and Sarrouilhe, 2002). On the other hand, Src- and Neu-transformation appears to induce internalization of Cx43 at the expense of membrane-associated Cx43 plaques (Lampe and Lau 2000, 2004).

GJIC is modulated by connexin phosphorylation. Some phosphorylation events are required for proper connexin membrane insertion and gap junctional communication (Musil and Goodenough 1991; Fig. 12.1E), while other phosphorylation events disrupt gap junctions and block intercellular communication (Lampe and Lau 2004). For example, protein kinase A activity can augment gap junctional communication by Cx43. In contrast, protein kinase C, MAPK, and Src kinase severely block GJIC by phosphorylating Cx43 on sites distinct from those recognized by PKA (Lampe et al. 2000; Lampe and Lau 2004).

While decreased GJIC has been associated with neoplastic transformation of many cells, the correlation is not that simple. Src transformation significantly decreases GJIC between some fibroblasts. However, specifically curtailing GJIC by Cx43 antisense technology does not affect cell growth (Goldberg et al. 1994). Moreover, site-directed mutations that render Cx43 unaffected by Src do not inhibit the growth of Src transformed cells (Warn-Cramer et al. 2003). Therefore, Src must augment cell growth by mechanisms that override, or do not involve, GJIC.

GJIC has long been implicated in the ability of nontransformed cells to normalize the growth of neighboring tumor cells. This phenomenon, called heterologous growth control, was first described by Stoker (Stoker et al. 1966) in the same year that Loewenstein and Kanno first noticed that Src decreases GJIC between cells. It was later reported that heterologous gap junctional communication between the transformed and nontransformed cells correlated with this phenomenon (Mehta et al. 1986; Bignami et al. 1988). More recently, it has been demonstrated that while Cx43-mediated GJIC may augment heterologous growth control (Goldberg et al. 1994), it is not essential for the normalization of Src-transformed cells by neighboring nontransformed cells (Alexander et al. 2004).

12.3.4
Direct Effects of Altering Connexin Gene Expression

The effect of gap junction proteins on growth has been more directly studied by analyzing both the loss of gap junctions and the forced expression of connexins. The loss of gap junctions in many cases correlates with an increase in growth rate in vitro and tumor formation in vivo (Yamasaki and Naus 1996;Vine and Bertram 2002). There is little evidence of connexin mutations associated with human cancer. Only very recently has Cx43 been shown to be mutated in human colon sporadic adenocarcinomas (Dubina et al. 2002). It remains to be seen if other connexin mutations exist in cancer. With regard to mice, the knockout of Cx32 resulted in increased susceptibility to induced liver cancer (Temme et al. 1997), as well as lung cancer (King and Lampe 2004b). These mice also display an increase in radiation-induced tumorigenesis in several tissues, including liver, lung, adrenal, lymph, and small intestine (King and Lampe 2004a). In this latter study, evidence suggested that MAPK-related pathways were preferentially activated in the absence of Cx32.

On the other hand, many transformed cells transfected with connexin cDNAs exhibit suppression of growth and/or tumorigenicity. The connexin family members shown to act as tumor suppressors include Cx43 (Mehta et al. 1991; Zhu et al. 1991; Rose et al. 1993; Mesnil et al. 1995; Zhang et al. 1998; Goldberg et al. 2000), Cx26 (Mesnil et al. 1995), Cx32 (Eghbali et al. 1991), and Cx30 (Princen et al. 2001). However, the ability of individual connexins to suppress transformed growth is highly dependent on cell type (Mesnil et al. 1995; Yamasaki and Naus 1996). In some cases, where connexins were able to suppress tumor cell growth, there was a strong correlation noted between growth suppression and forced expression of the connexin isoform that is expressed in the normal untransformed tissue (Yamasaki and Naus 1996).

12.4
Mechanisms of Connexin-Mediated Growth Suppression

For many years, the growth suppressive role of connexins focused primarily on the function of the gap junction channel. Several possible mechanisms have been considered for this avenue of growth control. In addition, numerous reports have now described growth suppression mediated by connexins in the possible absence of gap-junction channel formation. However, these data should be considered carefully since GJIC was assayed primarily by dye transfer, which may not accurately represent the extent to which these cells were coupled to each other (Goldberg et al. 1999; Goldberg et al. 2002).

12.4.1 Junction-Mediated Growth Suppression

The mechanism by which connexins act as growth inhibitors is unknown. Initial investigation of connexin-induced tumor suppression focused on its channel-forming ability, to allow for either the exchange of normalizing messages, or dispersion of growth-promoting signals (Loewenstein and Kanno 1966), as illustrated in Fig. 12.1A. In this regard, Goldberg and colleagues have attempted to identify transjunctional molecules which might control growth (Goldberg et al. 1998, 1999; Alexander and Goldberg 2003). They were able to identify ADP, ATP, glutamate, and glutathione as major transjunctional molecules which may be involved in growth control. They proposed the preferred passage of endogenous metabolites through Cx43 as opposed to Cx32 channels may underlie the selective ability of Cx43 to inhibit growth of C6 cells in vitro. While it is clear that gap junctions can selectively mediate intercellular exchange of ions and small molecules, to date there has been no identification of a direct intercellular mediator of growth control passing through gap junctions.

Therapeutically, there has been exploitation of the role of gap junctions in transmitting molecules affecting growth control. Gap junctions have been implicated in the so-called bystander effect observed in cell killing mediated by HSVtk expression and ganciclovir treatment (Mesnil et al. 1996; Dilber et al. 1997; Nicholas et al. 2003). This has led to a number of approaches to augment chemotherapy with enhanced gap junctional coupling (Touraine et al. 1998; Carystinos et al. 1999; Robe et al. 2000; Huang et al. 2001b; Robe et al. 2004). This "bystander effect" has also been shown to be involved in cisplatin-mediated tumor cell killing (Jensen and Glazer 2004) as well as radiation therapy (Azzam et al. 1998; Azzam et al. 2003).

With regard to tumor therapy, enhancement of apoptosis provides a clear therapeutic mechanism for exploitation (Ding and Fisher 2002). The role of gap junctions in apoptosis was suggested over 10 years ago (Trosko and Goodman 1994). In this regard, the "bystander effect" noted above has been shown to result in cell death via apoptosis (Samejima and Meruelo 1995). A more direct role for connexins in cell death has been demonstrated since increased Cx43 expression resulted in decreased expression of bcl-2 in conjunction with increased apoptosis following treatment with chemotherapeutic drugs (Huang et al. 2001b). In fact, several reports have demonstrated that enhanced expression of connexins alone induces apoptosis. Cx43 has been shown to enhance apoptosis in glioblastoma cells under low serum conditions (Huang et al. 2001a). More recently, adenoviral-mediated expression of Cx37 in endothelial cells resulted in enhanced apoptosis, while Cx40 and Cx43 had no effect (Seul et al. 2004). The Cx37-mediated apoptosis was abrogated by pharmacologically blocking gap junctions. These studies, as well as others, indicate that these effects, in many situations, are connexin- and cell type-specific (reviewed in Andrade-Rozental et al. 2000). While gap junctions have been implicated as mediators of enhanced apoptosis (Lin et al. 1998), Cx43 expression in the absence of gap junction formation has been shown to enhance cell survival (Lin et al. 2003). These authors postulated that cytoskeletal interactions and calcium homeostasis were involved in this process. Cell survival was also dependent upon the context of the injury. Thus, there remains substantial uncertainty around the role of connexins and gap junctions in apoptosis.

12.4.2 Nonjunctional Functions

Recent evidence challenges the view that functional gap junctions may not be necessary for the growth regulatory effects, suggesting that simply the expression of connexin genes may be sufficient. Indeed, several reports now challenge the view that connexins are only membrane-spanning, pore-forming proteins. For example, Cx43 inhibits the growth of both human glioblastoma and breast cancer cells in vivo, but there is no significant increase in coupling as assessed by dye transfer (Huang et al. 1998; Qin et al. 2003). Moreover, it has been shown that Cx43 appears to affect the cell cycle. Expression of Cx43 in dog kidney epithelial cells altered the expression of genes involved in regulating the cell cycle, specifically cyclin A, D1, D2, as well as CDK5 and CDK6 (Chen et al. 1995). Cx43 may also inhibit growth by delaying the G1-S phase transition of the cell cycle (Moorby and Patel 2001). Along this line, one report indicated that Cx43 promotes the degradation of S phase kinase-associated protein 2 (Skp2) which regulates the ubiquitination of p27, in turn increasing the level of p27 (Zhang et al. 2003). This effect was not influenced by gap-junction blockage, suggesting that it is communication-independent. Others have also shown that membrane localization of Cx43 is found not to be necessary for its growth suppressive effect (Olbina and Eckhart 2003). Furthermore, the carboxy-terminal tail (C-terminus) of Cx43 has been shown to be as effective as wild-type Cx43 in growth suppression. Deletion of the C-terminus in Cx43 abrogated the growth regulation (Moorby and Patel 2001), and transfection of the C-terminus of Cx43 alone was found to be sufficient to reduce the growth rate of tumor cells (Moorby and Patel 2001; Zhang et al. 2003). These truncated Cx43 proteins were present diffusely in the cytoplasm of transfected cells, confirming that membrane localization was not required for growth suppression. In addition, the C-terminal domain of Cx43 has been reported to localize to the nucleus in association with decreased cell proliferation, opening possibilities of direct mediation of gene expression (Dang et al. 2003).

Another nonjunctional aspect of connexins which must be briefly considered here relates to the existence and function of hemichannels, as illustrated in Fig. 12.1B. Substantial evidence supports the existence of hemichannels in vitro, where they may function as transmembrane ion channels (Dermietzel et al. 2003; Ebihara, 2003). They have also been implicated in the release of ATP to the extracellular space (Contreras et al. 2002; Stout et al. 2002). It is clear that deregulated opening of such hemichannels would be detrimental to cell survival, and could be exploited in the killing of tumor cells.

12.4.3 Connexin-Interacting Proteins

Many intracellular proteins have now been identified to interact with the connexin proteins and participate in signaling events (reviewed in Herve et al. 2004), as shown in Fig. 12.1F. The first interacting protein identified was Zonular occludens-1 (ZO-1), a peripheral membrane component of tight junctions and adherens junctions (Giepmans and Moolenaar 1998). It contains multiple protein interaction domains including an SH3 domain and three PDZ domains, the second of which interacts with Cx43. It has been proposed that ZO-1 may recruit signaling proteins to gap junction plaques (Giepmans and Moolenaar 1998).

The interaction of connexins with a number of cytoskeletal-associated proteins has suggested the existence of a gap junction complex. Colocalization of Cx43 with N-cadherin and p120catenin has been observed in neural crest cells (Xu et al. 2001), and β-catenin has also been shown to associate with Cx43 (Ai et al. 2000). The interaction α/β catenins with the cytoplasmic domain of cadherins and to other F-actin binding molecules such as vinculin and ZO-1 provides a structural link between adherens junctions, the actin cytoskeleton and the tight junction (Nagafuchi 2001). In addition, Cx43 also bind microtubules (Giepmans et al. 2001b) and the recent identification of Drebrin, an actin binding protein, as a novel Cx43-

interacting protein confirmed the tight relationship between gap junctions and the cytoskeleton (Butkevich et al. 2004). However, it is not clear whether Cx43 performs a role upstream in the cytoskeletal cascade by being part of a multi-protein complex that coordinately regulates the multi-membrane spanning assemblies, or if the connexons are actually being directed to certain cellular locations for channel forming/degradative pathways downstream of other receptors/signals. Indeed, Drebrin is necessary for Cx43 to be targeted to the membrane and the depletion of Drebrin causes Cx43 to be targeted to a degradative pathway (Butkevich et al. 2004).

Cx43 has been found to associate with other intracellular signaling molecules such as calmodulin (Van Eldik et al. 1985), and protein kinases including v-Src (Kanemitsu et al. 1997), c-Src (Giepmans et al. 2001a), PKCα and PKCε (Doble et al. 2000; Bowling et al. 2001), and DMPK (Mussini et al. 1999; Schiavon et al. 2002), as reviewed in detail by Herve et al. (2004). Most information about how these kinases affect GJIC by phosphorylation is available for Src, PKC, and MAPK (Cruciani and Mikalsen 2002). Much less is known for PKG, Ca^{2+}-calmodulin-dependent protein kinase, and casein kinase 1 (Cooper and Lampe 2002).

The interaction of Src with Cx43 is the most well known. Src tyrosine kinase contains SH2 and SH3 domains and is thought to affect Cx43 by at least two signaling routes. V-Src and c-Src can bind directly to and phosphorylate the Cx43 C-terminal tail via SH2/SH3-domain interactions (Kanemitsu et al. 1997; Zhou et al. 1999; Toyofuku et al. 2001) and might either modulate the gating of the gap junctions (reviewed by Lampe and Lau 2004; Warn-Cramer and Lau 2004), or the kinase might regulate the interaction of Cx with other proteins. Constitutively active c-Src disrupts the binding of ZO-1 and Cx43 by phosphorylating Tyr 265 of Cx43, thus provide a binding site for the SH2 domain of c-Src (Toyofuku et al. 2001).

Most research on gap junctions and cell growth has focused on Cx43, probably due to its widespread tissue distribution and its frequent expression in cells grown in culture. However, gap junction-associated proteins originally identified by their binding to Cx43 have also been discovered to associate with other members of the connexin family with tumor suppressive properties. Cx32 binds to calmodulin (Torok et al. 1997), which is implicated in regulating the gating of the gap junction channels (Peracchia et al. 2000). Cx32 also associates with ZO-1, Claudin-1, α-catenin and E-cadherin, and Cx26 interacts with occludin, α-catenin, E-cadherin (as reviewed in Herve et al. 2004). Although it is not clear how the interaction of connexins with kinases and cytoskeletal-associated proteins directly regulate growth, we can safely assume that phosphorylation of gap junctions by associating kinases is critical in regulating intercellular communication (Lampe and Lau 2000).

12.4.4 Direct Interaction of Cx43 with CCN3, a Growth Suppressor

More recently, the discovery of direct interaction of Cx43 with an antiproliferative factor CCN3 (NOV) suggests a more direct role played by connexins in growth control and tumor suppression (Fu et al. 2004; Gellhaus et al. 2004; Fig. 12.1F). Previously, we reported that transfection of Cx43 led to decreased proliferation of glioma cells (Zhu et al. 1991). As well, conditioned medium collected from transfected C6 cells (C6-Cx43) alone also possessed anti-proliferative properties (Zhu et al. 1992). Thus, the growth suppressive capacity of Cx43 appeared to be mediated, at least in part, by the secretion of soluble growth inhibitory factors into the surrounding media (Zhu et al. 1992;Goldberg et al. 2000). Using differential gene expression approaches, we have pursued the hypothesis that connexin gene expression might alter the expression of downstream growth regulatory genes (Naus et al. 2000). Among the most prominent differentially expressed genes were members of the CYR61/Connective Tissue Growth Factor/Nephroblastoma Overexpressed (CCN) family, which have been shown to play a role in tumorigenesis and growth control (Perbal 2004). One of these, CCN3, has known antiproliferative properties and led to our proposal that CCN3 serves as one of the soluble factors mediating Cx43-induced tumor suppression. CCN3 colocalizes with Cx43 at the plasma membrane in C6-Cx43 cells (Mcleod et al. 2001). Furthermore, CCN3 and Cx43 physically associate with each other in cultured cells (Fu et al. 2004; Gellhaus et al. 2004). Gellhaus and colleagues have confirmed these findings using gene chip analysis for choriocarcinoma cells (Gellhaus et al. 2004).

The CCN family of proteins encodes a family of multi-modular proteins that mediate diverse cellular functions such as adhesion, migration, and cellular growth (Brigstock 2003). The presence of a signal peptide indicates they exert their effects as secreted growth factors although recent data have also suggested alternate intracellular functions. CCN proteins are composed of a four-modular structure commonly found in other extracellular matrix proteins. Module 1 is a putative insulin growth factor binding domain, module 2 has similarity to von Willebrand type C while module 3 is likely to be thrombospondin 1 domain and, finally, the last domain is a cysteine-rich knot. It is proposed that the interaction of these modules, either in groups or independently, accounts for their specificity and diversity in biological functions. The CCN proteins appear to exert their cellular effects by binding to integrins.

CCN3 is thought to exist in two isoforms (Planque and Perbal 2003). An N-terminal truncated isoform containing the cysteine-rich module localizes to the nuclei and cytoplasm of tumor cells and leads to increased cell growth. In contrast, the full-length 48-kDa protein is secreted or remains at the cell membrane and inhibits proliferation (Joliot et al. 1992; Perbal 2001; Planque and Perbal 2003). Transfection of CCN3 cDNA into the human glioma cell line G59 significantly reduced their in vitro growth rate, as well as tumor size in vivo (Gupta et al. 2001). Following Cx43 transfection into C6 glioma cells, the redistribution of CCN3 from the cytoplasm to the plasma membrane coincided with the appearance of the 48-kDa isoform (Fu et al. 2004). The upregulation of full-length CCN3 could be the consequence of an increased level of mRNA transcript, reduced proteolytic degradation into the truncated isoforms, or a combination of both. Conditioned medium from C6-Cx43 causes a reduction in the growth rate of C6 cells in culture (Zhu et al. 1992). Thus, the secretion of CCN3 into the extracellular environment may account for the growth suppressive effects observed in the conditioned medium of C6 glioma cells transfected by Cx43.

The loss of Cx43 colocalization and co-immunoprecipitation with CCN3 in C6 cells transfected with a C-terminal truncated (trCT) Cx43 indicates that the C-terminus of Cx43 is necessary for its interaction with CCN3 (Fu et al. 2004). A similar study on Hs578T breast cancer cells shows inhibition of CCN3 translocation to the membrane in cells transfected with trCT Cx43 (Sin and Naus, unpubl. obs.). Further evidence that CCN3 might be a major player in the Cx43-dependent control of cell growth results from studies with another cell type, the choriocarcinoma cell line Jeg3 (Gellhaus et al. 2004). Co-immunoprecipitation experiments identified the region of interaction between amino acids 257 and 382 of the Cx43 C-terminus. This binding was apparently independent of the requirement for any cofactor molecule. However, this does not exclude the participation of other signaling molecules and the involvement of phosphorylation of Cx43 in this interaction process. The present evidence also does not rule out the potential ability of the truncated CCN3 isoform to interact with Cx43. CCN3 did not associate with Cx32 or Cx40 when these were expressed in C6 glioma cells (Fu et al. 2004). It is likely that the C-terminal of connexins, being the most variable among all connexin family members, confers cell-type specificity with regard to interactions with CCN3, and the resultant growth control.

12.5 Conclusions

In summary, there is accumulating evidence that connexin-mediated growth regulation not only depends on their channel properties, but also on their interaction with other proteins in an isoform- and cell type-specific manner. These various cellular mechanisms implicated in connexin function are outlined in Fig. 12.1. The differential ability of individual connexins to interact with downstream effector molecules is most likely to account for the unique, but overlapping C-terminus induced signal cascades. We and others have provided the first direct evidence that Cx43 may mediate their effects by associating directly with a protein with tumor-suppressive functions. With respect to CCN3, its biological properties have been proposed to depend on combinatorial events, resulting from interaction with various partners (Perbal 2001; Planque and Perbal 2003). Indeed, the multimodular structure of CCN3 suggests that it may lie at the hub of multiple signaling pathways. We discovered by immunocytochemistry that ZO-1 colocalizes with CCN3 and Cx43 at the plasma membrane (unpubl. observ.). By acting as a scaffolding protein, CCN3 could recruit different molecules such as Cx43 into a complex and coordinate cross-talk between different signals, ultimately contributing to growth regulation. The association of these C-terminal-associating proteins provides a mechanism for the possible modulation of the opening and closing of the gap junctions. A number of hypotheses can be generated in light of the emerging view of gap junctions as the anchorage of multi-protein complexes called a Nexus (Duffy et al. 2002; Herve et al. 2004).

Acknowledgements. We wish to thank Dr. P. Lampe for critical reading of this review, and Mr. Charles Lai for assistance with the illustration. This work was supported by a grant from the Canadian Institutes of Health Research MOP-14358 to CCN and grants from the United States National Institutes of Health CA88805 and EY014479 to GSG.

References

Ai Z, Fischer A, Spray DC, Brown AM, Fishman GI (2000) Wnt-1 regulation of connexin43 in cardiac myocytes. J Clin Invest 105:161–171

Alexander DB, Goldberg GS (2003) Transfer of biologically important molecules between cells through gap junction channels. Curr Med Chem 10:2045–2058

Alexander DB, Ichikawa H, Bechberger JF, Valiunas V, Ohki M, Naus CC, Kunimoto T, Tsuda H, Miller WT, Goldberg GS (2004) Normal cells control the growth of neighboring transformed cells independent of gap junctional communication and SRC activity. Cancer Res 64:1347–1358

Andrade-Rozental AF, Rozental R, Hopperstad MG, Wu JK, Vrionis FD, Spray DC (2000) Gap junctions: the "kiss of death" and the "kiss of life". Brain Res Brain Res Rev 32:308–315

Azarnia R, Mitcho M, Shalloway D, Loewenstein WR (1989) Junctional intercellular communication is cooperatively inhibited by oncogenes in transformation. Oncogene 4:1161–1168

Azzam EI, de Toledo SM, Gooding T, Little JB (1998) Intercellular communication is involved in the bystander regulation of gene expression in human cells exposed to very low fluences of alpha particles. Radiat Res 150:497–504

Azzam EI, de Toledo SM, Little JB (2003) Oxidative metabolism, gap junctions and the ionizing radiation-induced bystander effect. Oncogene 22:7050–7057
Bani-Yaghoub M, Felker JM, Sans C, Naus CC (2000) The effects of bone morphogenetic protein 2 and 4 (BMP2 and BMP4) on gap junctions during neurodevelopment. Exp Neurol 162:13–26
Bignami M, Rosa S, Falcone G, Tato F, Katoh F, Yamasaki H (1988) Specific viral oncogenes cause differential effects on cell-to-cell communication, relevant to the suppression of the transformed phenotype by normal cells. Mol Carcinog 1:67–75
Bowling N, Huang X, Sandusky GE, Fouts RL, Mintze K, Esterman M, Allen PD, Maddi R, McCall E, Vlahos CJ (2001) Protein kinase C-alpha and -epsilon modulate connexin-43 phosphorylation in human heart. J Mol Cell Cardiol 33:789–798
Brigstock DR (2003) The CCN family: a new stimulus package. J Endocrinol 178:169–175
Butkevich E, Hulsmann S, Wenzel D, Shirao T, Duden R, Majoul I (2004) Drebrin is a novel connexin-43 binding partner that links gap junctions to the submembrane cytoskeleton. Curr Biol 14:650–658
Carystinos GD, Katabi MM, Laird DW, Galipeau J, Chan H, Alaoui-Jamali MA, Batist G (1999) Cyclic-AMP induction of gap junctional intercellular communication increases bystander effect in suicide gene therapy. Clin Cancer Res 5:61–68
Chatterjee B, Meyer RA, Loredo GA, Coleman CM, Tuan R, Lo CW (2003) BMP regulation of the mouse connexin43 promoter in osteoblastic cells and embryos. Cell Commun Adhes 10:37–50
Chaumontet C, Droumaguet C, Bex V, Heberden C, Gaillard-Sanchez I, Martel P (1997) Flavonoids (apigenin, tangeretin) counteract tumor promoter-induced inhibition of intercellular communication of rat liver epithelial cells. Cancer Lett 114: 207–210
Chen SC, Pelletier DB, Ao P, Boynton AL (1995) Connexin43 reverses the phenotype of transformed cells and alters their expression of cyclin/cyclin-dependent kinases. Cell Growth Differ 6:681–690
Contreras JE, Sanchez HA, Eugenin EA, Speidel D, Theis M, Willecke K, Bukauskas FF, Bennett MV, Saez JC (2002) Metabolic inhibition induces opening of unapposed connexin 43 gap junction hemichannels and reduces gap junctional communication in cortical astrocytes in culture. Proc Natl Acad Sci USA 99:495–500
Cooper CD, Lampe PD (2002) Casein kinase 1 regulates connexin-43 gap junction assembly. J Biol Chem 277:44962–44968
Cruciani V, Mikalsen SO (2002) Connexins, gap junctional intercellular communication and kinases. Biol Cell 94:433–443
Dang X, Doble BW, Kardami E (2003) The carboxy-tail of connexin-43 localizes to the nucleus and inhibits cell growth. Mol Cell Biochem 242:35–38
Das PM, Singal R (2004) DNA methylation and cancer. J Clin Oncol 22:4632–4642
De Maio A, Vega VL, Contreras JE (2002) Gap junctions, homeostasis, and injury. J Cell Physiol 191:269–282
Defeijter AW, Matesic DF, Ruch RJ, Guan XJ, Chang CC, Trosko JE (1996) Localization and function of the connexin 43 gap-junction protein in normal and various oncogene-expressing rat liver epithelial cells. Mol Carcinogenesis 16:203–212
Dermietzel R, Meier C, Bukauskas F, Spray DC (2003) Following tracks of hemichannels. Cell Commun Adhes 10:335–340
Deschenes SM, Walcott JL, Wexler TL, Scherer SS, Fischbeck KH (1997) Altered trafficking of mutant connexin32. J Neurosci 17:9077–9084
Dilber M, Abedi MR, Christensson B, Bjorkstrand B, Kidder GM, Naus CCG, Gahrton G, Smith CIE (1997) Gap junctions promote the bystander effect of Herpes Simplex thymidine kinase in vivo. Cancer Res 57:1523–1528
Ding HF, Fisher DE (2002) Induction of apoptosis in cancer: new therapeutic opportunities. Ann Med 34:451–469

Doble BW, Ping P, Kardami E (2000) The epsilon subtype of protein kinase C is required for cardiomyocyte connexin-43 phosphorylation. Circ Res 86:293–301

Dubina MV, Iatckii NA, Popov DE, Vasil'ev SV, Krutovskikh VA (2002) Connexin 43, but not connexin 32, is mutated at advanced stages of human sporadic colon cancer. Oncogene 21:4992–4996

Duffy HS, Delmar M, Spray DC (2002) Formation of the gap junction nexus: binding partners for connexins. J Physiol Paris 96:243–249

Ebihara L (2003) New roles for connexons. News Physiol Sci 18:100–103

Echetebu CO, Ali M, Izban MG, MacKay L, Garfield RE (1999) Localization of regulatory protein binding sites in the proximal region of human myometrial connexin 43 gene. Mol Hum Reprod 5:757–766

Eghbali B, Kessler JA, Reid LM, Roy C, Spray DC (1991) Involvement of gap junctions in tumorigenesis: transfection of tumor cells with connexin 32 cDNA retards growth in vivo. Proc Natl Acad Sci USA 88:10701–10705

Enomoto T, Sasaki Y, Shiba Y, Kanno Y, Yamasaki H (1981) Tumor promoters cause a rapid and reversible inhibition of the formation and maintenance of electrical cell coupling in culture. Proc Natl Acad Sci USA 78:5628–5632

Esinduy CB, Chang CC, Trosko JE, Ruch RJ (1995) In vitro growth inhibition of neoplastically transformed cells by non-transformed cells: Requirement for gap junctional intercellular communication. Carcinogenesis 16:915–921

Fu CT, Bechberger JF, Ozog MA, Perbal B, Naus CC (2004) CCN3 (NOV) Interacts with Connexin43 in C6 Glioma Cells: Possible mechanism of connexin-mediated growth suppression. J Biol Chem 279:36943–36950

Gellhaus A, Dong X, Propson S, Maass K, Klein-Hitpass L, Kibschull M, Traub O, Willecke K, Perbal B, Lye SJ, Winterhager E (2004) Connexin43 interacts with NOV: A possible mechanism for negative regulation of cell growth in choriocarcinoma cells. J Biol Chem 279:36931–36942

Giepmans BN, Moolenaar WH (1998) The gap junction protein connexin43 interacts with the second PDZ domain of the zona occludens-1 protein. Curr Biol 8:931–934

Giepmans BN, Hengeveld T, Postma FR, Moolenaar WH (2001a) Interaction of c-Src with gap junction protein connexin-43. Role in the regulation of cell-cell communication. J Biol Chem 276:8544–8549

Giepmans BN, Verlaan I, Hengeveld T, Janssen H, Calafat J, Falk MM, Moolenaar WH (2001b) Gap junction protein connexin-43 interacts directly with microtubules. Curr Biol 11:1364–1368

Goldberg GS, Martyn KD, Lau AF (1994) A connexin 43 antisense vector reduces the ability of normal cells to inhibit the foci formation of transformed cells. Mol Carcinogen 11:106–114

Goldberg GS, Lampe PD, Sheedy D, Stewart CC, Nicholson BJ, Naus CCG (1998) Direct isolation and analysis of endogenous transjunctional ADP from Cx43 transfected C6 glioma cells. Exp Cell Res 239:82–92

Goldberg GS, Lampe PD, Nicholson BJ (1999) Selective transfer of endogenous metabolites through gap junctions composed of different connexins. Nat Cell Biol 1:457–459

Goldberg GS, Bechberger JF, Tajima Y, Merritt M, Omori Y, Gawinowicz MA, Narayanan R, Tan Y, Sanai Y, Yamasaki H, Naus CC, Tsuda H, Nicholson BJ (2000) Connexin43 suppresses MFG-E8 while inducing contact growth inhibition of glioma cells. Cancer Res 60:6018–6026

Goldberg GS, Moreno AP, Lampe PD (2002) Gap junctions between cells expressing connexin 43 or 32 show inverse permselectivity to adenosine and ATP. J Biol Chem 277:36725–36730

Govindarajan R, Zhao S, Song XH, Guo RJ, Wheelock M, Johnson KR, Mehta PP (2002) Impaired trafficking of connexins in androgen-independent human prostate cancer cell lines and its mitigation by alpha-catenin. J Biol Chem 277:50087–50097

Gupta N, Wang H, Mcleod TL, Naus CCG, Kyurkchiev S, Advani S, Yu J, Perbal B, Weichselbaum RR (2001) Inhibition of glioma cell growth and tumorigenic potential by CCN3 (NOV). J Clin Path Mol Path 54:293–299

Hayashi T, Nomata K, Chang CC, Ruch RJ, Trosko JE (1998) Cooperative effects of v-myc and c-Ha-ras oncogenes on gap junctional intercellular communication and tumorigenicity in rat liver epithelial cells. Cancer Lett 128:145–154

Herve JC, Sarrouilhe D (2002) Modulation of junctional communication by phosphorylation: protein phosphatases, the missing link in the chain. Biol Cell 94:423–432

Herve JC, Bourmeyster N, Sarrouilhe D (2004) Diversity in protein-protein interactions of connexins: emerging roles. Biochim Biophys Acta 1662:22–41

Hirai A, Yano T, Nishikawa K, Suzuki K, Asano R, Satoh H, Hagiwara K, Yamasaki H (2003) Down-regulation of connexin 32 gene expression through DNA methylation in a human renal cell carcinoma cell. Am J Nephrol 23:172–177

Hossain MZ, Bertram JS (1994) Retinoids suppress proliferation, induce cell spreading, and up- regulate connexin43 expression only in postconfluent 10T1/2 cells: Implications for the role of gap junctional communication. Cell Growth Differ 5:1253–1261

Huang RP, Fan Y, Hossain MZ, Peng A, Zeng ZL, Boynton AL (1998) Reversion of the neoplastic phenotype of human glioblastoma cells by connexin 43 (cx43). Cancer Res 58:5089–5096

Huang R, Liu YG, Lin Y, Fan Y, Boynton A, Yang D, Huang RP (2001a) Enhanced apoptosis under low serum conditions in human glioblastoma cells by connexin 43 (Cx43). Mol Carcinogenesis 32:128–138

Huang RP, Hossain MZ, Huang R, Gano J, Fan Y, Boynton AL (2001b) Connexin 43 (cx43) enhances chemotherapy-induced apoptosis in human glioblastoma cells. Int J Cancer 92:130–138

Jensen R, Glazer PM (2004) Cell-interdependent cisplatin killing by Ku/DNA-dependent protein kinase signaling transduced through gap junctions. Proc Natl Acad Sci USA 101:6134–6139

Joliot V, Martinerie C, Dambrine G, Plassiart G, Brisac M, Crochet J, Perbal B (1992) Proviral rearrangements and overexpression of a new cellular gene (nov) in myeloblastosis-associated virus type 1-induced nephroblastomas. Mol Cell Biol 12:10–21

Jou YS, Layhe B, Matesic DF, Chang CC, Defeijter AW, Lockwood L, Welsch CW, Klaunig JE, Trosko JE (1995) Inhibition of gap junctional intercellular communication and malignant transformation of rat liver epithelial cells by neu oncogene. Carcinogenesis 16:311–317

Kalimi GH, Hampton LL, Trosko JE, Thorgeirsson SS, Huggett AC (1992) Homologous and heterologous gap-junctional intercellular communication in v-raf-, v-myc-, and v-raf/v-myc-transduced rat liver epithelial cell lines. Mol Carcinogenesis 5:301–310

Kanemitsu MY, Loo LWM, Simon S, Lau AF, Eckhart W (1997) Tyrosine phosphorylation of connexin 43 by v-SRC is mediated by SH2 and SH3 domain interactions. J Biol Chem 272:22824–22831

Kapoor P, Saunders MM, Li Z, Zhou Z, Sheaffer N, Kunze EL, Samant RS, Welch DR, Donahue HJ (2004) Breast cancer metastatic potential: correlation with increased heterotypic gap junctional intercellular communication between breast cancer cells and osteoblastic cells. Int J Cancer 111:693–697

King TJ, Lampe PD (2004a) Mice deficient for the gap junction protein Connexin32 exhibit increased radiation-induced tumorigenesis associated with elevated mitogen-activated protein kinase (p44/Erk1, p42/Erk2) activation. Carcinogenesis 25:669–680

King TJ, Lampe PD (2004b) The gap junction protein connexin32 is a mouse lung tumor suppressor. Cancer Res 64:7191–7196

King TJ, Fukushima LH, Donlon TA, Hieber AD, Shimabukuro KA, Bertram JS (2000) Correlation between growth control, neoplastic potential and endogenous connexin43 expression in HeLa cell lines: implications for tumor progression. Carcinogenesis 21:311–315

Laing JG, Tadros PN, Westphale EM, Beyer EC (1997) Degradation of connexin43 gap junctions involves both the proteasome and the lysosome. Exp Cell Res 236:482–492

Lampe PD (1994) Analyzing phorbol ester effects on gap junctional communication: A dramatic inhibition of assembly. J Cell Biol 127:1895–1905

Lampe PD, Lau AF (2000) Regulation of gap junctions by phosphorylation of connexins. Arch Biochem Biophys 384:205–215

Lampe PD, Lau AF (2004) The effects of connexin phosphorylation on gap junctional communication. Int J Biochem Cell Biol 36:1171–1186

Lampe PD, Tenbroek EM, Burt JM, Kurata WE, Johnson RG, Lau AF (2000) Phosphorylation of connexin43 on serine368 by protein kinase C regulates gap junctional communication. J Cell Biol 149:1503–1512

Leithe E, Rivedal E (2004) Epidermal growth factor regulates ubiquitination, internalization and proteasome-dependent degradation of connexin43. J Cell Sci 117:1211–1220

Lin C, Weigel H, Cotrina ML, Liu S, Bueno E, Hansen AJ, Hansen TW, Goldman S, Nedergaard M (1998) Gap-junction-mediated propagation and amplification of cell injury. Nat Neurosci 1:494–500

Lin JH, Yang J, Liu S, Takano T, Wang X, Gao Q, Willecke K, Nedergaard M (2003) Connexin mediates gap junction-independent resistance to cellular injury. J Neurosci 23:430–441

Loewenstein WR, Kanno Y (1966) Intercellular communication and the control of tissue growth: lack of communication between cancer cells. Nature 209:1248–1249

Loncarek J, Yamasaki H, Levillain P, Milinkevitch S, Mesnil M (2003) The expression of the tumor suppressor gene connexin 26 is not mediated by methylation in human esophageal cancer cells. Mol Carcinogenesis 36:74–81

Mcleod TL, Bechberger JF, Naus CC (2001) Determination of a potential role of the CCN family of growth regulators in connexin43 transfected C6 glioma cells. Cell Commun Adhes 8:441–445

Mehta PP, Bertram JS, Loewenstein WR (1986) Growth inhibition of transformed cells correlates with their junctional communication with normal cells. Cell 44:187–196

Mehta PP, Bertram JS, Loewenstein WR (1989) The actions of retinoids on cellular growth correlate with their actions on gap junctional communication. J Cell Biol 108:1053–1065

Mehta PP, Hotz-Wagenblatt A, Rose B, Shalloway D (1991) Incorporation of the gene for a cell-cell channel protein into transformed cells leads to normalization of growth. J Membr Biol 124:207–225

Mesnil M (2002) Connexins and cancer. Biol Cell 94:493–500

Mesnil M, Krutovskikh V, Piccoli C, Elfgang C, Traub O, Willecke K, Yamasaki H (1995) Negative growth control of HeLa cells by connexin genes: Connexin species specificity. Cancer Res 55:629–639

Mesnil M, Piccoli C, Tiraby G, Willecke K, Yamasaki H (1996) Bystander killing of cancer cells by HSV-TK gene is mediated by connexins. Proc Natl Acad Sci USA 93:1831–1835

Moorby C, Patel M (2001) Dual functions for connexins: Cx43 regulates growth independently of gap junction formation. Exp Cell Res 271:238–248

Musil LS, Goodenough DA (1991) Biochemical analysis of connexin43 intracellular transport, phosphorylation, and assembly into gap junctional plaques. J Cell Biol 115:1357–1374

Musil LS, Le AC, VanSlyke JK, Roberts LM (2000) Regulation of connexin degradation as a mechanism to increase gap junction assembly and function. J Biol Chem 275:25207–25215

Mussini I, Biral D, Marin O, Furlan S, Salvatori S (1999) Myotonic dystrophy protein kinase expressed in rat cardiac muscle is associated with sarcoplasmic reticulum and gap junctions. J Histochem Cytochem 47:383–392
Nagafuchi A (2001) Molecular architecture of adherens junctions. Curr Opin Cell Biol 13:600–603
Naus CC, Bond SL, Bechberger JF, Rushlow W (2000) Identification of genes differentially expressed in C6 glioma cells transfected with connexin43. Brain Res Brain Res Rev 32:259–266
Nicholas TW, Read SB, Burrows FJ, Kruse CA (2003) Suicide gene therapy with Herpes simplex virus thymidine kinase and ganciclovir is enhanced with connexins to improve gap junctions and bystander effects. Histol Histopathol 18:495–507
Nishimura M, Saito T, Yamasaki H, Kudo R (2003) Suppression of gap junctional intercellular communication via 5' CpG island methylation in promoter region of E-cadherin gene in endometrial cancer cells. Carcinogenesis 24:1615–1623
Olbina G, Eckhart W (2003) Mutations in the second extracellular region of connexin 43 prevent localization to the plasma membrane, but do not affect its ability to suppress cell growth. Mol Cancer Res 1:690–700
Ozog MA, Bechberger JF, Naus CC (2002) Ciliary neurotrophic factor (CNTF) in combination with its soluble receptor (CNTFRα) increases connexin43 expression and suppresses growth of C6 glioma cells. Cancer Res 62:3544–3548
Ozog MA, Bernier SM, Bates DC, Chatterjee B, Lo CW, Naus CC (2004) The complex of CNTF-CNTFRα upregulates connexin43 and intercellular coupling in astrocytes via the JAK/STAT pathway. Mol Biol Cell (in press)
Peracchia C, Sotkis A, Wang XG, Peracchia LL, Persechini A (2000) Calmodulin directly gates gap junction channels. J Biol Chem 275:26220–26224
Perbal B (2001) NOV (nephroblastoma overexpressed) and the CCN family of genes: structural and functional issues. Mol Pathol 54:57–79
Perbal B (2004) CCN proteins: multifunctional signalling regulators. Lancet 363:62–64
Piechocki MP, Burk RD, Ruch RJ (1999) Regulation of connexin32 and connexin43 gene expression by DNA methylation in rat liver cells. Carcinogenesis 20:401–406
Planque N, Perbal B (2003) A structural approach to the role of CCN (CYR61/CTGF/NOV) proteins in tumourigenesis. Cancer Cell Int 3:15
Princen F, Robe P, Gros D, Jarry-Guichard T, Gielen J, Merville MP, Bours V (2001) Rat gap junction connexin-30 inhibits proliferation of glioma cell lines. Carcinogenesis 22:507–513
Qin H, Shao Q, Thomas T, Kalra J, Alaoui-Jamali MA, Laird DW (2003) Connexin26 regulates the expression of angiogenesis-related genes in human breast tumor cells by both GJIC-dependent and -independent mechanisms. Cell Commun Adhes 10:387–393
Reuss B, Unsicker K (1998) Regulation of gap junction communication by growth factors from non-neural cells to astroglia – a brief review. GLIA 24:32–38
Robe PA, Princen F, Martin D, Malgrange B, Stevenaert A, Moonen G, Gielen J, Merville M, Bours V (2000) Pharmacological modulation of the bystander effect in the herpes simplex virus thymidine kinase/ganciclovir gene therapy system: effects of dibutyryl adenosine 3',5'-cyclic monophosphate, alpha-glycyrrhetinic acid, and cytosine arabinoside. Biochem Pharmacol 60:241–249
Robe PA, Jolois O, N'Guyen M, Princen F, Malgrange B, Merville MP, Bours V (2004) Modulation of the HSV-TK/ganciclovir bystander effect by n-butyrate in glioblastoma: correlation with gap-junction intercellular communication. Int J Oncol 25:187–192
Rose B, Mehta PP, Loewenstein WR (1993) Gap-junction protein gene suppresses tumorigenicity. Carcinogenesis 14:1073–1075

Samejima Y, Meruelo D (1995) 'Bystander killing' induces apoptosis and is inhibited by forskolin. Gene Therapy 2:50–58

Saunders MM, Seraj MJ, Li Z, Zhou Z, Winter CR, Welch DR, Donahue HJ (2001) Breast cancer metastatic potential correlates with a breakdown in homospecific and heterospecific gap junctional intercellular communication. Cancer Res 61:1765–1767

Schiavon G, Furlan S, Marin O, Salvatori S (2002) Myotonic dystrophy protein kinase of the cardiac muscle: evaluation using an immunochemical approach. Microsc Res Tech 58:404–411

Segretain D, Decrouy X, Dompierre J, Escalier D, Rahman N, Fiorini C, Mograbi B, Siffroi JP, Huhtaniemi I, Fenichel P, Pointis G (2003) Sequestration of connexin43 in the early endosomes: an early event of Leydig cell tumor progression. Mol Carcinogenesis 38:179–187

Seul KH, Kang KY, Lee KS, Kim SH, Beyer EC (2004) Adenoviral delivery of human connexin37 induces endothelial cell death through apoptosis. Biochem Biophys Res Commun 319:1144–1151

Shevde LA, Samant RS, Goldberg SF, Sikaneta T, Alessandrini A, Donahue HJ, Mauger DT, Welch DR (2002) Suppression of human melanoma metastasis by the metastasis suppressor gene, BRMS1. Exp Cell Res 273:229–239

Simon AM, Goodenough DA (1998) Diverse functions of vertebrate gap junctions. Trends Cell Biol 8:477–483

Singal R, Tu ZJ, Vanwert JM, Ginder GD, Kiang DT (2000) Modulation of the connexin26 tumor suppressor gene expression through methylation in human mammary epithelial cell lines. Anticancer Res 20:59–64

Sohl G, Willecke K (2004) Gap junctions and the connexin protein family. Cardiovasc Res 62:228–232

Stoker MG, Shearer M, O'Neill C (1966) Growth inhibition of polyoma-transformed cells by contact with static normal fibroblasts. J Cell Sci 1:297–310

Stout CE, Costantin JL, Naus CC, Charles AC (2002) Intercellular calcium signaling in astrocytes via ATP release through connexin hemichannels. J Biol Chem 277:10482–10488

Swierenga SH, Yamasaki H (1992) Performance of tests for cell transformation and gap-junction intercellular communication for detecting nongenotoxic carcinogenic activity. IARC Sci Publ 165–193

Tan LW, Bianco T, Dobrovic A (2002) Variable promoter region CpG island methylation of the putative tumor suppressor gene connexin 26 in breast cancer. Carcinogenesis 23:231–236

Temme A, Buchmann A, Gabriel HD, Nelles E, Schwarz M, Willecke K (1997) High incidence of spontaneous and chemically induced liver tumors in mice deficient for connexin32. Curr Biol 7:713–716

Teunissen BE, Jansen AT, van Amersfoorth SC, O'Brien TX, Jongsma HJ, Bierhuizen MF (2003) Analysis of the rat connexin 43 proximal promoter in neonatal cardiomyocytes. Gene 322:123–136

Thomas T, Aasen T, Hodgins M, Laird DW (2003) Transport and function of cx26 mutants involved in skin and deafness disorders. Cell Commun Adhes 10:353–358

Torok K, Stauffer K, Evans WH (1997) Connexin32 of gap junctions contains two cytoplasmic calmodulin-binding domains. Biochem J 326:479–483

Touraine RL, Vahanian N, Ramsey WJ, Blaese RM (1998) Enhancement of the herpes simplex virus thymidine kinase/ganciclovir bystander effect and its antitumor efficacy in vivo by pharmacologic manipulation of gap junctions. Hum Gene Therapy 9:2385–2391

Toyofuku T, Akamatsu Y, Zhang H, Kuzuya T, Tada M, Hori M (2001) c-Src regulates the interaction between connexin-43 and ZO-1 in cardiac myocytes. J Biol Chem 276:1780–1788

Trosko JE, Goodman JI (1994) Intercellular communication may facilitate apoptosis: Implications for tumor promotion. Mol Carcinogenesis 11:8–12

Trosko JE, Chang CC (2001) Mechanism of up-regulated gap junctional intercellular communication during chemoprevention and chemotherapy of cancer. Mutat Res 480–481:219–229

Trosko JE, Ruch RJ (2002) Gap junctions as targets for cancer chemoprevention and chemotherapy. Curr Drug Targets 3:465–482

van der Heyden MA, Rook MB, Hermans MM, Rijksen G, Boonstra J, Defize LH, Destree OH (1998) Identification of connexin43 as a functional target for Wnt signalling. J Cell Sci 111:1741–1749

Van Eldik LJ, Hertzberg EL, Berdan RC, Gilula NB (1985) Interaction of calmodulin and other calcium-modulated proteins with mammalian and arthropod junctional membrane proteins. Biochem Biophys Res Commun 126:825–832

Vine AL, Bertram JS (2002) Cancer chemoprevention by connexins. Cancer Metastasis Rev 21:199–216

Warn-Cramer BJ, Lau AF (2004) Regulation of gap junctions by tyrosine protein kinases. Biochim Biophys Acta 1662:81–95

Warn-Cramer BJ, Lin R, Martyn K, Guyette CV, Lau AF (2003) Maintaining connexin43 gap junctional communication in v-Src cells does not alter growth properties associated with the transformed phenotype. Cell Commun Adhes 10:299–303

Xu X, Li WE, Huang GY, Meyer R, Chen T, Luo Y, Thomas MP, Radice GL, Lo CW (2001) Modulation of mouse neural crest cell motility by N-cadherin and connexin 43 gap junctions. J Cell Biol 154:217–230

Yamasaki H, Naus CCG (1996) Role of connexin genes in growth control. Carcinogenesis 17:1199–1213

Yano T, Ito F, Kobayashi K, Yonezawa Y, Suzuki K, Asano R, Hagiwara K, Nakazawa H, Toma H, Yamasaki H (2004) Hypermethylation of the CpG island of connexin 32, a candidate tumor suppressor gene in renal cell carcinomas from hemodialysis patients. Cancer Lett 208:137–142

Yu W, Dahl G, Werner R (1994) The connexin43 gene is responsive to oestrogen. Proc R Soc Lond [Biol] 255:125–132

Zhang YW, Kaneda M, Morita I (2003) The gap junction-independent tumor-suppressing effect of connexin 43. J Biol Chem 278:44852–44856

Zhang ZQ, Zhang W, Wang NQ, Bani-Yaghoub M, Lin ZX, Naus CC (1998) Suppression of tumorigenicity of human lung carcinoma cells after transfection with connexin43. Carcinogenesis 19:1889–1894

Zhou L, Kasperek EM, Nicholson BJ (1999) Dissection of the molecular basis of pp60(v-src) induced gating of connexin 43 gap junction channels. J Cell Biol 144:1033–1045

Zhu D, Caveney S, Kidder GM, Naus CC (1991) Transfection of C6 glioma cells with connexin 43 cDNA: analysis of expression, intercellular coupling, and cell proliferation. Proc Natl Acad Sci USA 88:1883–1887

Zhu D, Kidder GM, Caveney S, Naus CC (1992) Growth retardation in glioma cells cocultured with cells overexpressing a gap junction protein. Proc Natl Acad Sci USA 89:10218–10221

Subject Index

Printing: Krips bv, Meppel
Binding: Stürtz, Würzburg